5th Workshop on BioNLP Open Shared Tasks (BioNLP-OST 2019)

Hong Kong, China
4 November 2019

ISBN: 978-1-7138-0076-7

Additional copies of this publication are available from:

Curran Associates, Inc.
57 Morehouse Lane
Red Hook, NY 12571 USA
Phone: 845-758-0400
Fax: 845-758-2633
Email: curran@proceedings.com
Web: www.proceedings.com

EMNLP-IJCNLP 2019

BioNLP-OST 2019

**Proceedings of
the 5th Workshop on BioNLP Open Shared Tasks**

November 4, 2019
Hong Kong, China

Introduction

In the era of machine learning and AI, the importance of data, either for training or for evaluating the learning-based models, is more and more evident. The performance of any AI device is greatly affected by the data it is trained on. Also, it is common that an evaluated performance of an AI device based on one benchmark data set is critically different when evaluated on another data set. This is particularly true for text mining where the performance strongly depends on various factors: the goal or the type and the language of the documents. These examples dictates the importance of sharing benchmark datasets, so that evaluation results can be comparable to each other. It is a key for efficient advancement of the technology.

BioNLP Open Shared Tasks is organized to promote the sharing of computational tasks of biomedical text mining and also solutions to them. Here sharing a task means sharing benchmark datasets and evaluation systems. It is a continuation of the previous efforts organized around the BioNLP Shared Task (BioNLP-ST) workshop series (2009, 2011, 2013, 2016).

This year, six tasks are contributed by voluntary task organizers. Two tasks, the Bacteria-Biotope (BB) and SeeDev tasks, are a continuation of their previous editions. BB targets the extraction of information about bacterial biotopes and phenotypes, while SeeDev focuses on extracting events of genetic and molecular mechanisms involved in plant seed development. The PharmaCoNER task is a named entity recognition task for pharmacological substances, compounds and proteins. Particularly it targets Spanish texts, which brings the new challenge of dealing with multilingualism. The CRAFT task is presented as a highly challenging task, aiming at annotating texts with rich semantics, and a full stack of linguistic structures. AGAC proposes to extract compositional concepts for drug repurposing. Finally RDoc is an Information Retrieval task in the field of neuroscience.

For the six tasks, a total of 45 teams participated. For the workshop, paper submissions were open exclusively to the teams that had completed at least one task as well as the task organizers. 43 reviewers in the Program Committee selected 30 papers to be presented for the workshop out of 38 submitted papers. We are happy to present the papers and we believe it to be a rare chance to compare various tasks of biomedical text mining, and also various solutions to them.

BioNLP-OST Organizers
- Jin-Dong Kim, DBCLS
- Claire Nédellec, INRA
- Robert Bossy, INRA
- Louise Deléger, INRA

Organizing Committee:

Jin-Dong Kim, DBCLS, Japan
Claire Nédellec, INRA, France
Robert Bossy, INRA, France
Louise Deléger, INRA, France

Program Committee:

Eiji Aramaki, NAIST, Japan
Mouhamadou Ba, INRA, France
Michael Bada, University of Colorado Denver, USA
William Baumgartner, University of Colorado Denver, USA
Robert Bossy, INRA, France
Tiffany Callahan, University of Colorado Denver, USA
Thierry Charnois, LIPN Université Paris 13, France
Kevin Cohen, Univ. of Colorado, USA
Nigel Collier, Univ. of Cambridge, UK
Louise Deléger, INRA, France
Dina Demner-Fushman, NLM, USA
Richard Eckart de Castilho, UKP Technische Universität Darmstadt, Germany
Arnaud Ferré, INRA, France
Juliane Fluck, Fraunhofer SCAI, Germany
Cyril Grouin, LIMSI CNRS, France
Tianyong Hao, South China Normal Univ., China
Antonio Jimeno Yepes, IBM research, Australia
Indika Kahanda, Montana State University, USA
Jin-Dong Kim, DBCLS, Japan
Sun Kim, NCBI, USA
Martin Krallinger, Barcelona Supercomputing Center, Spain
Anastasia Krithara, Demokritos, Greece
Robert Leaman, NCBI, USA
Jake Lever, Stanford University, USA
Zhiyong Lu, NCBI, USA
Claire Nédellec, INRA, France
Mariana Neves, German Federal Institute for Risk Assessment, Germany
Arzucan Özgür, Bogazici University, Turkey
Fabio Rinaldi, Univ of Zurich, Switzerland
Kirk Roberts, University of Texas, USA
Min Song, Yonsei Univ., Korea
Xavier Tannier, LIMICS Sorbonne Université, France
Yuka Tateishi, NBDC, Japan
Karin Verspoor, Univ. of Melbourne, Australia
Chih-Hsuan Wei, NCBI, USA
Davy Weissenbacher, UPenn, USA
Jingbo Xia, Huazhong Agricultural Univ., China
Kaiyin Zhou, Huazhong Agricultural Univ., China

Table of Contents

BioNLP-OST 2019 RDoC Tasks: Multi-grain Neural Relevance Ranking Using Topics and Attention Based Query-Document-Sentence Interactions
Pankaj Gupta, Yatin Chaudhary and Hinrich Schütze 227

Conference Program

November 4, 2019

09:00–09:10 *Opening*

Session 1: PharmaCoNER

09:10–09:30 *PharmaCoNER: Pharmacological Substances, Compounds and proteins Named Entity Recognition track*
Aitor Gonzalez Agirre, Montserrat Marimon, Ander Intxaurrondo, Obdulia Rabal, Marta Villegas and Martin Krallinger

09:30–09:40 *When Specialization Helps: Using Pooled Contextualized Embeddings to Detect Chemical and Biomedical Entities in Spanish*
Manuel Stoeckel, Wahed Hemati and Alexander Mehler

09:40–09:50 *VSP at PharmaCoNER 2019: Recognition of Pharmacological Substances, Compounds and Proteins with Recurrent Neural Networks in Spanish Clinical Cases*
Víctor Suárez-Paniagua

09:50–10:00 *IxaMed at PharmacoNER Challenge 2019*
Xabier Lahuerta, Iakes Goenaga, Koldo Gojenola, Aitziber Atutxa Salazar and Maite Oronoz

10:00–10:10 *NLNDE: Enhancing Neural Sequence Taggers with Attention and Noisy Channel for Robust Pharmacological Entity Detection*
Lukas Lange, Heike Adel and Jannik Strötgen

10:10–10:20 *A Deep Learning-Based System for PharmaCoNER*
Ying Xiong, Yedan Shen, Yuanhang Huang, Shuai Chen, Buzhou Tang, Xiaolong Wang, Qingcai Chen, Jun Yan and Yi Zhou

10:30–11:00 *Coffee Break*

Session 2: PharmaCoNER and AGAC

Poster Presentations

12:20–14:00   *RACAI's System at PharmaCoNER 2019*
Radu Ion, Vasile Florian Păiş and Maria Mitrofan

12:20–14:00   *Transfer Learning in Biomedical Named Entity Recognition: An Evaluation of BERT in the PharmaCoNER task*
Cong Sun and Zhihao Yang

12:20–14:00   *A Multi-Task Learning Framework for Extracting Bacteria Biotope Information*
Qi Zhang, Chao Liu, Ying Chi, Xuansong Xie and Xiansheng Hua

12:20–14:00   *YNU-junyi in BioNLP-OST 2019: Using CNN-LSTM Model with Embeddings for SeeDev Binary Event Extraction*
junyi li, Xiaobing Zhou, Yuhang Wu and Bin Wang

12:20–14:00   *Using Snomed to recognize and index chemical and drug mentions.*
Pilar López Úbeda, Manuel Carlos Díaz Galiano, L. Alfonso Urena Lopez and Maite Martin

12:30–14:00   *Lunch Break*

Session 3: BB and SeeDev

14:00–14:20   *Bacteria Biotope at BioNLP Open Shared Tasks 2019*
Robert Bossy, Louise Deléger, Estelle Chaix, Mouhamadou Ba and Claire Nédellec

14:20–14:40   *Linguistically Informed Relation Extraction and Neural Architectures for Nested Named Entity Recognition in BioNLP-OST 2019*
Pankaj Gupta, Usama Yaseen and Hinrich Schütze

14:40–14:50   *An ensemble CNN method for biomedical entity normalization*
Pan Deng, Haipeng Chen, Mengyao Huang, Xiaowen Ruan and Liang Xu

14:50–15:00   *BOUN-ISIK Participation: An Unsupervised Approach for the Named Entity Normalization and Relation Extraction of Bacteria Biotopes*
İlknur Karadeniz, Ömer Faruk Tuna and Arzucan Özgür

PharmaCoNER: Pharmacological Substances, Compounds and proteins Named Entity Recognition track

Aitor Gonzalez-Agirre
Centro Nacional de
Investigaciones Oncológicas (CNIO)
Barcelona Supercomputing
Center (BSC)
aitor.gonzalez@bsc.es

Montserrat Marimon
Centro Nacional de
Investigaciones Oncológicas (CNIO)
Barcelona Supercomputing
Center (BSC)
montserrat.marimon@bsc.es

Ander Intxaurrondo
Centro Nacional de
Investigaciones Oncológicas (CNIO)
Barcelona Supercomputing
Center (BSC)
ander.intxaurrondo@bsc.es

Obdulia Rabal
Center for Applied
Medical Research (CIMA)
University of Navarra
orabal@unav.es

Marta Villegas
Centro Nacional de
Investigaciones Oncológicas (CNIO)
Barcelona Supercomputing
Center (BSC)
marta.villegas@bsc.es

Martin Kralligner
Centro Nacional de
Investigaciones Oncológicas (CNIO)
Barcelona Supercomputing
Center (BSC)
martin.krallinger@bsc.es

Abstract

One of the biomedical entity types of relevance for medicine or biosciences are chemical compounds and drugs. The correct detection these entities is critical for other text mining applications building on them, such as adverse drug-reaction detection, medication-related fake news or drug-target extraction. Although a significant effort was made to detect mentions of drugs/chemicals in English texts, so far only very limited attempts were made to recognize them in medical documents in other languages. Taking into account the growing amount of medical publications and clinical records written in Spanish, we have organized the first shared task on detecting drug and chemical entities in Spanish medical documents. Additionally, we included a clinical concept-indexing sub-track asking teams to return SNOMED-CT identifiers related to drugs/chemicals for a collection of documents. For this task, named PharmaCoNER, we generated annotation guidelines together with a corpus of 1,000 manually annotated clinical case studies. A total of 22 teams participated in the sub-track 1, (77 system runs), and 7 teams in the sub-track 2 (19 system runs). Top scoring teams used sophisticated deep learning approaches yielding very competitive results with F-measures above 0.91. These results indicate that there is a real interest in promoting biomedical text mining efforts beyond English. We foresee that the PharmaCoNER annotation guidelines, corpus and participant systems will foster the development of new resources for clinical and biomedical text mining systems of Spanish medical data.

1 Introduction

Efficient access to mentions of drugs, medications and chemical entities contained in clinical texts, scientific articles, patents or even the web is a pressing need shared by biomedical researchers and clinicians (Krallinger et al., 2017). Biomedical text mining is one of the most prolific application domains of natural language processing technologies (Zweigenbaum et al., 2007). The recognition of pharmaceutical drugs/chemical entities is a critical step required for the subsequent detection of relations with other biomedically relevant entities such as genes/proteins, diseases or adverse reactions (Vazquez et al., 2011). Text mining and information extraction systems were published that tried to find protein-drug relations (including ligand-protein interactions and pharmacogenomics information), medication-related al-

Proceedings of the 5th Workshop on BioNLP Open Shared Tasks, pages 1–10
Hong Kong, China, November 4, 2019. ©2019 Association for Computational Linguistics

lergies, chemical metabolic reactions, drug-drug interactions (Herrero-Zazo et al., 2013), disease-drug relations, as well as drug safety-related issues. The correct identification of drug mentions is also needed for other complex relation types like drug dosage recognition, duration of medical treatments or drug repurposing.

The importance of chemical and drug name recognition motivated several-shared tasks in the past, such as the CHEMDNER tracks (Krallinger et al., 2015) or the i2b2 medication challenge (Uzuner et al., 2010b,a), with a considerable number of participants and impact (Doan et al., 2010; Yang, 2010).

Currently, most of the biomedical and clinical NLP research, is done on English documents, while only few tasks were carried out using non-English texts, or were multilingual. Nonetheless, it is important to highlight that there is a considerable amount of biomedically relevant content published in other languages than English, and particularly clinical texts are entirely written in the native language of each country.

Spanish is a language spoken by more than 572 million people in the world today, either as a native, second or foreign language. It is the second language in the world by number of native speakers with more than 477 million people. According to results derived from WHO statistics, just in Spain there are over 180 thousand practicing physicians, more than 247 thousand nursing and midwifery personnel or 55 thousand pharmaceutical personnel. These facts, and the extrapolation to other Spanish speaking countries explains why a considerable subset of the PubMed database records corresponds to Spanish medical articles. Moreover, PubMed does only contain a part of the medical literature originally published in Spanish, which is also stored in other resources such as MEDES, SciELO, IBECS or CUIDEN.

Following the outline of previous chemical/drug NER efforts, in particular the BioCreative CHEMDNER tracks, we have carried out the first task on chemical and drug mention recognition from Spanish medical texts, namely from a corpus of Spanish clinical case studies. Thus, this track addressed the automatic extraction of chemical, drug, gene/protein mentions from clinical case studies written in Spanish. The main aim was to promote the development of named entity recognition tools of practical relevance, that is, chemi-

cal and drug mentions in non-English content, determining the current-state-of-the art, identifying challenges and comparing the strategies and results to those published for English data.

2 Methods

2.1 Track Description

The PharmaCoNER track was one of the six tracks of the BioNLP-OST 2019 / EMNLP-IJCNLP workshop[1]. It was the first community challenge track devoted to the recognition of pharmaceutical drugs and chemical entities in medical texts in Spanish.

For this track, two scenarios or sub-tracks were proposed:

- *NER offset and entity classification*. The first sub-track focused on the recognition and classification of entities.

- *Concept indexing*. The second sub-track consisted of concept indexing, where, for each document, the participating teams had to generate the list of the unique SNOMED-CT concept identifiers, which were compared to the manually annotated concept IDs corresponding to the pharmaceutical drugs and chemical entities.

2.2 Track data

We prepared a manually classified collection of clinical case report sections derived from open access Spanish medical publications, named the Spanish Clinical Case Corpus (SPACCC)[2]. The corpus contained a total of 1,000 clinical cases / 396,988 words. It is noteworthy that this kind of narrative shows properties of both the biomedical and medical literature, as well as clinical records. Case reports are considered as the scientific paper of a single clinical observation. Moreover, the clinical cases were not restricted to a single medical discipline, covering a variety of medical disciplines, including oncology, urology, cardiology, pneumology or infectious diseases. This is key to cover a diverse set of chemicals and drugs.

The PharmaCoNER corpus had a total of 7,624 entity mentions, corresponding to four different mention types[3]. Figure 1 shows a screenshot of a

---

[1]https://2019.bionlp-ost.org/

[2]https://github.com/PlanTL-SANIDAD/SPACCC

[3]For a detailed description of the mentions types, see (Rabal et al., 2018).

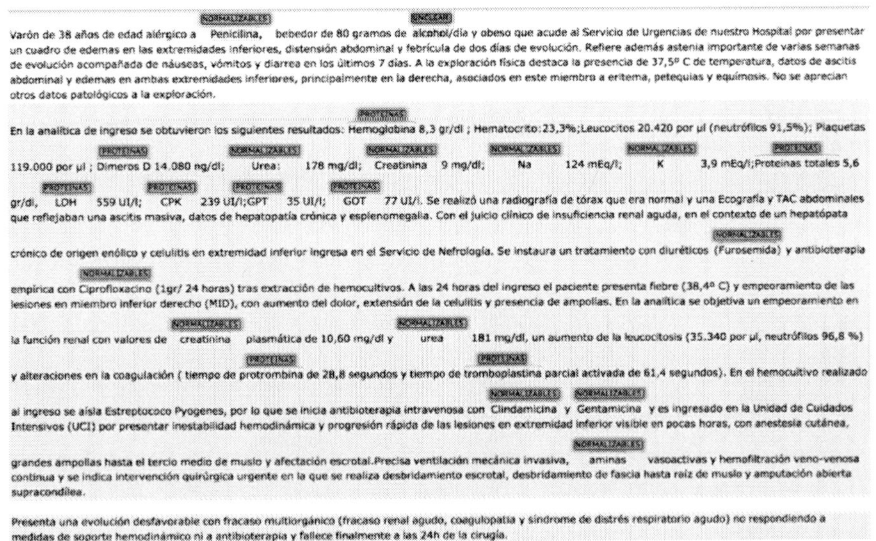

Figure 1: PharmaCoNER annotation example.

clinical case annotated using the BRAT tool. The overall annotation statistics were:

- NORMALIZABLES (normalizable): 4,398 mentions of chemicals that could be manually normalized to a unique concept identifier (primarily SNOMED-CT).

- NO_NORMALIZABLES (not normalizable): 50 mentions of chemicals that could not be normalized manually to a unique concept identifier.

- PROTEINAS (proteins): 3,009 mentions of proteins and genes following an adaptation of the BioCreative GPRO track annotation guidelines. This class included also peptides, peptide hormones and antibodies.

- UNCLEAR: 167 cases of general substance class mentions of clinical or biomedical relevance, including certain pharmaceutical formulations, general treatments, chemotherapy programs, vaccines and a predefined set of general substances (e.g.: Estragn, Silimarina, Bromelana, Melanina, Vaselina, Lanolina, Alcohol, Tabaco, Marihuana, Cannabis, Opio and Gluten)[4].

The annotation process of the PharmaCoNER corpus was inspired by previous annotation schemes and corpora used for the BioCreative CHEMDNER (Krallinger et al., 2015) and GPRO tracks (Pérez-Pérez et al., 2017), translating the guidelines used for these tracks into Spanish and adapting them to the characteristics and needs of clinically oriented documents by modifying the annotation criteria and rules to cover medical information needs. This adaptation was carried out in collaboration with practicing physicians and medicinal chemistry experts. The adaptation, translation and refinement of the guidelines (Rabal et al., 2018) was done on a sample set of the SPACCC corpus and linked to an iterative process of annotation consistency analysis through inter-annotator agreement (IAA) studies until a high annotation quality in terms of IAA was reached. The final, IAA measure obtained for this corpus was calculated on a set of 50 records that were double annotated (blinded) by two different expert annotators, reaching a pairwise agreement of 93% on the exact entity mention comparison level and 76% agreement when also the entity concept normalization was taken into account. Entity normalization was carried out primarily against the SNOMED-CT knowledge base. Note that there is a SNOMED-CT version directly released by the Spanish Ministry of Health twice a year.

The PharmaCoNER corpus was randomly sampled into three subsets: the train set (500 clinical cases), and the development and test sets (250 clinical cases each). These clinical cases were

---

[4]Mentions of this class were not part of the entities evaluated by this track, but served as additional annotations of medical relevance.

manually annotated using a customized version of
AnnotateIt. Then, the BRAT annotation toolkit
(Stenetorp et al., 2012) was used to correct errors
and add missing annotations. The statistics of the
number of label for each datasets are shown in Table 1.

Table 1: Distribution of labels in the PharmaCoNER
datasets.

Label	Train	Dev	Test	Overall
NORMALIZABLES	2,304	1,121	973	4,398
NO_NORMALIZABLES	24	16	10	50
PROTEINAS	1,405	745	859	3,009
UNCLEAR	89	44	34	167

Together with the test set, we released an additional collection of 3,501 documents (background
set[5]) to make sure that participating teams were
not able to do manual corrections and also to promote that these systems would potentially be able
to scale to larger data collections.

Moreover, we provided also the following resources: (1) Spanish medical text tokenizer, sentence splitter, lemmatizer and POS tagger; (2)
Dictionary of chemicals, compounds and drugs
in Spanish; (3) Sense inventory of Spanish medical abbreviation and their long forms; (4) Spanish
drug naming file with prefixes and suffixes rules;
and (5) a large background set of medical and
health documents in Spanish.

2.3 Evaluation metrics

We released an evaluation script that supported the
evaluation of the predictions of the participating
teams. For both sub-tracks, the primary evaluation
metrics used consisted of standard measures from
the NLP community, namely micro-averaged precision, recall, and balanced F-score, the last one
being the official evaluation measure:

$$\text{Precision: } P = \frac{TP}{TP+FP}$$

$$\text{Recall: } R = \frac{TP}{TP+FN}$$

$$\text{F-score: } F1 = 2 * \frac{(P*R)}{(P+R)}$$

where TP = true positives, FP = false positive
and FN = false negative.

Teams could submit up to five prediction files
(or system runs) in a predefined prediction format: BRAT, for sub-track 1, and TSV files, for
sub-track 2.

3 Participation and Results

3.1 Participation

To participate in the PharmaCoNER track it was
necessary to register both on the official website[6]
and in the CodaLab competition[7]. Training and
development sets were made available for download on the official website[8], and the evaluation
script was uploaded to GitHub[9], to ensure a transparent evaluation.

As we already said, submissions had to be provided in a predefined prediction format: BRAT, for
sub-track 1, and TSV files, for sub-track 2. Additionally we plan to release the corpus also in the
popular PubAnnotation format (Kim and Wang,
2012).

The participants had a period of almost two
months to develop their system. In the middle of
this period, the test and background sets were released with the 3,751 documents that the participants had to process and label, although the final
evaluation was done only on the 250 documents
corresponding to the test set. The intention was
to use the background set to enable the construction of participant-generated Silver Standard corpus. As we have mentioned, the participants could
submit a maximum of 5 system runs, and, once
the submission deadline expired, we published the
Gold Standard annotations of the test set, in order to ensure a transparent evaluation process and
help participants to carry out a more detailed error
analysis.

A total of 22 teams participated in the sub-track
1, submitting a total of 77 systems, and 7 teams
in the sub-track 2, submitting a total of 19 runs.
Teams from eleven different nationalities participated in the track: seven teams from Spain, three
from China, and one team from each: Finland,
France, India, Japan, Romania, Russia, United
Kingdom and the United States. Three participants belong to a commercial institution. Table

---

[5]The background set included the training, development
and test sets, and an additional collection of 2,751 unlabeled
clinical cases (total of 3,751 clinical cases).

[6]http://temu.bsc.es/pharmaconer/
[7]https://competitions.codalab.org/
competitions/23159
[8]http://temu.bsc.es/pharmaconer/index.
php/data/
[9]https://github.com/PlanTL-SANIDAD/
PharmaCoNER-CODALAB-Evaluation-Script

Table 2: Overview of Team Participation in the PharmaCoNER track.

Username	Organization/Institution/Company	Members	Country	Comm.
alily	Carlos III University of Madrid	3	Spain	No
ayan7246	Unaffiliated	1	India	No
chaanim	University of Turku	2	Finland	No
CongSun	Dalian University of Technology	3	China	No
Edson	University of Côte d'Azur	4	France	No
foxlf823	UMASS Lowell	3	United States	No
FSL	Unaffiliated	2	Spain	No
ghada.alfattni	University of Manchester	3	United Kingdom	No
ixamed	University of the Basque Country	5	Spain	No
JoyHan	-	-	-	-
lluisp	Universitat Politècnica de Catalunya	1	Spain	No
lukas.lange	Bosch Center for Artificial Intelligence	3	Germany	Yes
m-stoeckel	Goethe University Frankfurt	2	Germany	No
m.domrachev	Unaffiliated	1	Russia	No
naiven	JD	1	China	Yes
plubeda	Universidad de Jan	4	Spain	No
raduion	Research Institute for AI "Mihai Draganescu"	3	Romania	No
rriveraz	Carlos III University of Madrid	3	Spain	No
sohrab	National Institute of Advanced Industrial Science and Technology	4	Japan	No
tEarth	-	-	-	-
uyaseen	Siemens AG	2	Germany	Yes
VSP	Carlos III University of Madrid	1	Spain	No
xiongying	Harbin Institute of Technology	4	China	Yes

2 summarizes the most relevant information about the participants (we lack the information from two of the teams, because they registered at CodaLab, but not at our website).

3.2 Baseline system

We produced three baseline systems for the track: The first one is a very simple baseline based on vocabulary transfer, and the other two baseline systems are competitive baselines based on the PharmaCoNER Tagger (Armengol-Estapé et al., 2019), a deep learning-based tool for automatically finding chemicals and drugs in Spanish medical texts.

In the vocabulary transfer approach, each annotation from the train and development datasets was transferred to the test dataset using strict string matching. For those cases where the text was the same, but the entity type was different, we decided to annotate all entity types that matched that text.

In the two baselines based on the Pharma-CoNER Tagger, we used the default parameters, a hidden layer of size 300, and early stop (best model at epoch 35). The models were trained using the GloVe embeddings (Pennington et al., 2014) from SBWC[10] (from now on *baseline-glove*) and the Medical Word Embeddings for Spanish (Soares et al., 2019) (from now on *baseline-med*). The corpus was tokenized using spaCy.

<hr>

[10]https://github.com/dccuchile/spanish-word-embeddings

3.3 Results

Table 3 shows the results for sub-track 1 (*NER offset and entity type classification*), ordered by team performance (first column), then system performance (second column).

The top scoring system was submitted by *xiongying*, with an F-score of 0.91052, being relatively close to the next two participants: *FSL*, ranked 2nd with a F-score of 0.90968, and *m-stoeckel*, ranked 3rd with a F-score of 0.89888. Participant *Edson* submitted five systems that scored almost zero. Once he noticed the error, he submitted two fixed submissions. These submissions were made after the publication of the results but before the release of the test set with GS annotations. These late submissions of *Edson* are marked with an asterisks in the table, including the hypothetical ranking of his team/systems.

Note that all of the teams were well above the baseline based on vocabulary transfer, which would rank last if we ignored the submission with errors. The competitive baseline trained with the GloVe embeddings would rank 16, and the one trained with embeddings that are specific for clinical texts in Spanish would rank 13. It is remarkable that 12 teams out of 20 managed to beat a very competitive baseline based on a well known Deep Learning tool.

Table 4 shows the results for sub-track 2 (*Concept Indexing*), ordered by team performance (first

Table 3: Results for sub-track 1: *NER offset and entity type classification.*

Team Rank	System Rank	User	Precision	Recall	F1
	1		0.91226	0.90879	0.91052
	2		0.91589	0.90445	0.91013
1	3	xiongying	0.91008	0.90662	0.90835
	4		0.90751	0.90554	0.90652
	5		0.90205	0.90988	0.90595
2	3	FSL	0.90625	0.91314	0.90968
	7		0.90708	0.89082	0.89888
3	8	m-stoeckel	0.89297	0.89685	0.89491
	13		0.88839	0.86369	0.87586
	9		0.90463	0.88056	0.89243
4	10	CongSun	0.90704	0.87405	0.89024
	14		0.89183	0.85939	0.87531
	17		0.88732	0.85071	0.86863
5	11	naiven	0.90315	0.87079	0.88668
	12		0.88950	0.88274	0.88610
	27		0.85162	0.87242	0.86189
6	28	lukas.lange	0.86307	0.85885	0.86095
	31		0.85078	0.86048	0.85560
	32		0.85520	0.85288	0.85404
7	15	chaanim	0.87568	0.87188	0.87378
	16		0.88098	0.85993	0.87033
8	22	foxlf823	0.87218	0.85939	0.86574
	23		0.87674	0.85342	0.86492
	18		0.90222	0.83659	0.86817
	21		0.90088	0.83388	0.86608
9	42	ixamed	0.82981	0.85233	0.84092
	49		0.81914	0.80402	0.81151
	50		0.81914	0.80402	0.81151
	19		0.86881	0.86645	0.86763
	26		0.87079	0.85613	0.86340
10	39	sohrab	0.85320	0.83931	0.84620
	41		0.83665	0.84528	0.84094
	46		0.88483	0.77579	0.82673
	20		0.90581	0.83008	0.86629
	24		0.90482	0.82573	0.86347
11	25	uyaseen	0.90482	0.82573	0.86347
	33		0.84644	0.85885	0.85260
	37		0.88941	0.81650	0.85140
12	29	m.domrachev	0.87073	0.84473	0.85754
	30		0.87073	0.84473	0.85754
-	-	*baseline-med*	*0.87020*	*0.83713*	*0.85335*
	34		0.88538	0.82193	0.85248
13	35	rriveraz	0.88538	0.82193	0.85248
	36		0.88538	0.82193	0.85248
	38		0.90189	0.80347	0.84984
	40		0.90043	0.79533	0.84462
14	47	raduion	0.89327	0.76330	0.82319
	48		0.78281	0.84528	0.81284
	52		0.92530	0.71281	0.80527
	43		0.88882	0.78990	0.83645
	44		0.89176	0.78719	0.83622
15	45	lluisp	0.88991	0.78556	0.83449
	53		0.81160	0.76710	0.78872
	61		0.73211	0.73887	0.73548
-	-	*baseline-glove*	*0.83259*	*0.80999*	*0.82113*
16	51	ghada.alfattni	0.85039	0.77144	0.80900
	55		0.82776	0.72530	0.77315
	54		0.88507	0.69815	0.78058
	56		0.85992	0.69653	0.76965
17	60	plubeda	0.92602	0.61835	0.74154
	62		0.84404	0.64929	0.73397
18	57	alily	0.86034	0.68893	0.76515
	59		0.86981	0.67101	0.75759
19	58	VSP	0.81621	0.71607	0.76287
	63		0.74668	0.61129	0.67224
	67		0.43812	0.48046	0.45831
20	68	ayan7246	0.36910	0.47991	0.41728
	69		0.33333	0.48046	0.39360
	70		0.52283	0.19273	0.28163
	64		0.88519	0.54098	0.67155
21	65	JoyHan	0.52523	0.52666	0.52594
	66		0.88350	0.37193	0.52349
-	-	*baseline-vt*	*0.67330*	*0.60641*	*0.63810*
	71		0.00280	0.00163	0.00206
	72		0.00008	0.00163	0.00015
22	73	Edson	0.00007	0.00217	0.00014
	74		0.00007	0.00217	0.00014
	75		0.00002	0.00054	0.00004
20*	60*	Edson	0.80660	0.68920	0.74330
	70*		0.63350	0.14930	0.24160

Table 4: Results for sub-track 2: *Concept Indexing*.

Team Rank	System Rank	User	Precision	Recall	F1
1	1	FSL	0.91108	0.92083	0.91593
2	2	ixamed	0.87964	0.82882	0.85347
	3		0.87623	0.82810	0.85149
	9		0.82374	0.83666	0.83015
	10		0.81232	0.80884	0.81058
3	4	xiongying	0.82835	0.85021	0.83914
	5		0.83809	0.83809	0.83809
	6		0.82202	0.84665	0.83415
	7		0.82032	0.84665	0.83327
	8		0.81699	0.84379	0.83018
4	11	sohrab	0.87532	0.73609	0.79969
	12		0.88003	0.73252	0.79953
5	13	plubeda	0.85207	0.63267	0.72616
	14		0.82887	0.61840	0.70833
	15		0.87879	0.55849	0.68295
	16		0.83350	0.57846	0.68295
6	17	VSP	0.66502	0.55215	0.60335
7	18	rriveraz	0.50000	0.49287	0.49641
	19		0.48641	0.49786	0.49207

Table 5: Results by category for sub-track 1.

	NORMALIZABLES			NO_NORMALIZABLES			PROTEINAS		
	Precision	Recall	F1	Precision	Recall	F1	Precision	Recall	F1
Min	0.31976	0.17986	0.28618	0.00000	0.00000	0.00000	0.32377	0.12224	0.19981
Mean	0.87217	0.81754	0.83880	0.19844	0.03276	0.04984	0.81654	0.76428	0.78494
Median	0.90977	0.86434	0.87922	0.00000	0.00000	0.00000	0.85626	0.82421	0.83445
Maximum	0.95924	0.94142	0.94253	1.00000	0.40000	0.38095	0.89831	0.89406	0.88709
Std Dev	0.12742	0.12967	0.13065	0.38104	0.06854	0.09247	0.11328	0.15302	0.13668
Best team	raduion	FSL	xiongying	m-stoeckel sohrab xiongying	FSL	FSL	plubeda	xiongying	xiongying

column), then system performance (second column). The top scoring system for sub-track 2 was submitted by *FSL*, with a F-score of 0.91593, showing a significantly better result when compared to the second best submission (more than 6 points) provided by *ixamed*, with a F-score of 0.85347. The third team was *xiongying*, the best participant in the sub-track 1, with a F-score of 0.83914.

Some statistics of the results are shown in Table 6. There was a high variability among the systems, with a difference of 6 point between the best system and the median for sub-track 1, and of 10 points for sub-track 2. The difference between the best system and the mean of all system was still higher. This proved that the task, was quite difficult.

As additional analysis, results by category, including the best teams for category and metric, are shown in Table 5. The performance of the systems was systematically better for the NORMALIZABLES category, 4-9 points better in respect with the PROTEINAS category. Surprisingly, the

Table 6: Statistics by track.

Track	Measure	Precision	Recall	F1
1	Minimum	0.33333	0.19273	0.28163
	Mean	0.84211	0.77916	0.80493
	Median	0.88417	0.82791	0.85248
	Maximum	0.92602	0.91314	0.91052
	Std Dev	0.12071	0.13840	0.12819
2	Minimum	0.48641	0.49287	0.49207
	Mean	0.80152	0.72885	0.75936
	Median	0.82887	0.80884	0.81058
	Maximum	0.91108	0.92083	0.91593
	Std Dev	0.11975	0.14059	0.12057

median for the NO_NORMALIZABLES category was 0, suggesting that at least half of the systems ignored this category.

3.4 Combination of systems

In this section, we present an experiment we performed to combine the systems submitted to the track to see if we could improve the results. We combined the systems using a voting scenario: we accepted as good the annotations that had been predicted by N systems.

The first system accepted all the annotations

predicted by, at least, one of the systems, while the last one accepted only the annotations that were predicted by, at least, N systems. The results of this experiment are shown in Table 7.

As expected, as the value of N increased (the number of required votes was increased), the recall got worse and the precision improved. Based on the maximum value of F-score for sub-track 1 on the train and development sets we selected 20 as the optimum value for combining systems (F-score of 0.98408). We used this value for N on the test set, obtaining an F-score of 0.92355, 1.3 points better than the best system. This score was lower than the best one that could be obtained for the test set (0.92426, with N = 18), but the difference was (in practice) negligible.

The combined systems did not improve the results for sub-track 2. The maximum value of F-score on the train and development sets was obtained combining 6-7 systems (F-score of 0.97352 in the Dev set for N = 6). This scored 0.87073 in the test set, 4.5 points below the best system. This was probably a consequence of amount of systems and the performance gap between the best systems and the others. For the future, we will combine the system using more sophisticated approaches.

4 Discussion and Conclusions

The results of the first chemical and drug named entity recognition track from clinical case reports in Spanish are very encouraging, both in terms of the number of participants, not only from Spanish-speaking countries, as well as in terms of the obtained system results, which are already reaching a level of performance that would make the resulting tools very valuable resources for processing the vast amount of medical data generated worldwide in Spanish.

We had structured the track into two sub-tracks to cover different practical aspects of the resulting systems. The named entity recognition track of chemicals/drugs had the aim of serving as a building block task for future down-stream text mining of more complex information types, including the detection of medication duration, dosage, drug-drug-interactions, therapeutic target relations and drug/chemical induced adverse effects. The concept-indexing sub-track was more concerned with the development of sophisticated semantic retrieval engines and the exploitation of high impact normative terminologies such as SNOMED CT.

Table 7: Combining systems using a voting scheme.

Track	#	Train	Dev	Test
	1	0.77448	0.64485	0.36036
	2	0.89285	0.78679	0.71539
	3	0.94173	0.85545	0.78403
	4	0.95583	0.88711	0.82505
	5	0.96638	0.91222	0.85261
	6	0.97523	0.92725	0.87124
	7	0.98024	0.93859	0.88286
	8	0.98452	0.94902	0.89519
	9	0.98792	0.95772	0.90438
	10	0.98989	0.96386	0.90828
	11	0.99160	0.96906	0.91038
	12	0.99319	0.97280	0.91436
	13	0.99386	0.97431	0.91615
	14	0.99412	0.97760	0.91880
	15	0.99505	0.97808	0.92124
	16	0.99518	0.97856	0.92253
	17	0.99558	0.98162	0.92320
	18	0.99598	0.98368	0.92426
	19	0.99571	0.98362	0.92418
	20	**0.99571**	**0.98408**	**0.92355**
	21	0.99585	0.98244	0.92372
	22	0.99598	0.98182	0.92074
	23	0.99638	0.97740	0.91872
	24	0.99638	0.97569	0.91641
1	25	0.99651	0.96952	0.91202
	26	0.99610	0.96665	0.90815
	27	0.99516	0.96294	0.90392
	28	0.99421	0.95715	0.90152
	29	0.99217	0.95135	0.89164
	30	0.99025	0.94474	0.88598
	31	0.98669	0.93575	0.88197
	32	0.98641	0.92965	0.87602
	33	0.98462	0.92197	0.86712
	34	0.98198	0.91708	0.85794
	35	0.98073	0.91311	0.85002
	36	0.97738	0.90290	0.84011
	37	0.97285	0.89545	0.82827
	38	0.97058	0.88458	0.81402
	39	0.96829	0.87545	0.80040
	40	0.96397	0.86574	0.78042
	41	0.95860	0.85169	0.75992
	42	0.95258	0.83427	0.73265
	43	0.94455	0.81401	0.69613
	44	0.93128	0.79135	0.66063
	45	0.91042	0.76815	0.57034
	46	0.88054	0.72273	0.50117
	47	0.84009	0.66146	0.46420
	48	0.75949	0.57517	0.41148
	49	0.34449	0.26516	0.20968
	1	0.65485	0.63039	0.57822
	2	0.81233	0.80333	0.70237
	3	0.90781	0.91254	0.80749
	4	0.92741	0.93277	0.81967
	5	0.97599	0.96739	0.84716
	6	**0.98009**	**0.97352**	**0.87073**
	7	0.98298	0.97020	0.87106
	8	0.97983	0.97061	0.87719
	9	0.97207	0.94607	0.86444
2	10	0.96992	0.93864	0.86435
	11	0.95120	0.90585	0.85479
	12	0.94721	0.88803	0.84603
	13	0.88231	0.80623	0.80337
	14	0.87566	0.78624	0.79056
	15	0.85368	0.73964	0.75661
	16	0.81724	0.68416	0.72358
	17	0.43734	0.38884	0.37916
	18	0.43231	0.39216	0.36889
	19	0.40541	0.36462	0.33152

Surprisingly we had a considerably higher number of participants for the NER sub-track when compared to the concept-indexing sub-track. Future evaluation efforts should potentially consider also an entity grounding/normalization of chemical and drug mentions in clinical case reports.

Most of the participating systems were based on the use of sophisticated deep learning and neural net approaches, which are becoming the state of the art methods for named entity recognition tasks also in specialized domains such as biomedicine or for non-English data.

When analyzing the more difficult mention types for participating teams, it is still clear that very short abbreviations (1-2 letters) are cumbersome to recognize correctly, due to their high level of implicit ambiguity. Solving such cases would probably require larger manually annotated corpora or the generation of other complementary resources specifically suited for the recognition and resolution of short abbreviations. We did not observe any particular issues related to the clinical disciplines of the case reports, thus it seems that drug NER systems should work well across all medical specialties. It is important to place the very competitive results obtained for Pharma-CoNER into its context, in terms data collections used. When compared to the biomedical literature or medicinal chemistry patents, clinical case reports show a lower degree of variability in terms of the chemicals and drug mentions used, as in the clinic only a limited number of medications and chemical entities are being used for treatment, biochemical testing or explored in clinical settings and analysis.

The construction of high quality Gold Standard manually annotated corpora can be considered one of the major bottlenecks for the development of biomedical named entity recognition systems. During this task, we have promoted the collaborative generation of a larger Silver Standard corpus generated through the predictions of all the participating teams. A more detailed examination of this resource and approaches on how to optimally merge/combine multiple annotations and in turn train new systems using this silver standard dataset might give new insights on how to speed up the creation of new NER tools/annotated datasets.

One of the difficulties we have also encountered during this task was due to the use of a very popular third party platform for organizing online shared tasks on data mining tasks, including text mining and NLP. The explored resource, Codalab, had a server crash, and no proper up to date backup system in place (including user registration info, as well as data collections). Thus, the use of resources with a more focused support for biomedical text mining datasets, corpora, services and shared task organization, such as PubAnnotation would have been a better choice for hosting all the relevant data and predictions for biomedical shared tasks.

Acknowledgments

We acknowledge the Encargo of Plan TL (SEAD) to CNIO and BSC for funding, and the scientific committee for their valuable comments and guidance. We would also like to thank Siamak Barzegar for his help in setting up PharmaCoNER at CodaLab.

References

Jordi Armengol-Estapé, Felipe Soares, Montserrat Marimon, and Martin Krallinger. 2019. Pharmaconer tagger: a deep learning-based tool for automatically finding chemicals and drugs in spanish medical texts. *Genomics Inform*, 17(2):e15–.

Son Doan, Lisa Bastarache, Sergio Klimkowski, Joshua C Denny, and Hua Xu. 2010. Integrating existing natural language processing tools for medication extraction from discharge summaries. *Journal of the American Medical Informatics Association*, 17(5):528–531.

María Herrero-Zazo, Isabel Segura-Bedmar, Paloma Martínez, and Thierry Declerck. 2013. The ddi corpus: An annotated corpus with pharmacological substances and drug–drug interactions. *Journal of biomedical informatics*, 46(5):914–920.

Jin-Dong Kim and Yue Wang. 2012. Pubannotation: A persistent and sharable corpus and annotation repository. In *Proceedings of the 2012 Workshop on Biomedical Natural Language Processing*, BioNLP '12, pages 202–205, Stroudsburg, PA, USA. Association for Computational Linguistics.

Martin Krallinger, Obdulia Rabal, Florian Leitner, Miguel Vazquez, David Salgado, Zhiyong Lu, Robert Leaman, Yanan Lu, Donghong Ji, Daniel M Lowe, et al. 2015. The chemdner corpus of chemicals and drugs and its annotation principles. *Journal of cheminformatics*, 7(1):S2.

Martin Krallinger, Obdulia Rabal, Analia Lourenco, Julen Oyarzabal, and Alfonso Valencia. 2017. Information retrieval and text mining technologies for chemistry. *Chemical reviews*, 117(12):7673–7761.

Jeffrey Pennington, Richard Socher, and Christopher D. Manning. 2014. Glove: Global vectors for word representation. In *Empirical Methods in Natural Language Processing (EMNLP)*, pages 1532–1543.

Martin Pérez-Pérez, Obdulia Rabal, Gael Pérez-Rodríguez, Miguel Vazquez, Florentino Fdez-Riverola, Julen Oyarzabal, Alfonso Valencia, Anália Lourenço, and Martin Krallinger. 2017. Evaluation of chemical and gene/protein entity recognition systems at biocreative v. 5: the cemp and gpro patents tracks. In *BC V.5 - Workshop Proceedings*.

Obdulia Rabal, Ander Intxaurrondo, and Martin Krallinger. 2018. Guías de anotación y normalización de compuestos químicos.

Felipe Soares, Marta Villegas, Aitor Gonzalez-Agirre, Martin Krallinger, and Jordi Armengol-Estapé. 2019. Medical word embeddings for Spanish: Development and evaluation. In *Proceedings of the 2nd Clinical Natural Language Processing Workshop*, pages 124–133, Minneapolis, Minnesota, USA. Association for Computational Linguistics.

Pontus Stenetorp, Sampo Pyysalo, Goran Topić, Tomoko Ohta, Sophia Ananiadou, and Jun'ichi Tsujii. 2012. Brat: A web-based tool for nlp-assisted text annotation. In *Proceedings of the Demonstrations at the 13th Conference of the European Chapter of the Association for Computational Linguistics*, EACL '12, pages 102–107, Stroudsburg, PA, USA. Association for Computational Linguistics.

Özlem Uzuner, Imre Solti, and Eithon Cadag. 2010a. Extracting medication information from clinical text. *Journal of the American Medical Informatics Association*, 17(5):514–518.

Özlem Uzuner, Imre Solti, Fei Xia, and Eithon Cadag. 2010b. Community annotation experiment for ground truth generation for the i2b2 medication challenge. *Journal of the American Medical Informatics Association*, 17(5):519–523.

Miguel Vazquez, Martin Krallinger, Florian Leitner, and Alfonso Valencia. 2011. Text mining for drugs and chemical compounds: methods, tools and applications. *Molecular Informatics*, 30(6-7):506–519.

Hui Yang. 2010. Automatic extraction of medication information from medical discharge summaries. *Journal of the American Medical Informatics Association*, 17(5):545–548.

Pierre Zweigenbaum, Dina Demner-Fushman, Hong Yu, and Kevin B Cohen. 2007. Frontiers of biomedical text mining: current progress. *Briefings in bioinformatics*, 8(5):358–375.

When Specialization Helps: Using Pooled Contextualized Embeddings to Detect Chemical and Biomedical Entities in Spanish

Manuel Stoeckel

Goethe University Frankfurt

Text Technology Lab

`manuel.stoeckel@stud.uni-frankfurt.de`

Wahed Hemati

Goethe University Frankfurt

Text Technology Lab

`hemati@em.uni-frankfurt.de`

Alexander Mehler

Goethe University Frankfurt

Text Technology Lab

`mehler@em.uni-frankfurt.de`

Abstract

The recognition of pharmacological substances, compounds and proteins is an essential preliminary work for the recognition of relations between chemicals and other biomedically relevant units. In this paper, we describe an approach to Task 1 of the PharmaCoNER Challenge, which involves the recognition of mentions of chemicals and drugs in Spanish medical texts. We train a state-of-the-art BiLSTM-CRF sequence tagger with stacked Pooled Contextualized Embeddings, word and sub-word embeddings using the open-source framework FLAIR. We present a new corpus composed of articles and papers from Spanish health science journals, termed the *Spanish Health Corpus*, and use it to train domain-specific embeddings which we incorporate in our model training. We achieve a result of 89.76% F1-score using pre-trained embeddings and are able to improve these results to 90.52% F1-score using specialized embeddings.

1 Introduction

Efficient access to information on chemicals and pharmaceutical units has become increasingly important for researchers in various chemical disciplines. However, manual annotation of these units to create knowledge bases is a laborious process given the ever-increasing number of papers and patents in bio/chemical and pharmaceutical research. Thus, *Natural Language Processing* (NLP) can be employed to detect such entities and their relations from the relevant literature. Previous work has been successful in detecting and classifying chemical substances or in extracting complex relations between chemical substances (Krallinger et al., 2015; Hemati and Mehler, 2019).

While most NLP research is conducted on English datasets, there are a considerable number of non-English biomedically relevant texts written in other languages, e.g. clinical texts. In order to advance the further development of biomedical and pharmaceutical entity recognition facing this linguistic diversity, the PharmaCoNER task challenges participants with *Named Entity Recognition* (NER) for pharmacological substances, compounds and proteins on a Spanish corpus (Gonzalez-Agirre et al., 2019b). The PharmaCoNER task belongs to the *BioNLP Open Shared Tasks 2019* (BioNLP-OST 2019) Workshop and distinguishes two tracks: the first track focuses on NER offset and entity classification, while the second task deals with concept indexing.

In this paper we present an architecture for NER of chemical and pharmacological units in Spanish texts that produces an F-score of up to 90%. Source code and instructions for reproducing these results are available on GitHub[1] and we are offering an interactive web service for testing our models.[2] The article is organized as follows: First, we describe the resources used to train our model and explain our methodical approach. This includes a detailed description of the PharmaCoNER dataset and the kind of preprocessing we performed on the input texts. Afterwards, we give a thorough description of our architecture. Finally, we discuss our results and give our conclusions.

2 Materials and Methods

2.1 Datasets

In this section, we describe the datasets used in our experiments and the architecture of our NER tagger.

---

[1] `www.git.io/JenqE`
[2] `espharmaner.texttechnologylab.org`

Proceedings of the 5th Workshop on BioNLP Open Shared Tasks, pages 11–15

Hong Kong, China, November 4, 2019. ©2019 Association for Computational Linguistics

PharmaCoNER The corpus accompanying the PharmaCoNER task, that is, the *Spanish Clinical Case Corpus* (SPACCC), contains 1 000 manually classified clinical cases and comprises 396 988 token (Gonzalez-Agirre et al., 2019a). The corpus was derived from open access Spanish medical publications and (according to the creators) shows properties of both biomedical and medical literature as well as clinical records.

The SPACCC corpus is given in brat stand-off format[3] as two separate files per document, one containing the plain text, the other containing the annotations with character level offsets on the raw text. We converted the corpus into a CoNLL2003 compatible format, applying common whitespace tokenization and splitting tokens on non-alphanumeric characters, as this increased the performance of our model.

Spanish Health Corpus In this section, we describe the Spanish Health Corpus, a collection of 7353 diverse Spanish health and science journal articles and papers. The corpus was obtained from SciELO[4] by means of an automated crawler.[5] The content of the articles in this corpus was downloaded as embedded text from the respective websites and stripped of any structural elements, like HTML tags. Then, the raw text was split into sentences using DEEP-EOS, a neural network sentence boundary detection tool created by Stefan Schweter which is publicly available on GitHub.[6]

We trained a Spanish DEEP-EOS LSTM model on 100 000 Spanish Wikipedia sentences extracted from the Leipzig Corpora Collection (Goldhahn et al., 2012). Our DEEP-EOS model achieves an accuracy of 99.65% on separate 100 000 test sentences. The resulting sentences were then tokenized based on the procedure mentioned in the previous section. This resulted in a set of 957 648 sentences containing 32 346 137 words in total. We used this corpus to train special word embeddings for our system that we believe have a positive impact on the performance of our models.

2.2 System Architecture

Our system was built with FLAIR (Akbik et al., 2019a), an easy to use open-source NLP framework that is able to produce state-of-the-art results for sequence tagging tasks (eg. Akbik et al., 2018, 2019b). We follow the approach of Akbik et al., using FLAIR to stack (i.e. concatenate) character and word embeddings to improve recognition rates. We further expand this model by adding sub-word embeddings to the stacked embeddings. These stacked embeddings serves as input for a BiLSTM-CRF sequence tagger (Huang et al., 2015).

For our best performing model, we used two different token-level embeddings, a WANG2VEC-based embedding (Ling et al., 2015) and a FAST-TEXT-based embedding (Bojanowski et al., 2017), a single byte-pair sub-word embedding (Heinzerling and Strube, 2018) and one context sensitive character-level language model (Akbik et al., 2019b). Figure 1 gives a visual depiction of our best performing model. The following paragraphs describe the used embeddings in more detail.

Pooled Contextualized Embeddings *Contextualized String Embeddings* (Akbik et al., 2018, CSEs) use pre-trained character-level language models from which hidden states at the start and end character positions of each word are extracted to create embeddings for any string in sentence contexts. This model is further developed by Akbik et al. (2019b) who introduce an expansion to CSEs in terms of *Pooled Contextualized Embeddings* (PCEs).

PCEs tackle the problem of embedding rare words by applying a pooling operation on different contextual embeddings of the word. The authors follow the idea that words which occur in underspecified contexts should be familiar to the reader from previous mentions. So when a word is processed during the training of a character-level language model, all previous contextualized instances of the word are pooled and concatenated with the current instance to create a "global" word representation (Akbik et al., 2019b). The authors experiment with three pooling operations (*min, max* and *mean*). In this way, they are able to achieve state-of-the-art results in four major NER tasks (Akbik et al., 2019b).

In our architecture, we employ pre-trained Spanish Pooled Contextualized Embeddings.[7]

wang2vec Embeddings This model, proposed by Ling et al. (2015), is an extension of the token-

---

[3]`brat.nlplab.org/standoff.html`
[4]`www.scielo.org`
[5]See our GitHub repository for the list of documents.
[6]`www.github.com/stefan-it/deep-eos`

[7]These models were trained by Yihwa Kim (`www.github.com/iamyihwa`).

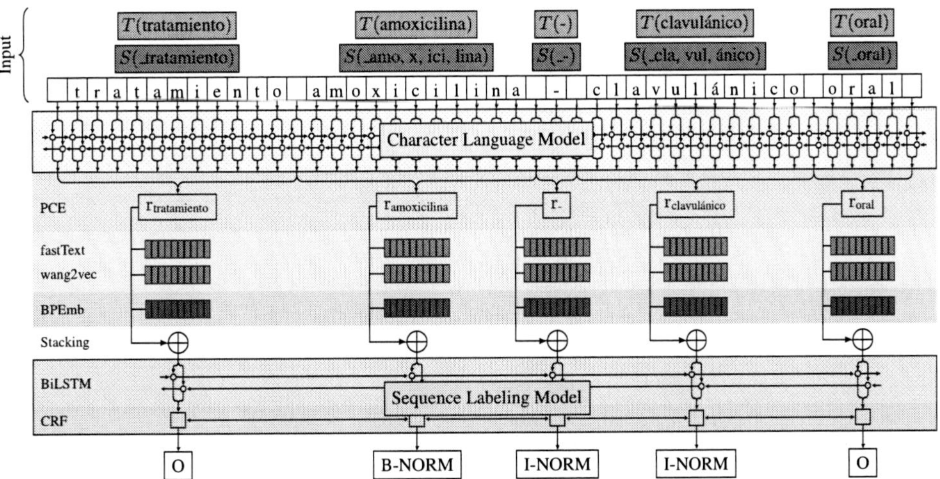

Figure 1: The architecture of the best performing model in our experiments. The PCEs are generated from C character features, while FASTTEXT and WANG2VEC embeddings are trained on T tokens, and BPEMB uses S syllable input. The embeddings are stacked and serve as input for a BiLSTM-CRF Sequence Labeling Model.

level WORD2VEC model of Mikolov et al. (2013). During training, WANG2VEC makes a prediction for each neighboring position of the target word instead of making a single prediction for all neighbours. Thus, the resulting embeddings are better at capturing syntactic, positional information (Ling et al., 2015).

We trained 300 dimensional WANG2VEC-based embeddings based on 100 iterations using default parameters on the Spanish Health Corpus.

fastText Embeddings Unlike WORD2VEC or WANG2VEC, FASTTEXT (Bojanowski et al., 2017) models words as sets of character n-grams, where all n-grams from sizes 3-6 are used during training. FASTTEXT can thus represent rare words that were not present in the vocabulary of the training files if their skip-grams were observed during training. Before the words are split into n-grams, special boundary symbols are added. The embeddings are thus also able to learn information about word prefixes and suffixes (Bojanowski et al., 2017). We used pre-trained 300 dimensional Spanish FASTTEXT embeddings from Grave et al. (2018) in our initial submission to the Pharma-CoNER task.[8]

We replaced them with our own 300 dimensional embeddings trained on the Spanish Health Corpus with standard parameter settings during our experimental phase.

Byte-Pair Embeddings. Similar to FASTTEXT, Byte-Pair embeddings (Heinzerling and Strube, 2018, BPEMB) are trained on a pre-processed corpus that contains sub-word entities. But in contrast to FASTTEXT, words in the training corpus are represented as combinations of *syllables* instead of skip-grams. These syllables or *subword units* are learned from the corpus prior to the segmentation using Byte-Pair-Encoding (Sennrich et al., 2016) for a predefined number.

In our experiments, we used pre-trained 300 dimensional Spanish Byte-Pair embeddings made available by Heinzerling and Strube (2018) with a syllable vocabulary size of $100\,000$.[9]

2.3 Experiments

We conducted extensive experiments to optimize our models. The ease of use of FLAIR enables us to swap embeddings and optimizers on the fly and perform a state-of-the-art hyper-parameter search. Following the "best known configurations" for NER tasks in English, German and Dutch according Akbik et al.'s GitHub repository,[10] we trained the BiLSTM-CRF sequence tagger with a hidden size of 256, a single LSTM layer (unless stated otherwise) and no dropout. We used common *Stochastic Gradient Descent* (SDG) with a learning rate of 0.1, mini-batch size of 32, an annealing rate of 0.5 with a patience of 3 and default parame-

---

[8]www.fasttext.cc/docs/en/ crawl-vectors.html

[9]www.github.com/bheinzerling/bpemb
[10]www.github.com/zalandoresearch/flair

ters otherwise. The training takes about 80 epochs with these settings.

We performed a parameter search with FLAIR's wrapper of the hyper parameter selection tool HY-PEROPT (Bergstra et al., 2013). We chose our initial search parameters similar to the search conducted by Akbik et al. (2019b), which includes a learning rate $\in \{0.01, 0.05, 0.1\}$ and mini-batch size $\in \{8, 16, 32\}$. Using this parameter set we were unable to improve our models performance over the performance using the suggested ones. In addition, we ran a sparse parameter search with a different array of possible choices: hidden size $\in \{256, 512\}$, dropout $\in [0, 0.5]$, number of RNN layers $\in \{1, 2\}$ and learning rate $\in \{0.05, 0.1, 0.15\}$. While all of the trained models performed very well, we were unable to outperform our previous best model.

All experiments were performed either on a NVIDIA GTX 1660 with 6 GiB VRAM available or on a NVIDIA GTX 1080 Ti with 11 GiB VRAM available.

3 Evaluation

Results　Table 1 compares the scores of our systems. All scores were computed using the official evaluation script provided by the organizers of the PharmaCoNER task on the gold standard test data, which was released after the end of the challenge phase. After establishing a baseline using mean-pooled PCEs only, we added pre-trained Byte-Pair embeddings (BPEMB-PRE) and pre-trained FAST-TEXT (FT-PRE) embeddings. While Byte-Pair embeddings alone were able to increase the performance of the model by $+3.61\%$ F1-score, further adding pre-trained FASTTEXT embeddings only increased the systems performance about $+0.11\%$ for a total of $+3.72\%$ against our baseline. This confirms the observations of Akbik et al. (2019b) according to which stacking token-level embeddings on PCEs can improve the performance of the model significantly. Adding a second LSTM layer to the BiLSTM sequence tagger decreased the models F1-score by 0.54% as can be seen in the second entry in row 3 of table 1.

After the challenge phase, we replaced the pre-trained FASTTEXT embeddings with our self-trained, specialized embeddings (FT^S) and added the specialized WANG2VEC ($w2v^S$) embeddings. This increased the performance of the system to 90.34% F1-score. The choice of mean-pooled

Model	F1-Score	Precision	Recall
PCE-PRE (BSE)	86.04	88.59	83.64
BSE + BPEMB-PRE[†] (SBM)	89.65	90.45	88.86
SBM + FT-PRE			
1 LSTM layer[†‡]	89.76	90.69	88.85
2 LSTM layers[†]	89.22	89.10	89.34
SBM + FT^S + $w2v^S$			
min-pooled[*]	90.31	90.02	89.71
max-pooled[*]	90.34	**90.97**	89.71
mean-pooled[*]	**90.52**	90.79	**90.30**

Table 1: All scores in %. BSE denotes our baseline, while SBM denotes our first submission model. The notation "X + Y" is to be read as "X stacked with Y". Legend: [†] indicates challenge submissions, [‡] indicates the best challenge submission, [S] indicates self-trained specialized embeddings, [*] indicates models built after the challenge deadline.

PCEs in favor of min-pooled PCEs resulted in a further increase in performance to 90.52% F1-score, representing a total increase of $+4.48\%$ over our baseline and $+0.76\%$ over our best result during the challenge phase, while choosing max-pooled PCEs results in the highest precision score of all our models (90.97%).

4 Conclusion and Future Work

Our experiments show that with current frameworks like FLAIR it is possible to achieve very good test results with little time spent on system development or implementation. Good results can be achieved with pre-trained models and embeddings that are available in many languages thanks to the NLP community's ongoing efforts.

Our experiments confirm our expectations regarding the usability of special embeddings. The embeddings that are trained on the Spanish Health Corpus contribute to significantly increasing the performance of our system, even with such a small training corpus. Our results show that the use of domain-specific embeddings can significantly improve the performance of sequence tagging models even in the case of small corpora.

We will be continuing our experiments in due time, using larger corpora for our training. In the mean time all our results, datasets and code necessary to reproduce our experiments have been made publicly available on GitHub and can be tested with an interactive web service.

Acknowledgements

We would like to thank the anonymous reviewers for their fair opinions and the organisers for their patience and help with problems during the submission of the workshop papers.

References

Alan Akbik, Tanja Bergmann, Duncan Blythe, Kashif Rasul, Stefan Schweter, and Roland Vollgraf. 2019a. FLAIR: An easy-to-use framework for state-of-the-art NLP. In *Proceedings of the 2019 Conference of the North American Chapter of the Association for Computational Linguistics (Demonstrations)*, pages 54–59, Minneapolis, Minnesota. Association for Computational Linguistics.

Alan Akbik, Tanja Bergmann, and Roland Vollgraf. 2019b. Pooled contextualized embeddings for named entity recognition. In *NAACL 2019, 2019 Annual Conference of the North American Chapter of the Association for Computational Linguistics*, page 724–728.

Alan Akbik, Duncan Blythe, and Roland Vollgraf. 2018. Contextual string embeddings for sequence labeling. In *COLING 2018, 27th International Conference on Computational Linguistics*, pages 1638–1649.

J. Bergstra, D. Yamins, and D. D. Cox. 2013. Making a science of model search: Hyperparameter optimization in hundreds of dimensions for vision architectures. In *Proceedings of the 30th International Conference on International Conference on Machine Learning - Volume 28*, ICML'13, pages I–115–I–123. JMLR.org.

Piotr Bojanowski, Edouard Grave, Armand Joulin, and Tomas Mikolov. 2017. Enriching word vectors with subword information. *Transactions of the Association for Computational Linguistics*, 5:135–146.

Dirk Goldhahn, Thomas Eckart, and Uwe Quasthoff. 2012. Building large monolingual dictionaries at the leipzig corpora collection: From 100 to 200 languages. In *In Proceedings of the Eight International Conference on Language Resources and Evaluation (LREC'12*.

Aitor Gonzalez-Agirre, Ander Intxaurrondo, and Jose Antonio Lopez-Martin. 2019a. Description of the corpus for the PharmaCoNER challenge. http://temu.bsc.es/pharmaconer/index.php/description-of-the-corpus/, [Accessed: 01.08.2019].

Aitor Gonzalez-Agirre, Montserrat Marimon, Ander Intxaurrondo, Obdulia Rabal, Marta Villegas, and Martin Krallinger. 2019b. Pharmaconer: Pharmacological substances, compounds and proteins named entity recognition track. In *Proceedings of the BioNLP Open Shared Tasks (BioNLP-OST)*, pages 1–X, Hong Kong, China. Association for Computational Linguistics.

Edouard Grave, Piotr Bojanowski, Prakhar Gupta, Armand Joulin, and Tomas Mikolov. 2018. Learning word vectors for 157 languages. In *Proceedings of the International Conference on Language Resources and Evaluation (LREC 2018)*.

Benjamin Heinzerling and Michael Strube. 2018. BPEmb: Tokenization-free Pre-trained Subword Embeddings in 275 Languages. In *Proceedings of the Eleventh International Conference on Language Resources and Evaluation (LREC 2018)*, Miyazaki, Japan. European Language Resources Association (ELRA).

Wahed Hemati and Alexander Mehler. 2019. LSTMVoter: chemical named entity recognition using a conglomerate of sequence labeling tools. *Journal of Cheminformatics*, 11(1):7.

Zhiheng Huang, Wei Xu, and Kai Yu. 2015. Bidirectional lstm-crf models for sequence tagging. Cite arxiv:1508.01991.

Martin Krallinger, Florian Leitner, Obdulia Rabal, Miguel Vazquez, Julen Oyarzabal, and Alfonso Valencia. 2015. Chemdner: The drugs and chemical names extraction challenge. *Journal of cheminformatics*, 7(1):S1.

Wang Ling, Chris Dyer, Alan Black, and Isabel Trancoso. 2015. Two/too simple adaptations of word2vec for syntax problems. In *Proceedings of the 2015 Conference of the North American Chapter of the Association for Computational Linguistics: Human Language Technologies*. Association for Computational Linguistics.

Tomas Mikolov, Ilya Sutskever, Kai Chen, Greg Corrado, and Jeffrey Dean. 2013. Distributed representations of words and phrases and their compositionality. In *Neural and Information Processing System (NIPS)*.

Rico Sennrich, Barry Haddow, and Alexandra Birch. 2016. Neural machine translation of rare words with subword units. In *Proceedings of the 54th Annual Meeting of the Association for Computational Linguistics*, pages 1715–1725. Association for Computational Linguistics (ACL).

VSP at PharmaCoNER 2019: Recognition of Pharmacological Substances, Compounds and Proteins with Recurrent Neural Networks in Spanish Clinical Cases

Víctor Suárez-Paniagua
Computer Science Department,
Carlos III University of Madrid.
Leganés 28911, Madrid, Spain.
vspaniag@inf.uc3m.es

Abstract

This paper presents the participation of the VSP team for the PharmaCoNER Tracks from the BioNLP Open Shared Task 2019. The system consists of a neural model for the Named Entity Recognition of drugs, medications and chemical entities in Spanish and the use of the Spanish Edition of SNOMED CT term search engine for the concept normalization of the recognized mentions. The neural network is implemented with two bidirectional Recurrent Neural Networks with LSTM cells that creates a feature vector for each word of the sentences in order to classify the entities. The first layer uses the characters of each word and the resulting vector is aggregated to the second layer together with its word embedding in order to create the feature vector of the word. In addition, a Conditional Random Field layer classifies the vector representation of each word in one of the mention types. The system obtains a performance of 76.29%, and 60.34% in F1 for the classification of the Named Entity Recognition task and the Concept indexing task, respectively. This method presents good results with a basic approach without using pretrained word embeddings or any hand-crafted features.

1 Introduction

Nowadays, the task of finding the essential data about the patients in medical records is very difficult because of the highly increasing amount of unstructured documents generated by the doctors. Thus, the automatic extraction of the mentions related with drugs, medications and chemical entities in the clinical case studies can reduces the time of healthcare professionals expend reviewing these medical documents in order to retrieve the most relevant information.

Previously, some Natural Language Processing (NLP) shared tasks were organized in order to promote the develop of automatic systems given the importance of this task. The i2b2 shared task was the first NLP challenge for identifying Protected Health Information in the clinical narratives (Özlem Uzuner et al., 2007). The CHEMDNER task was focused on the Named Entity Recognition (NER) of chemical compounds and drug names in PubMed abstracts and chemistry journals (Krallinger et al., 2015).

The goal of the BioNLP Open Shared Task 2019 is to create NLP challenges for developing systems in order to extract information from biomedical corpora. Concretely, the PharmaCoNER Task is focusing on the recognition of pharmacological substance, compound and protein mentions from Spanish medical texts.

Currently, deep learning approaches overcome traditional machine learning systems on the majority of NLP tasks, such as text classification (Kim, 2014), language modeling (Mikolov et al., 2013) and machine translation (Cho et al., 2014). Moreover, these models have the advantage of automatically learn the most relevant features without defining rules by hand. Concretely, the LSTM-CRF Model proposed by (Lample et al., 2016) improves the performance of a CRF with hand-crafted features for different biomedical NER tasks (Habibi et al., 2017). The main idea of this system is to create a word vector representation using a bidirectional Recurrent Neural Network with LSTM cells (BiLSTM) with character information encoded in another BiLSTM layer in order to classify the tag of each word in the sentences with a CRF classifier. Following this approach, the system proposed in (Dernoncourt et al., 2016) uses a BiLSTM-

Proceedings of the 5th Workshop on BioNLP Open Shared Tasks, pages 16–20
Hong Kong, China, November 4, 2019. ©2019 Association for Computational Linguistics

CRF Model with character and word levels for the de-identification of patient notes using the i2b2 dataset that overcomes the previous systems in this task.

This paper presents the participation of the author, as VSP team, at the tasks proposed by PharmaCoNER about the classification of pharmacological substances, compounds and proteins and the Concept Indexing of the recognized mentions from clinical cases in Spanish. The proposed system follows the same approaches of (Lample et al., 2016) and (Dernoncourt et al., 2016) for the NER task with some modifications for the Spanish language implemented with NeuroNER tool (Dernoncourt et al., 2017) because the architecture obtains good performance for the recognition of biomedical entities. In addition, a simple SNOMED CT term search engine is implemented for the concept normalization.

2 Dataset

The corpus of the PharmaCoNER task contains 1,000 clinical cases derived from the Spanish Clinical Case Corpus (SPACCC)[1] with manually annotated mentions such as pharmacological substances, compounds and proteins by clinical documentalists. The documents are randomly divided into the training, validation and test sets for creating, developing and ranking the different systems, respectively.

The corpus contains four different entity types:

- *NORMALIZABLES*: they are chemicals that can be normalized to a unique concept identifier.

- *NO_NORMALIZABLES*: they are chemicals that cannot be normalized. These mentions were used for training the system, but they were not taken into consideration for the results in the task of NER or Concept Indexing.

- *PROTEINAS*: this entity type refers to mentions of proteins and genes following the annotation schema of BioCreative GPRO (Pérez-Pérez et al., 2017).

- *UNCLEAR*: these mentions are cases of general substances, such as pharmaceutical formulations, general treatments, chemotherapy programs, vaccines and a predefined set of general substances.

Additionally, all mentions without the *NO_NORMALIZABLES* tag are annotated with its corresponding SNOMED CT normalization concept.

3 Method

This section presents the Neural architecture for the classification of the entity types and the concept normalization method in Spanish clinical cases. Figure 1 presents the process of the NER task using two BiLSTMs for the character and token levels in order to create each word representation until its classification by a CRF.

3.1 Data preprocessing

The first step is a preprocessing of the sentences in the corpus, which prepares the inputs for the neural model. Firstly, the clinical cases are separated into sentences using a sentence splitter and the words of these sentences are extracted by a tokenizer, both were adapted for the Spanish language. For the experiments, the previous processes were performed by the spaCy tool in Python (Explosion AI, 2017). Once the sentences were divided into word, the BIOES tag schema encodes each token with an entity type (B tag is the beginning token, I tag is the inside token, E tag is the ending token, S tag is the single token and O tag is the outside token). In many previous NER tasks, using this codification is better than the BIO tag scheme (Ratinov and Roth, 2009), but the number of labels increases because there are two additional tags for each class. Thus, the number of possible classes are the 4 tags times the 4 entity types and the O tag for the PharmaCoNER corpus.

3.2 BiLSTM layers

RNNs are very effective in feature learning when the inputs are sequences. Concretely, the Long Short-Term Memory cell (LSTM) (Hochreiter and Schmidhuber, 1997) defines four gates for creating the representation of

---

[1]`https://doi.org/10.5281/zenodo.2560316`

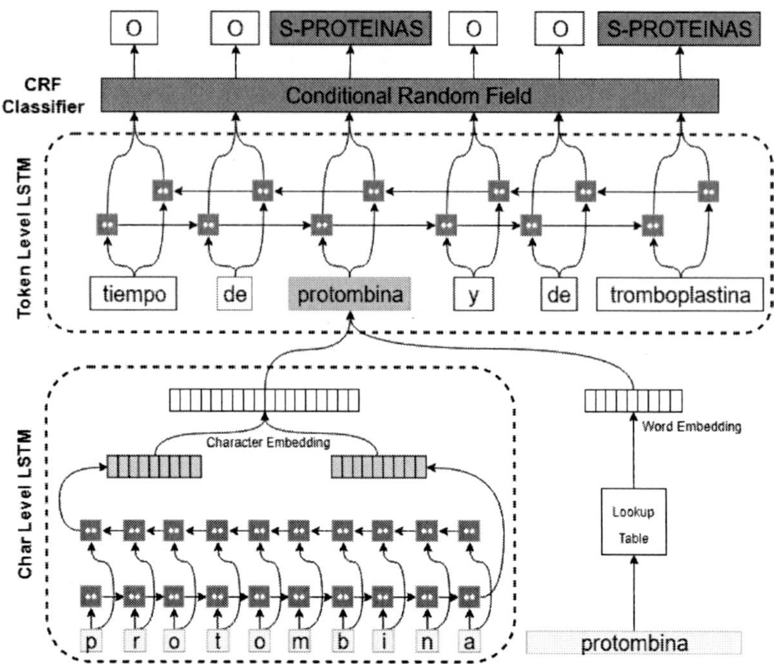

Figure 1: Neural model for the recognition of mentions in Spanish clinical cases using the PharmaCoNER task 2019 corpus.

each input taking the information of the current and previous cells. Thus, each output is a combination of the current and the previous cell states. Furthermore, another LSTM can be applied in the other direction from the end of the sequence to the start in order to extract the relevant features of each input in both directions.

3.2.1 Character level

The first layer takes each word of the sentences individually. These tokens are decomposed into characters that are the input of the BiLSTM. Once all the inputs are computed by the network, the last output vectors of both directions are concatenated in order to create the vector representation of the word according to its characters.

3.2.2 Token level

The second layer takes the embedding of each word in the sentence and concatenates them with the outputs of the first BiLSTM with the character representation. In addition, a Dropout layer is applied to the word representation in order to prevent overfitting in the training phase. In this case, the outputs of

each direction in one token are concatenated for the classification layer.

3.3 Contional Random Field Classifier

CRF (Lafferty et al., 2001) is the sequential version of the Softmax that aggregates the label predicted in the previous output as part of the input. In NER tasks, CRF shows better results than Softmax because it adds a higher probability to the correct labelled sequence. For instance, the I tag cannot be before a B tag or after a E tag by definition. For the proposed system, the CRF classifies the output vector of the BiLSTM layer with the token information in one of the classes.

3.4 Concept Indexing

After the NER task, the concept indexing is applied to all recognized entities in the sentences for the term normalization. To this end, the Spanish Edition of the SNOMED CT International Browser[2] searches each mention and gives its normalization term. Moreover, The Spanish Medical Abbreviation DataBase

---

[2]https://prod-browser-exten.ihtsdotools.org/

(AbreMES-DB)[3] is used in order to disambiguate the acronyms and the resulting term is searched in the SNOMED CT International Browser. In the cases where there are more than one normalization concept for a term, a very naive approach is followed where the first node in the term list is chosen as the final output.

4 Results and Discussion

The architecture was trained over the training set during 100 epochs with shuffled mini-batches and choosing the best performance over the validation set via stopping criteria. The values of the two BiLSTM and CRF parameters for generating the prediction of the test set are presented in Table 1. Additionally, a gradient clipping keeps the weight of the network in a low range preventing the exploding gradient problem. The embeddings of the characters and words are randomly initialized and learned during the training of the network. The main goal of this work is to test the performance of the proposed neural model on this dataset without using pretrained word embeddings or any hand-crafted features. In future work, the impact of different pretrained word embeddings will be covered.

Table 1: The parameters of the neural model and their values used for the PharmaCoNER results.

Parameter	Value
Character embeddings dimension	25
Character-level LSTM hidden units	25
Word embeddings dimension	300
Word-level LSTM hidden units	256
Optimizer	SGD
Learning rate	0.001
Dropout rate	0.5
Gradient clipping	5

The results were measured with precision (P), recall (R) and F-measure (F1) using the True Positives (TP), False Positives (FP) and False Negatives (FN) for its calculation. Table 2 presents the results of the system over the test set of the PharmaCoNER tasks. The performance for the entity type classification and the performance for the Concept Indexing task are 76.29% and 60.34% in F1, respectively.

[3]https://zenodo.org/record/2207130#.XHPEFYUo85k

Table 2: Official results of the neural Model for the two tasks of the PharmaCoNER.

Task	R	P	F1
NER	71.61%	81.62%	76.29%
Concept Indexing	55.22%	66.5%	60.34%

Table 3 presents the results of the NER task for each entity type independently. It can be observed that the number of FN is higher than FP in all the classes giving better results in Precision than in Recall. The performance of the classes are directly proportional of the number of instances in the training set. In order to alleviate this problem, the use of oversampling techniques will be tackled in future works to increase the number of examples of the less representative classes and making this dataset more balanced.

5 Conclusions and Future work

This paper presents a model where a neural model classifies mentions from clinical texts in Spanish and the Concept Indexing uses the SNOMED CT search engine for their normalization. The neural architecture is based on RNNs in both direction of the sentences using LSTM for the computation of the outputs. Finally, a CRF classifier performs the classification for tagging the entity types. The results shows a performance of 76.29% in F1 for the classification of the pharmacological substances, compounds and proteins in the PharmaCoNER corpus and the normalization system reaches to 60.34% in F1. In spite of the basic approaches, the results are very promising in both tasks. As future work, it is proposed to pretrain the word embeddings with collections of biomedical documents and the aggregation of other embeddings such as Part-of-Speech tags, syntactic parse trees or semantic tags, that could increase the representation of each word in order to improve its classification. Moreover, fine-tuning the parameters of the model according to the PharmaCoNER corpus will be useful in order to increase the performance of the method. Furthermore, adding more layers to each BiLSTM is proposed to be included in the architecture. In addition, other complex concept indexing rules could be applied to chose the best nor-

Table 3: Performance of the neural model for each category in the Named Entity Recognition Task of the PharmaCoNER.

Label	TP	FN	FP	R	P	F1
NORMALIZABLES	707	266	94	72.66%	88.26%	79.71%
PROTEINAS	612	247	203	71.25%	75.09%	73.12%
UNCLEAR	20	14	6	58.82%	76.92%	66.67%

malization term in the cases that they are multiple possibilities.

References

Kyunghyun Cho, Bart van Merrienboer, Caglar Gulcehre, Dzmitry Bahdanau, Fethi Bougares, Holger Schwenk, and Yoshua Bengio. 2014. Learning phrase representations using RNN encoder–decoder for statistical machine translation. In *Proceedings of the 2014 Conference on Empirical Methods in Natural Language Processing (EMNLP)*, pages 1724–1734, Doha, Qatar. Association for Computational Linguistics.

Franck Dernoncourt, Ji Young Lee, and Peter Szolovits. 2017. NeuroNER: an easy-to-use program for named-entity recognition based on neural networks. In *Proceedings of the 2017 Conference on Empirical Methods in Natural Language Processing: System Demonstrations*, pages 97–102, Copenhagen, Denmark. Association for Computational Linguistics.

Franck Dernoncourt, Ji Young Lee, Ozlem Uzuner, and Peter Szolovits. 2016. De-identification of patient notes with recurrent neural networks. *Journal of the American Medical Informatics Association : JAMIA*, 24.

Explosion AI. 2017. spaCy - Industrial-strength Natural Language Processing in Python.

Maryam Habibi, Leon Weber, Mariana Neves, David Luis Wiegandt, and Ulf Leser. 2017. Deep learning with word embeddings improves biomedical named entity recognition. *Bioinformatics*, 33(14):i37–i48.

Sepp Hochreiter and Jürgen Schmidhuber. 1997. Long short-term memory. *Neural computation*, 9(8):1735–1780.

Y. Kim. 2014. Convolutional neural networks for sentence classification. In *Proceedings of the 2014 Conference on Empirical Methods in Natural Language Processing (EMNLP)*, pages 1746–1751.

Martin Krallinger, Florian Leitner, Obdulia Rabal, Miguel Vazquez, Julen Oyarzabal, and Alfonso Valencia. 2015. Chemdner: The drugs and chemical names extraction challenge. *J. Cheminformatics*, 7(S-1):S1.

John D. Lafferty, Andrew McCallum, and Fernando C. N. Pereira. 2001. Conditional random fields: Probabilistic models for segmenting and labeling sequence data. pages 282–289.

Guillaume Lample, Miguel Ballesteros, Sandeep Subramanian, Kazuya Kawakami, and Chris Dyer. 2016. Neural architectures for named entity recognition. In *Proceedings of the 2016 Conference of the North American Chapter of the Association for Computational Linguistics: Human Language Technologies*, pages 260–270, San Diego, California. Association for Computational Linguistics.

Tomas Mikolov, Ilya Sutskever, Kai Chen, Greg S Corrado, and Jeff Dean. 2013. Distributed representations of words and phrases and their compositionality. In *Advances in neural information processing systems*, pages 3111–3119.

Martin Pérez-Pérez, Obdulia Rabal, Gael Pérez-Rodríguez1, Miguel Vazquez, Florentino Fdez-Riverola, Julen Oyarzabal, Alfonso Valencia, Anália Lourenço, and Martin Krallinger. 2017. Evaluation of chemical and gene/protein entity recognition systems at biocreative v.5: the cemp and gpro patents tracks. In *Proceedings of the BioCreative V.5 Challenge Evaluation Workshop*, page 11–18.

Lev Ratinov and Dan Roth. 2009. Design challenges and misconceptions in named entity recognition. In *Proceedings of the Thirteenth Conference on Computational Natural Language Learning (CoNLL-2009)*, pages 147–155. Association for Computational Linguistics.

Özlem Uzuner, Yuan Luo, and Peter Szolovits. 2007. Evaluating the state-of-the-art in automatic de-identification. *Journal of the American Medical Informatics Association*, 14(5):550 – 563.

IxaMed at PharmacoNER Challenge 2019

Xabier Lahuerta and **Iakes Goenaga** and **Koldo Gojenola** and
Aitziber Atutxa and **Maite Oronoz**
IXA NLP Group
UPV/EHU University of the Basque Country
Donostia, Basque Country
xlahuerta001@ikasle.ehu.eus, iakes.goenaga,
koldo.gojenola, aitziber.atucha, maite.oronoz{*@ehu.eus*}

Abstract

The aim of this paper is to present our approach (IxaMed) on the PharmacoNER 2019 task. The task consists of identifying chemical, drug, and gene/protein mentions from clinical case studies written in Spanish. The evaluation of the task is divided in two scenarios: one corresponding to the detection of named entities and one corresponding to the indexation of named entities that have been previously identified. In order to identify named entities we have made use of a Bi-LSTM with a CRF on top in combination with different types of word embeddings. We have achieved our best result (86.81 F-Score) combining pretrained word embeddings of Wikipedia and Electronic Health Records (50M words) with contextual string embeddings of Wikipedia and Electronic Health Records. On the other hand, for the indexation of the named entities we have used the Levenshtein distance obtaining a 85.34 F-Score as our best result.

1 Introduction

The aim of this paper is to present our approach in the PharmacoNER 2019 task (Gonzalez et al., 2019), on Medical Entity Recognition and Concept Indexing. The task consists of identifying different types of entities in the clinical domain in Spanish. The evaluation of the task is divided in two scenarios: the detection of medical entities, and the linking of each entity with its corresponding Concept Unique Identifier, a task called Concept Indexing.

The training corpus contains a manually classified collection of clinical cases derived from Open access Spanish medical publications (SPACCC) (Intxaurrondo, 2018). It contains a total of 1,000 clinical cases (396,988 words). This kind of narrative shows properties of both the biomedical and medical literature as well as clinical records.

In order to carry out the tasks, for Named Entity Recognition we have made use of a Recurrent Neural Network (RNN) to identify named entities feeding it with different types of embeddings, combining pretrained word embeddings and contextualized character-level word embeddings or contextual string embeddings. Furthermore, for Concept Indexing task we have opted to use a simple but effective Levenshtein distance method. We have achieved a F-score of 86.81 identifying named entities and 85.34 in Concept Indexing.

2 Related work

The SemEval 2014 Task 7 (Pradhan et al., 2014) was similar to the present competition, except for the number and types of entities to be identified (diseases and others) and the fact that discontinuous entities were also included. Task 7 in SemEval 2014 also comprised two subtasks, medical entity recognition and concept indexation. To tackle the first subtask, different teams used approaches as MaxEnt, SVM or CRF in combination with the extraction of syntactic and semantic attributes. The authors in (Tang et al., 2014) obtained the best results in strict F-Score with 78.5 on the development set and 81.3 on the test set. Their results were 4.7 points higher than those of the second ranked team (Kaewphan et al., 2014).

For the second subtask, namely Concept Indexation, the solutions proposed were very similar among the different teams. As in the NER task, the winner was (Tang et al., 2014) with an accuracy of 74.1 on the test set. Their solution was based on the cosine similarity using Vector Space Model (VSM). The team in (Ghiasvand and Kate, 2014) assigned the Concept Unique Identifier (CUI) code by comparing candidate strings with the terms obtained from the training set and the contents in the Unified Medical Language System (UMLS). They also proposed a method based on edit distance, more precisely Levenshtein distance (Levenshtein, 1966). The second best team (Kaewphan et al., 2014) employed word embeddings, word2vec (Mikolov et al., 2013), for word representation and the cosine similarity to find the closest standard term in UMLS. As a novelty, they implemented a binary classification based on Support Vector Machines (SVMs).

In SemEval 2015 (task 14), the evaluation was the only difference compared to SemEval 2014 (task 7). Besides strict evaluation (correct CUI and complete entity identification), relaxed evaluation was also pursued (successful CUI assignment and partly successful entity identification). In this case, the winning team was (Pathak et al., 2015), which obtained in the strict evaluation an F-score of 75.7, and in the relaxed one an F-score of 78. The methods used were similar to those used in SemEval-2014. In this case, a CRF was used to detect entities and a SVM classifier to determine if these were joined or not (and thus catch discontinuous entities). Regarding Concept Indexing, they used basically customized look-ups, like Dictionary look-up (exact match of entity word permutations, LVG), Customized Dictionary look-up (split UMLS entities by function words), and Customized Dictionary look-up (list of possible UMLS spans and application of Levenshtein distance). The second highest ranked team (Leal et al., 2015) obtained, for strict evaluation, an F-score of 74 and in the relaxed one 76.5. They employed a CRF to identify entities (also discontinuous entities), and for Concept Indexing they applied exact match on the terminology content of the Systematized Nomenclature of Medicine - Clinical Terms (SNOMED-CT) enriching it with an abbreviation dictionary built on the training set. They also implemented a comparison method exploiting SNOMED-CT tree structure, Lucene index and Levenshtein average after splitting each recognized entity and each SNOMED-CT candidate.

Besides these competitions in recent years, improvements have been made mostly in the entity recognition subtask using neural networks such as Bi-LSTM + CRFs (Lample et al., 2016). (Casillas et al., 2019) used the tool for the detection of entities in clinical texts in Spanish, obtaining improvements with respect to previous works (Perez et al., 2017), from an F1-Score of 70.30

to 72.01. Employing a similar system (Goenaga et al., 2018) obtained the first position at the last IberEval shared task (Hermenegildo Fabregat and Araujo, 2018).

3 Resources

Apart from the tools we will present in the following sections, we made use of external data with the intention of completing the information the system extracts from the corpus provided by the organization. For this purpose we employed word-embeddings (Mikolov et al., 2013) that we have calculated (window length = 1, dimensions = 300, algorithm = SkipNgram) from Electronic Health Records (50M words), together with pretrained word-embeddings (window=5, dimensions=300, algorithm= Skip-gram) that have been calculated with Wikipedia2Vec (Yamada et al., 2018).

On the other hand, we have also used contextual string embeddings (Akbik et al., 2018) we have calculated from Electronic Health Records (number of layers=1, hidden size=2,048, sequence length=250, mini batch size=32) and Wikipedia (number of layers=1, hidden size=1,024, sequence length=250, mini batch size=100).

4 Methods

In this section we will explore the different methods we have used to perform the two sub-tasks of the shared task.

4.1 Track 1: NER Offset and Entity Classification

In this section we present our approach in order to extract named entities in track 1 of the shared task. For this purpose we employed a neural network based architecture, more precisely an specific Bi-LSTM (a RNN subclass, (Hochreiter and Schmidhuber, 1997)) with a CRF on top of it (Lample et al., 2016; Ma and Hovy, 2016) using as input raw text and the word-embeddings we have mentioned in section 3. This kind of neural network is widely used to pursue sequence to sequence tagging (Ma and Hovy, 2016; Jagannatha and Yu, 2016). One of the advantages of using Bi-LSTM in contrast to other machine learning techniques such as SVM, Perceptron or CRFs is that the size of the context is automatically learned by the LSTM and there is no need to perform any complicated text preprocessing to obtain features to feed the tool.

One of the strengths of our approach is that it combines different types of embeddings based on different types of corpus. On one hand, we use embeddings that have been calculated on a general domain corpus (Wikipedia) and embeddings that have been calculated on a medical domain corpus (EHRs). On the other hand, we stack pre-trained word embeddings, character-level embeddings and contextual string embeddings and we feed the neural network with them. While the pre-trained word embeddings and character-level embeddings are well known by the scientific community, the contextual string embeddings have been introduced recently (Akbik et al., 2018). This type of embeddings is based on recent advances in neural Language Modeling (LM) that have allowed a language to be modeled as distributions over sequences of characters instead of words (Sutskever et al., 2014), (Graves, 2013), (Kim et al., 2016).

Recent work has shown that by learning to predict the next character on the basis of previous characters, such models learn internal representations that capture syntactic and semantic properties: even though trained without an explicit notion of word and sentence boundaries, they have been shown to generate grammatically correct text, including words, subclauses, quotes and sentences (Sutskever et al., 2014), (Graves, 2013), (Karpathy et al., 2015).

The main features of these contextual string embeddings or contextualized character-level word embeddings are the following:

- They can be pre-trained on large unlabeled corpora.

- They are able to capture the meaning of the words in context and are able to produce different embeddings for polysemous words depending on their usage.

- They model words and contexts as sequences of characters, to both better handle rare and misspelled words as well as model subword structures such as prefixes and endings.

Lastly, we have sent two runs for named entity recognition (track 1): one run with the setup mentioned above, Bi-LSTM + CRF stacking pre-trained and contextual embeddings, and one run with the same setup and using the development corpus for training for a few epochs (fine-tuning) as a last step.

4.2 Track 2: Concept Indexing

The normalization of given named entities consists in linking named entities to concepts in standardized medical terminologies, allowing generalization across contexts. The task consists in assigning, to each term, its corresponding Concept Unique Index. For example, "corticoide", "corticoides" and "cortecostiroides" are all normalized to the same Concept (B-255877006). In our work, we made use of a Text Similarity based mapping from the given terms to different sets:

- The terms present in the training set. This set is limited but gives an account of standard and non-standard terms present in spontaneously written health records.

- SNOMED-CT terms that can be considered a standard terminology.

We tried approximate searching to guarantee a matching, by a string-based similarity measure, as the well-known Levenshtein distance, a standard soft-matching approach in text normalization. We computed the Levenshtein distance between the input string and the set of terms that served as reference. Edit distance is used to quantify similarity between two strings, counting the minimum number of operations required to transform one string into another. The most common metric is the Levenshtein Distance (Levenshtein, 1966) in where the basic edit operations are removal, insertion and substitution of a single character. This metric finds the minimum distance for each spontaneous diagnostic term (SpoDT) with respect to all standard Diagnostic Terms (DictDT), obtaining the best candidate match (see equation 1).

$$minLev(SpoDT, DictDTs) \qquad (1)$$

Hence, strings were searched in the reference-set and ranked according to this distance.

Exact matching of spontaneous expressions in standard dictionaries is not a good option, because it obtains a low accuracy. By contrast, matching with respect to previously classified non-standard expressions is well-worthy. However, the results show a considerable boost when using as reference the set of spontaneous terms and the standard reference (SNOMED CT).

We also tried a different approach using a sequence-to-sequence approach that, although it

NER	
Basic	**Fine-tuned**
86.60	86.81

Table 1: The results we have obtained for NER task. Basic = Combination of word embeddings and contextual string embeddings as an input of a Bi-LSTM with a CRF on top. Fine-tuned = Basic setup + fine-tuning on development set.

Concept Indexing	
Levenshtein Dist. 1	**Levenshtein Dist. 2**
85.14	85.34

Table 2: Results for Concept Indexing task. Levenshtein Distance 1 = Levenshtein distance applied to the entities extracted by the basic setup. Levenshtein Distance 2 = Levenshtein distance applied to the entities extracted by the fine-tuned setup.

gave promising results (an F-Score around 65% for concept Indexing), it was around 20 absolute points below the simplest option of using the Levenshtein distance. We think that this could be interesting to examine the strengths and weaknesses of each approach, and try to combine their positive aspects in a single combined or ensemble system, but we leave it as future work.

5   Results

In this section we present the results we have achieved for both tracks, NER and Concept indexing respectively. For this purpose we have compiled all the results in tables 1 and 2. If we observe the results obtained for both tracks we see a logical correlation between F-Score obtained for NER and the F-Score obtained for Concept Indexing. In other words, the better is the result for NER the better is the result for Concept Indexing. This is due to the fact that we use the output of the NER system as input of the Concept Indexing system.

Furthermore, if we analyze the results for each track we can observe we surpass the F-Score of 85.00 in all cases, thus confirming the robustness of our approaches. For NER, applying a Bi-LSTM with a CRF on top and feeding this neural network with stacked pretrained and contextual embeddings we have achieved a F-Score of 86.60. In contrast, fine-tuning on development set the previously mentioned neural network we outperform this result by 0.21. Although the improvement is not significant we have met our goal, that is to say, we have outperformed the basic setup avoiding overfitting.

Moreover, we have applied Levenshtein distance in order to assign a concept index to named entities that have been identified by NER system. We have achieved a 85.14 of F-Score when the input for the Concept Indexing system are named entities extracted by the basic NER system and a 85.34 of F-Score when the input are the named entities extracted by the fine-tuned NER system.

6   Conclusions

The purpose of this work was to evaluate the feasibility of different approaches to medical entity detection and concept indexing. Entity detection was dealt with a sequential tagger that uses word embeddings and contextual string embeddings acquired from electronic health records and Wikipedia. Concept normalization was approached by Text Similarity techniques. Surprisingly, the Levenshtein-based system obtained relatively good results, and this aspect deserves a further study of the strengths and weaknesses of each approach.

Acknowledgements

This work has been partially funded by:

- The Spanish ministry (projects PROSA-MED: TIN2016-77820-C3-1-R, DOMINO: PGC2018-102041-B-I00, both from MCIU/AEI/FEDER, UE).

- The Basque Government (projects DE-TEAMI: 2014111003).

We gratefully acknowledge the support of NVIDIA Corporation with the donation of the Titan X Pascal GPU used for this research.

References

Alan Akbik, Duncan Blythe, and Roland Vollgraf. 2018. Contextual string embeddings for sequence labeling. In *Proceedings of the 27th International Conference on Computational Linguistics*, pages 1638–1649.

Arantza Casillas, Nerea Ezeiza, Iakes Goenaga, Alicia Perez, and Xabier Soto. 2019. Measuring the Effect of Different Types of Unsupervised Word Representations on Medical Named Entity Recognition. *International Journal of Medical Informatics (https://doi.org/10.1016/j.ijmedinf.2019.05.022)*.

Omid Ghiasvand and Rohit J Kate. 2014. UWM: Disorder Mention Extraction from Clinical Text Using CRFs and Normalization Using Learned Edit Distance Patterns. In *SemEval@ COLING*, pages 828–832.

Iakes Goenaga, Aitziber Atutxa, Koldo Gojenola, Arantza Casillas, Arantza Daz de Ilarraza, Nerea Ezeiza, Maite Oronoz, Alicia Perez, and Olatz Perez de Viñaspre. 2018. A Hybrid Approach For Automatic Disability Annotation. In *Proceedings of the Third Workshop on Evaluation of Human Language Technologies for Iberian Languages (IberEval 2018). ISSN 1613-073. Vol-2150. Pages 31-36.*

Aitor Gonzalez, Montserrat Marimon, Ander Intxaurrondo, Obdulia Rabal, Marta Villegas, and Martin Krallinger. 2019. PharmaCoNER: Pharmacological Substances, Compounds and proteins Named Entity Recognition track. In *Proceedings of the BioNLP Open Shared Tasks (BioNLP-OST)*, pages 1–X, Hong Kong, China. Association for Computational Linguistics.

Alex Graves. 2013. Generating sequences with recurrent neural networks. *arXiv preprint arXiv:1308.0850.*

Juan Martinez-Romo Hermenegildo Fabregat and Lourdes Araujo. 2018. Overview of the DIANN Task: Disability Annotation Task. In *Proceedings of the Third Workshop on Evaluation of Human Language Technologies for Iberian Languages (IberEval 2018).*

Sepp Hochreiter and Jürgen Schmidhuber. 1997. Long short-term memory. *Neural Comput.*, 9(8):1735–1780.

Ander Intxaurrondo. 2018. SPACCC (Version 2019-02-01) [Data set]. Zenodo.

Abhyuday N Jagannatha and Hong Yu. 2016. Structured prediction models for RNN based sequence labeling in clinical text. In *Proceedings of the Conference on Empirical Methods in Natural Language Processing. Conference on Empirical Methods in Natural Language Processing*, volume 2016, page 856.

Suwisa Kaewphan, Kai Hakala, and Filip Ginter. 2014. UTU: Disease Mention Recognition and Normalization with CRFs and Vector Space Representations. In *SemEval@COLING*, pages 807–811.

Andrej Karpathy, Justin Johnson, and Li Fei-Fei. 2015. Visualizing and understanding recurrent networks. *arXiv preprint arXiv:1506.02078.*

Yoon Kim, Yacine Jernite, David Sontag, and Alexander M Rush. 2016. Character-aware neural language models. In *Thirtieth AAAI Conference on Artificial Intelligence.*

Guillaume Lample, Miguel Ballesteros, Sandeep Subramanian, Kazuya Kawakami, and Chris Dyer. 2016. Neural architectures for named entity recognition. *arXiv preprint arXiv:1603.01360.*

André Leal, Bruno Martins, and Francisco Couto. 2015. ULisboa: Recognition and normalization of medical concepts. In *Proceedings of the 9th International Workshop on Semantic Evaluation (SemEval 2015)*, pages 406–411. Association for Computational Linguistics.

Vladimir I Levenshtein. 1966. Binary codes capable of correcting deletions, insertions, and reversals. In *Soviet physics doklady*, volume 10, pages 707–710.

Xuezhe Ma and Eduard H. Hovy. 2016. End-to-end Sequence Labeling via Bi-directional LSTM-CNNs-CRF. In *ACL (1)*. The Association for Computer Linguistics.

Tomas Mikolov, Ilya Sutskever, Kai Chen, Greg S Corrado, and Jeff Dean. 2013. Distributed representations of words and phrases and their compositionality. In *Advances in neural information processing systems*, pages 3111–3119.

Parth Pathak, Pinal Patel, Vishal Panchal, Sagar Soni, Kinjal Dani, Amrish Patel, and Narayan Choudhary. 2015. ezDI: A Supervised NLP System for Clinical Narrative Analysis. In *Proceedings of the 9th International Workshop on Semantic Evaluation (SemEval 2015)*, pages 412–416. Association for Computational Linguistics.

Alicia Perez, Rebecka Weegar, Arantza Casillas, Koldo Gojenola, Maite Oronoz, and Hercules Dalianis. 2017. Semi-supervised medical entity recognition: A study on Spanish and Swedish clinical corpora. *Journal of biomedical informatics*, 71:16–30.

Sameer Pradhan, Noémie Elhadad, Wendy W Chapman, Suresh Manandhar, and Guergana Savova. 2014. Semeval-2014 task 7: Analysis of clinical text. In *SemEval@ COLING*, pages 54–62.

Ilya Sutskever, Oriol Vinyals, and Quoc V Le. 2014. Sequence to sequence learning with neural networks. In *Advances in neural information processing systems*, pages 3104–3112.

Yaoyun Zhang1 Jingqi Wang1 Buzhou Tang, Yonghui Wu1 Min Jiang, and Yukun Chen3 Hua Xu. 2014. UTH_CCB: a report for semeval 2014–task 7 analysis of clinical text. *SemEval 2014*, page 802.

Ikuya Yamada, Akari Asai, Hiroyuki Shindo, Hideaki Takeda, and Yoshiyasu Takefuji. 2018. Wikipedia2Vec: An Optimized Tool for Learning Embeddings of Words and Entities from Wikipedia. *arXiv preprint 1812.06280.*

NLNDE: Enhancing Neural Sequence Taggers with Attention and Noisy Channel for Robust Pharmacological Entity Detection

Lukas Lange[1,2,3]   **Heike Adel**[1]   **Jannik Strötgen**[1]

[1] Bosch Center for Artificial Intelligence, Renningen, Germany
[2] Spoken Language Systems (LSV), Saarland University, Saarbrücken, Germany
[3] Saarbrücken Graduate School of Computer Science, Saarbrücken, Germany
{Lukas.Lange,Heike.Adel,Jannik.Stroetgen}@de.bosch.com

Abstract

Named entity recognition has been extensively studied on English news texts. However, the transfer to other domains and languages is still a challenging problem. In this paper, we describe the system with which we participated in the first subtrack of the PharmaCoNER competition of the BioNLP Open Shared Tasks 2019. Aiming at pharmacological entity detection in Spanish texts, the task provides a non-standard domain and language setting. However, we propose an architecture that requires neither language nor domain expertise. We treat the task as a sequence labeling task and experiment with attention-based embedding selection and the training on automatically annotated data to further improve our system's performance. Our system achieves promising results, especially by combining the different techniques, and reaches up to 88.6% F1 in the competition.

1 Introduction

The detection and classification of pharmacological and biomedical entities in texts is especially challenging due to the domain's nature with long and complex entity names, which usually requires the design and usage of handcrafted rules and features. Natural language processing (NLP) research focused on this topic for quite a while on English texts, e.g., the drugs and chemical names extraction challenge (CHEMDNER) (Krallinger et al., 2015) or tracks for chemical entity recognition at BioCreative (Pérez-Pérez et al., 2017). Following these tasks, the Pharmacological Substances, Compounds and Proteins and Named Entity Recognition track (PharmaCoNER) is the first competition on this topic on Spanish data (Gonzalez-Agirre et al., 2019).

Named entity recognition (NER) and classification is the first subtrack of Pharma-

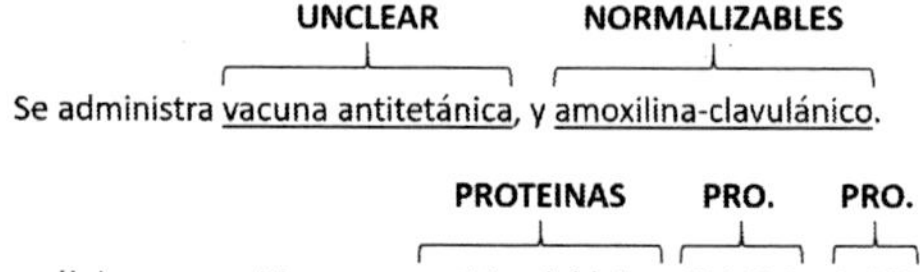

Figure 1: Annotated sample sentences (PRO. is short for PROTEINAS).

CoNER and aims at distinguishing four entity types: PROTEINAS, NORMALIZABLES, NO-NORMALIZABLES, and UNCLEAR. Our model was trained on all four entity types, although the NO-NORMALIZABLES type was not considered during the official evaluation due to its ambiguous definition. Two annotated sample sentences from the training data are shown in Figure 1.

In this paper, we describe our submissions to and their results in the first subtrack of PharmaCoNER. We address this task as a sequence-labeling problem and implement a system that relies **N**either on **L**anguage **N**or on **D**omain **E**xpertise (NLNDE). For this, we use a combination of different state-of-the-art approaches from NLP to tackle its challenges without the need for handcrafted features.

We train recurrent neural networks with conditional random field (CRF) output layers which are state of the art for different sequence labeling tasks, such as named entity recognition (Lample et al., 2016), part-of-speech tagging (Kemos et al., 2019) and de-identification (Liu et al., 2017). In our different runs, we further explore the advantages of domain-specific fastText embeddings that have been pre-trained on SciELO and Wikipedia articles (Soares et al., 2019) to investigate the impact of domain knowledge. Note that the training of these embeddings requires only a collection of domain-specific text but no human domain expertise. Based on these models, we train an attention-

Proceedings of the 5th Workshop on BioNLP Open Shared Tasks, pages 26–32
Hong Kong, China, November 4, 2019. ©2019 Association for Computational Linguistics

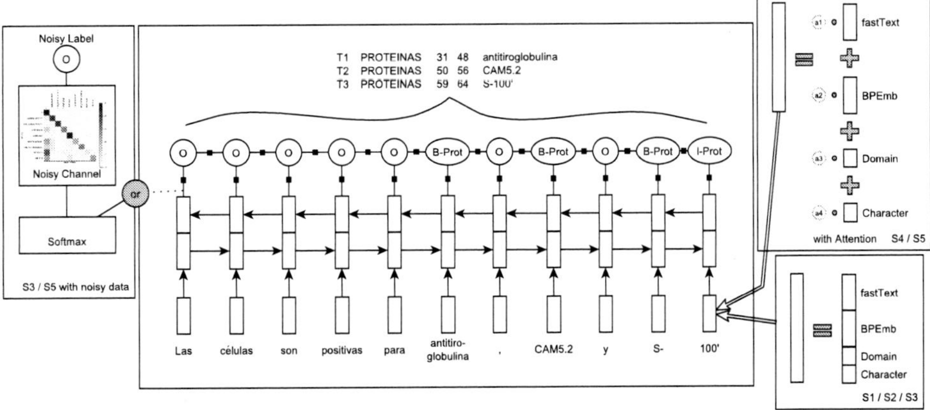

Figure 2: Architecture of our models. The label prefixes "B-" and "I-" show how we address the task as a sequence-labeling task. The word representations are either the concatenated embeddings (in runs S1–S3) or the attention-based weighted embeddings (in runs S4–S5).

based embedding selection function in order to leverage multiple different word embeddings effectively. Finally, we extend the training data with automatically annotated data, which was sampled from the same domain and annotated with information from Wikidata.[1]

2 Methods

In this section, we present our system, the attention function for embedding selection, and the noisy channel model.

2.1 NLNDE System

In Figure 2, the architecture of our models is depicted, which we explain in the following.

Input Embeddings. We tokenize the input with the tokenizer provided by the shared task organizers (Intxaurrondo, 2019). We noticed that the tokenizer sometimes merges multi-word expressions into a single token joined with underscores for contiguous words. As a result, some tokens cannot be aligned with the corresponding entity annotations. To address this, we split those tokens into their components in a postprocessing step. Then, we represent each token with the following embeddings (see bottom right box of Figure 2):

- *Character embeddings*: We use the concatenated last forward and backward hidden states of a bidirectional long short-term

memory (BiLSTM) network (Hochreiter and Schmidhuber, 1997) over character embeddings (50 dimensions, randomly initialized, fine-tuned during training (Lample et al., 2016)).

- *Domain-independent fastText embeddings* (100 dimensions, pre-trained on Spanish text (Grave et al., 2018)).

- *Domain-specific fastText embeddings* (100 dimensions, pre-trained on Spanish SciELO and Wikipedia articles (Soares et al., 2019)).

- *Byte-pair encoding embeddings* (300 dimensions, vocabulary size of 200,000, pre-trained on Spanish text (Heinzerling and Strube, 2018)).

Note that except for the character embeddings, we do not fine-tune any of the embeddings. All embeddings are concatenated into a single word representation vector.

Word Features. We also experiment with extending the input representations with the following features:

- *Part-of-speech (POS)*: The POS tags are generated by the POS-tagger provided by the shared task organizers (Intxaurrondo, 2019). The tags are embedded into a 20-dimensional randomly initialized embedding and learned during training. The embedded vector is used as the representation for the POS tag.

---

[1]https://www.wikidata.org/

- *Length*: For each word, we encode its length in a one-hot vector. Words with more than nine characters share the same vector (10 dimensions).

- *Frequency*: We consider the relative frequency f of each word and bin the frequencies into ten groups. The first group contains the most frequent words that have relative frequencies above 1% ($f > 1\%$). The remaining bins are constructed in the following manner: $f > 0.5\%$, $f > 0.1\%$, $f > 0.05\%$, etc. (one-hot encoded, 10 dimensions).

- *Word shape*: We distinguish between uppercased, lowercased, starts with capital letter, numeric, mostly numeric, punctuation, mostly punctuation, only letters, alphanumeric and other (one-hot encoded, 10 dimensions).

All features are concatenated into a single feature vector f of 50 dimensions.

BiLSTM-CRF Layers. The input representation is fed into a BiLSTM with a conditional random field (CRF) output layer, similar to the model of Lample et al. (2016). The CRF output layer is a linear-chain CRF, i.e., it learns transition scores between the output classes. For training, the forward algorithm is used to sum the scores for all possible sequences. During decoding, the Viterbi algorithm is applied to obtain the sequence with the maximum score.

Hyperparameters and Training. The hyperparameters are the same across all runs. We use a BiLSTM hidden size of 256 and train the network with the NADAM optimizer (Dozat, 2016) using a learning rate of 0.002 and a batch size of 32. For regularization, we employ early stopping on the development set and apply dropout with probability 0.5 on the input representations.

2.2 Attention for Embedding Selection

As we are combining different word embeddings, some of them may be more beneficial for certain words than others, e.g., domain-specific embeddings for in-domain words. Kiela et al. (2018) used an attention mechanism for weighting and selecting the best embeddings for each word. We extend this idea and propose the following attention function to weight the embeddings depending on additional word features.

For the attention-based models, all n embeddings e are mapped to the same size using a linear mapping $Q_i \in \mathbb{R}^{E \times E_i}$ without bias, with $x_i \in \mathbb{R}^E$ being the i-th embedding e_i mapped from their original size E_i to the maximal embedding size $E = \max_m(E_m)$.

$$x_i = Q_i \cdot e_i \tag{1}$$

In order to allow the model to make an informed decision which embeddings to focus on, we use the word features described in Section 2.1 as an additional input to the attention function. The vector $f \in \mathbb{R}^F$ representing the features for each word is concatenated to each embedding x_i. The attention weight a_i for each embedding x_i is computed with the softmax function, by feeding x_i and f into a fully-connected hidden layer of size H with the parameters $W \in \mathbb{R}^{H \times E}, U \in \mathbb{R}^{H \times F}, V \in \mathbb{R}^{1 \times H}$.

$$a_i = \frac{\exp(V \cdot \tanh(W x_i + U f))}{\sum_{l=1}^{n} \exp(V \cdot \tanh(W x_l + U f))} \tag{2}$$

Finally, the embeddings x_i are weighted using the attention weights a_i resulting in the word representation:

$$e = \sum_i a_i \cdot x_i \tag{3}$$

Then, this word representation $e \in \mathbb{R}^E$ is fed into the BiLSTM-CRF. Compared to a concatenation of the different embeddings, this results in a lower-dimensional word representation and, thus, requires fewer parameters in the BiLSTM layer. The attention-based embedding selection is shown in the upper right box of Figure 2.

2.3 Training on Noisy Data

As it was shown in multiple low-resource settings (Dgani et al., 2018; Fang and Cohn, 2016; Mnih and Hinton, 2012; Paul et al., 2019; Yang et al., 2018), the performance of NER and other NLP systems can be substantially improved by training on additional noisy data which is labeled in a distantly supervised manner (Mintz et al., 2009). With this approach, the noisy data is cheap to create, but also error-prone and can even decrease performance if used as training data without noise handling as shown by Hedderich and Klakow (2018).

Extraction of Noisy Data. We create gazetteers for the different entity types by extracting names and aliases of possible entities from Wikidata for the following categories and their subclasses:[2]

- PROTEINAS: enzyme, gene, hormone, protein.

- NORMALIZABLES: allotropy, alloy, amino acid, antibody, carbohydrate, diagnostic procedure, dye, lipid, mineral, nucleotide, oil, reagent, chemical compounds, peptide, plant, polymer, vaccine.

- UNCLEAR and NO-NORMALIZABLES: The gazetteer was constructed from entity mentions in the training data that appeared at least twice and examples from the annotation guidelines.

Then, we retrieve unlabeled documents from the same domain from the SciELO archive (Packer, 1998). Finally, we use the extracted gazetteers to automatically annotate the SciELO data with the method from Lange et al. (2019). We use case-insensitive string matching for PROTEINAS and strict string matching for the other types. This allows to create additional training instances, but at the same time introduces noise into the system. To avoid that the noisy labels result in a decrease of performance, we train on the noisy data with a special noise handling method adapted from Goldberger and Ben-Reuven (2016), which will be described in the following.

Noisy Channel and Confusion Matrix. First, we annotate each word of the training data using the same method as for generating the noisy data. Thus, each word in the training data has a clean, true label y and a noisy label $\hat{y}$ from which we can model the noise distribution $p(\hat{y} = j|y = i)$ with a confusion matrix, as shown in Figure 3. We transform the distribution of the predicted (clean) labels to the noisy label distribution through a so-called noisy channel (Goldberger and Ben-Reuven, 2016):

$$p(\hat{y} = j|x) = \sum_{i=1}^{k} p(\hat{y} = j|y = i)p(y = i|x) \quad (4)$$

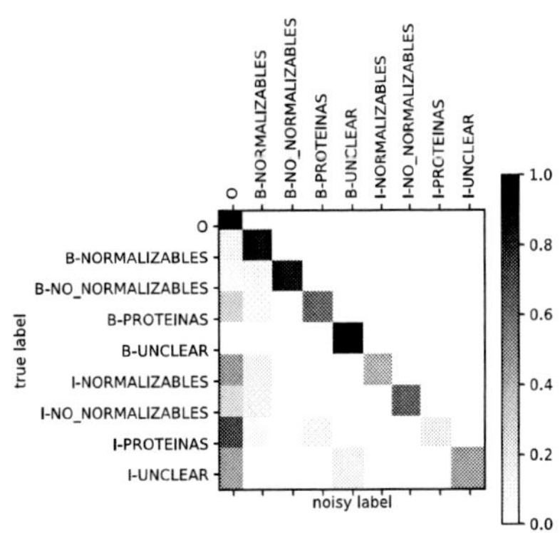

Figure 3: Confusion matrix for the automatic annotation on the training data used for the noisy channel initialization.

where k is the number of classes and $p(y = i|x)$ is the probability of a label y having a specific class i given the feature x.

We initialize the noisy channel weights using the learned confusion matrix on the training set, for which clean and noisy labels are available.

Training with Confusion Matrix. The sequence tagging model is then trained alternately on the clean data with the CRF output layer and on the noisy data with the noisy channel layer, as shown in Figure 2. The number of noisy training instances is constantly decreased by 5% after every training epoch to at least 100 sentences, as we observed that the noisy data helps in particular for the first epochs, but decreases performance if the amount is not reduced. Note that we shuffle the noisy data after each training epoch. Thus, the model is trained on new samples of noisy sentences in every epoch.

3 Submissions

We submitted five runs to the PharmaCoNER competition. All of them are based on the architecture described in Section 2.1.

S1 (*Base*): Our first run, the base system for all of the following runs, uses a concatenation of three embeddings (character, BPEmb, fastText) which were all trained on Wikipedia. Thus, this run does not include any form of domain knowledge, and it uses neither noisy data nor attention for embedding selection.

S2 (*Domain*): Our second run uses the three embeddings from S1 plus domain-specific fast-

---

[2]WikiData identifiers used for the extraction: Q8047, Q7187, Q11364, Q8054, Q81915, Q37756, Q8066, Q79460, Q11358, Q177719, Q189720, Q11367, Q7946, Q28745, Q42962, Q2356542, Q47154513, Q172847, Q756, Q81163, Q134808.

S_{ID}	Development			Test		
	P	R	F1	P	R	F1
S1	86.3	85.0	85.6	85.5	85.3	85.4
S2	87.1	85.8	86.4	86.3	85.9	86.1
S3	87.5	87.5	87.5	85.1	86.0	85.6
S4	88.0	**88.0**	88.0	85.2	87.2	86.2
S5	**89.1**	87.5	**88.3**	**89.0**	**88.3**	**88.6**

Table 1: Precision (P), Recall (R) and F1 for Task 1.

Text embeddings to incorporate knowledge about word distributions within the domain.

S3 (*Noise*): Our third run extends the model of S2 with training on additional noisy data (cf. Section 2.3). Moreover, we use the feature vector as an additional input, which is different from runs S1 and S2.

S4 (*Attention*): The fourth run uses the attention function for word embedding selection (cf. Section 2.2). Apart from that, the model is identical to S2 and only trained on clean data.

S5 (*Attention+Noise*): Our last run has the same architecture as S4 but is trained on the noisy data in addition. It thus combines domain-independent and domain-specific word embeddings, attention-based embedding selection, and training on noisy data.

4 Results and Analysis

This section describes our results and analysis.

4.1 Experimental Results

In Table 1, we report the results on the Pharma-CoNER development and test sets using the official shared task evaluation metrics.

Adding domain-knowledge (S2) to the base model S1 improves the performance on the development and the test set. The training on noisy data (S3) and the attention function alone (S4) do not lead to strong improvements on the test set; the noise model S3 even decreases performance. The combination of all proposed methods (run S5 *Attention+Noise*) outperforms all other models.

While we are able to see the improvements step by step introduced by our methods on the development set, such improvements are not observable one-to-one on the test set. We assume that model S5 performs best at generalizing to unseen words due to the training on additional data and the attention function based on basic word properties like word length or frequency. The other models seem

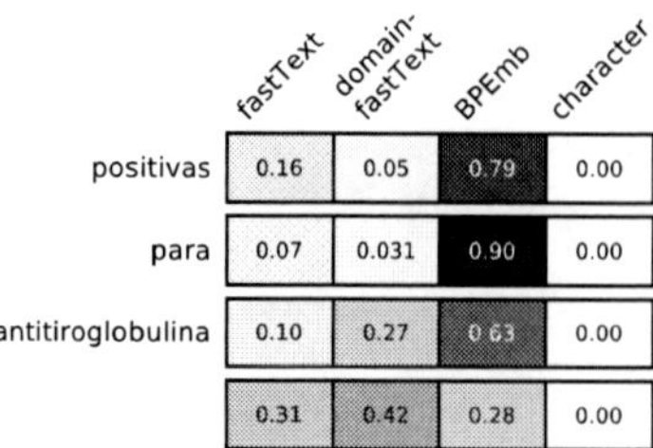

Figure 4: The attention weights of our model for the four embeddings. Darker color indicates higher weight.

to overfit on the development set, even though this set was never used for training but only for early stopping.

4.2 Analysis of Attention Weights

The attention-based models learn to focus mostly on the byte-pair-encoding embeddings, as shown in Figure 4. In particular, for words from the general domain (positivas) and stopwords (para), our model focuses on these embeddings. For domain-specific words (antitiroglobulina, CAM5.2), the model learns to focus more on the fastText embeddings and especially the domain-specific embeddings. Interestingly, the character embeddings are never assigned a noticeable weight. This may be attributed to the fact that the other embeddings are all subword embeddings and that they are able to generate meaningful vectors for out-of-vocabulary words. Moreover, the character embeddings were randomly initialized and had to be learned during training while the other models were pretrained.

5 Conclusions

In this paper, we described our system for the first subtrack of the PharmaCoNER competition. We trained a bi-directional long short-term memory network and explored different input representations. We proposed to use a feature-based attention function for embedding selection and training on noisy data, which in combination increased performance by more than 3 F1 points up to 88.6%. This shows that we can successfully extract these special types of entities without the need for domain or language-specific model architectures.

Acknowledgments

We would like to thank the anonymous reviewers for their helpful suggestions and comments.

References

Y. Dgani, H. Greenspan, and J. Goldberger. 2018. Training a neural network based on unreliable human annotation of medical images. In *2018 IEEE 15th International Symposium on Biomedical Imaging (ISBI 2018)*, pages 39–42.

Timothy Dozat. 2016. Incorporating nesterov momentum into adam. In *Proceedings of the Workshop of the International Conference on Learning Representations (ICLR)*.

Meng Fang and Trevor Cohn. 2016. Learning when to trust distant supervision: An application to low-resource POS tagging using cross-lingual projection. In *Proceedings of The 20th SIGNLL Conference on Computational Natural Language Learning*, pages 178–186, Berlin, Germany. Association for Computational Linguistics.

Jacob Goldberger and Ehud Ben-Reuven. 2016. Training deep neural-networks using a noise adaptation layer. In *Proceedings of the International Conference on Learning Representations (ICLR)*.

Aitor Gonzalez-Agirre, Montserrat Marimon, Ander Intxaurrondo, Obdulia Rabal, Marta Villegas, and Martin Krallinger. 2019. Pharmaconer: Pharmacological substances, compounds and proteins named entity recognition track. In *Proceedings of the BioNLP Open Shared Tasks (BioNLP-OST)*, Hong Kong, China. Association for Computational Linguistics.

Edouard Grave, Piotr Bojanowski, Prakhar Gupta, Armand Joulin, and Tomas Mikolov. 2018. Learning word vectors for 157 languages. In *Proceedings of the Eleventh International Conference on Language Resources and Evaluation (LREC-2018)*, Miyazaki, Japan. European Languages Resources Association (ELRA).

Michael A. Hedderich and Dietrich Klakow. 2018. Training a neural network in a low-resource setting on automatically annotated noisy data. In *Proceedings of the Workshop on Deep Learning Approaches for Low-Resource NLP*. Association for Computational Linguistics.

Benjamin Heinzerling and Michael Strube. 2018. BPEmb: Tokenization-free pre-trained subword embeddings in 275 languages. In *Proceedings of the Eleventh International Conference on Language Resources and Evaluation (LREC-2018)*, Miyazaki, Japan. European Languages Resources Association (ELRA).

Sepp Hochreiter and Jürgen Schmidhuber. 1997. Long short-term memory. *Neural computation*, 9(8):1735–1780.

Ander Intxaurrondo. 2019. SPACCC (spanish clinical case corpus) tokenizer.

Apostolos Kemos, Heike Adel, and Hinrich Schütze. 2019. Neural semi-Markov conditional random fields for robust character-based part-of-speech tagging. In *Proceedings of the 2019 Conference of the North American Chapter of the Association for Computational Linguistics: Human Language Technologies, Volume 1 (Long and Short Papers)*, pages 2736–2743, Minneapolis, Minnesota. Association for Computational Linguistics.

Douwe Kiela, Changhan Wang, and Kyunghyun Cho. 2018. Dynamic meta-embeddings for improved sentence representations. In *Proceedings of the 2018 Conference on Empirical Methods in Natural Language Processing*, pages 1466–1477, Brussels, Belgium. Association for Computational Linguistics.

Martin Krallinger, Florian Leitner, Obdulia Rabal, Miguel Vazquez, Julen Oyarzabal, and Alfonso Valencia. 2015. CHEMDNER: The drugs and chemical names extraction challenge. *Journal of cheminformatics*, 7(1):S1.

Guillaume Lample, Miguel Ballesteros, Sandeep Subramanian, Kazuya Kawakami, and Chris Dyer. 2016. Neural architectures for named entity recognition. In *Proceedings of the 2016 Conference of the North American Chapter of the Association for Computational Linguistics: Human Language Technologies*, pages 260–270, San Diego, California. Association for Computational Linguistics.

Lukas Lange, Michael A. Hedderich, and Dietrich Klakow. 2019. Feature-dependent confusion matrices for low-resource ner labeling with noisy labels. In *Proceedings of 2019 Conference on Empirical Methods in Natural Language Processing and 9th International Joint Conference on Natural Language Processing*, Hong Kong, China. Association for Computational Linguistics.

Zengjian Liu, Buzhou Tang, Xiaolong Wang, and Qingcai Chen. 2017. De-identification of clinical notes via recurrent neural network and conditional random field. *Journal of Biomedical Informatics*, 75(S):S34–S42.

Mike Mintz, Steven Bills, Rion Snow, and Daniel Jurafsky. 2009. Distant supervision for relation extraction without labeled data. In *Proceedings of the Joint Conference of the 47th Annual Meeting of the ACL and the 4th International Joint Conference on Natural Language Processing of the AFNLP*, pages 1003–1011, Suntec, Singapore. Association for Computational Linguistics.

Volodymyr Mnih and Geoffrey Hinton. 2012. Learning to label aerial images from noisy data. In *Proceedings of the 29th International Coference on International Conference on Machine Learning*, ICML'12, pages 203–210, USA. Omnipress.

Abel Laerte Packer. 1998. Scielo: uma metodologia para publicação eletrônica. *Ciência da informação*, 27(2).

Debjit Paul, Mittul Singh, Michael A. Hedderich, and
Dietrich Klakow. 2019. Handling noisy labels for
robustly learning from self-training data for low-
resource sequence labeling. *CoRR*, abs/1903.12008.

Martín Pérez-Pérez, Obdulia Rabal, Gael Pérez-
Rodríguez, Miguel Vazquez, Florentino Fdez-
Riverola, Julen Oyarzábal, A Valencia, Anália
Loureno, and Martin Krallinger. 2017. Evaluation
of chemical and gene/protein entity recognition sys-
tems at biocreative v. 5: the CEMP and GPRO
patents tracks. In *Proceedings of the BioCreative
Workshop*, volume 5, pages 3–11.

Felipe Soares, Marta Villegas, Aitor Gonzalez-Agirre,
Martin Krallinger, and Jordi Armengol-Estapé.
2019. Medical word embeddings for Spanish: Development and evaluation.

Yaosheng Yang, Wenliang Chen, Zhenghua Li,
Zhengqiu He, and Min Zhang. 2018. Distantly supervised NER with partial annotation learning and
reinforcement learning. In *Proceedings of the 27th
International Conference on Computational Linguistics*, pages 2159–2169, Santa Fe, New Mexico,
USA. Association for Computational Linguistics.

A Deep Learning-Based System for PharmaCoNER

Ying Xiong[1], Yedan Shen[1], Yuanhang Huang[1], Shuai Chen[1], Buzhou Tang[1,2*] Xiaolong Wang[1],
Qingcai Chen[1,2], Jun Yan[3], Yi Zhou[4*]
[1]Department of Computer Science, Harbin Institute of Technology, Shenzhen, China, 518055
[2]Peng Cheng Laboratory
[3]Yidu Cloud (Beijing) Technology Co., Ltd, Beijing
[4]Sun YAT-SEN UNIVERSITY
{xiongying0929, hyhang7, chenshuai726, tangbuzhou, qingcai.chen}@gmail.com
shenyedan@stu.hit.edu.cn, wangxl@insun.hit.edu.cn, Jun.YAN@Yiducloud.cn, zhouyi@sysu.edu.cn
* Corresponding author

Abstract

The Biological Text Mining Unit at BSC and CNIO organized the first shared task on chemical & drug mention recognition from Spanish medical texts called PharmaCoNER (Pharmacological Substances, Compounds and proteins and Named Entity Recognition track) in 2019. The shared task includes two tracks: one for NER offset and entity classification (track 1) and the other one for concept indexing (track 2). We developed a pipeline system based on deep learning methods for this shared task, specifically, a subsystem based on BERT (Bidirectional Encoder Representations from Transformers) for NER offset and entity classification and a subsystem based on Bpool (Bi-LSTM with max/mean pooling) for concept indexing. Evaluation conducted on the shared task data showed that our system achieves a micro-average F1-score of 0.9105 on track 1 and a micro-average F1-score of 0.8391 on track 2.

1 Introduction

Efficient access to mentions of clinical entities is very important for using clinical text. The way to extract clinical entities embedded in the text is natural language processing (NLP). In the last decades, clinical entity extraction has attracted plenty of attention of researchers, clinicians, and enterprises in the clinical domain. The development of technology for clinical entity extraction mainly benefits from related NLP challenges including tasks of biomedical entity recognition and normalization, such as the BioCreative (Critical Assessment of Information Extraction systems in Biology) challenges (e.g., the CHEMDNER (Chemical compound and drug name recognition) track (Leaman et al., 2013)),

the i2b2 (the Center of Informatics for Integrating Biology and Bedside) challenges (Uzuner et al., 2011), SemEval (Semantic Evaluation) challenges (Elhadad et al., 2015) and the ShARe/CLEF eHealth Evaluation Lab shared tasks (Kelly et al., 2016). A large number of various kinds of methods have been proposed for biomedical entity recognition and normalization. Lots of machine learning methods such as conditional random fields (CRF) (Lafferty et al., 2001), structured support vector machines (SSVM) (Tsochantaridis et al., 2005) and bidirectional long-short-term memory with conditional random fields (BiLSTM-CRF) (Huang et al., 2015) have been applied for biomedical entity recognition, support vector machines (SVM) (Grouin et al., 2010) and ranking based on convolutional neural network (CNN) (Li et al., 2017) for clinical entity normalization. Although there have been a few promising results, most of them focus on the clinical text in English. Recently, clinical entity extraction for clinical text in other languages has also begun to receive much attention. For example, in 2016, NTCIR organized the first challenge about information extraction from clinical documents in Japanese (Morita et al., 2013). In 2017, CCKS organized the first challenge about information extraction from clinical records in Chinese (Hu et al., 2017).

To accelerate development of techniques of information extraction from clinical text in Spanish, Martin Krallinger et al. organized a shared task particular for chemical & drug mention recognition from Spanish medical texts called PharmaCoNER in 2019 (Gonzalez-Agirre, Aitor et al., 2019), which includes two tracks: track 1 for NER offset and entity classification and track 2 for concept indexing. The organizers provided an annotated corpus of 1000 clinical cases, 500 cases out of which were used as the

Proceedings of the 5th Workshop on BioNLP Open Shared Tasks, pages 33–37
Hong Kong, China, November 4, 2019. ©2019 Association for Computational Linguistics

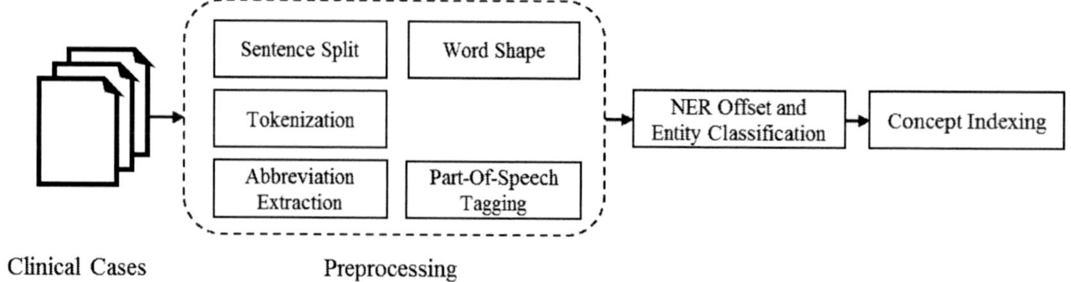

Figure 1: Overview architecture of our system for the PharmaCoNER task

training set, 250 cases as the development set and 250 cases as the test set. We participated in this shared task and developed a pipeline system based on two latest deep learning methods: BERT (Bidirectional Encoder Representations from Transformers) (Devlin et al., 2019) and Bpool (Bi-LSTM with max/mean pooling) (Conneau et al., 2017). The system developed on the training and development sets achieved a micro-average F1-score of 0.9105 on track 1 and a micro-average F1-score of 0.8391 on track 2 on the independent test set.

2 Material and Methods

As shown in Figure 1, We first developed a preprocessing module to split clinical cases into sentences, tokenized the sentences and extracted some features for each token, then a BERT-based subsystem for NER offset and entity classification, and finally a Bpool-based system for concept indexing. All of them were individually presented in the following sections in detail.

2.1 Dataset

The PharmaCoNER organizers asked medical experts to annotate a corpus of 1000 clinical cases with chemical & drug mentions for the shared task according to a pre-defined guideline. The corpus was divided into a training set, a development set and a test set. The test set was hidden in a background set of 3751 clinical cases when testing during the competition. The statistics of the corpus, including the number of documents, chemical & drug mentions in different types are listed in Table 1, where "UNK" denotes unknown. It should be noted that the chemical & drug mentions annotated with UNCLEAR were not considered during the competition.

2.2 Preprocessing

We split each clinical case into sentences using ';', '?', '!', '\n' or '.' which is not in numbers, and further split each sentence into tokens using the method proposed by Liu (Liu et al., 2015), which was specially designed for clinical text. We adopted Ab3P tools[1] to extract full names of abbreviations, and SPACCC_POS-TAGGER tool[2] for POS tagging and lemmatization. Besides, we used the same way as Liu (Liu et al., 2015) to get each word's word shape.

2.3 NER offset and entity classification

NER offset and entity is a typical NER problem usually recognized as a sequence labeling problem. In this study, we adopted "BIO" tagging schema to represent chemical & drug mentions, where 'B', 'I' and 'O' represent beginning, inside and outside of a chemical & drug mentions respectively, and developed a system based on BERT. First, character-level representation, POS tagging representation and word shape representation of each word were concatenated into the word representation of BERT, and then a CRF layer was appended to BERT for chemical & drug mentions recognition.

2.4 Concept Indexing

After chemical & drug mentions were recognized, we first constructed <mention, standard terminology> pairs as candidates for matching, and then built a Bpool-based matching model (Conneau et al., 2017) according to the candidates. Standard terminologies were selected into candidates in the following two ways:

---

[1] (https://github.com/ncbi-nlp/Ab3P)

[2] (https://github.com/PlanTL-SANIDAD/SPACCC_POS-TAGGER)

Statistic	#Training	#Development	#Test	#Background
DOCUMENT	500	250	250	3751
NORMALIZABLES	2304	1121	973	UNK
NO_NORMALIZABLES	24	16	10	UNK
PROTEINAS	1405	745	859	UNK
UNCLEAR	89	44	0	UNK

Table 1. Statistics of the PharmaCoNER Corpus.

1) Top n terminologies ranked by Levenshtein distance[3] with a given mention at char-level and at token-level.

2) Terminologies selected by 1) and the given mention's synonyms appearing in the standard terminology vocabulary.

After the terminology selection, a Bpool-based matching model at character-level was utilized to judge whether two mentions were matching or not.

2.5 Evaluation

The performance of our system was measured by micro-average precision (P), recall (R), and F1-score (F1), which were calculated by the official tool provided by the PharmaCoNER organizers[4].

2.6 Experiments Setup

In this study, for track1, we first optimized model on the development set and then fine-tuned the model on the training and development sets for 5 more epochs. For standard terminology selection, we optimized n from 10 to 50 with step 10, and finally set it to 40. For track2, we optimized the model on the training and development sets via 10-fold cross validation. The hyper-parameters and parameter estimation algorithm used for model training were listed in Table 2. The pre-trained BERT[5] was used as the initial neural language model and fine-tuned on all datasets provided by the shared task organizers. The embeddings of character, POS and word shape were randomly initialized from a uniform distribution. It is worth noting that in the BERT model, the update of the parameters

included in the BERT used the learning rate of 2e-5, and the parameter update of other features used a learning rate of 0.003.

Hyper-parameter	Value
Dimension of character representation	BERT:30; Bpool:50
Dimension of POS representation	30
Dropout probability	0.1
Learning rate	BERT: 2e-5; Bpool: 1e-3
Training epochs	Bert:15; Bpool:20
Parameter estimation algorithm	BERT: adam with warmup; Bpool: adam

Table 2. Hyper-parameters and parameter estimation algorithm used for deep learning methods.

3 Results

The highest micro-average precisions, recalls and F1-scores of our system on the two tracks were listed in Table 3. Our system achieved a micro-average precision of 0.9123, recall of 0.9088 and F1-score of 0.9105 on track1, and a micro-average precision of 0.8284, recall of 0.8502 and F1-score of 0.8391 on track2. Among three types of chemical & drug mentions considered in the shared task, our system performed best on NORMALIZABLES and worst on NO_NORMALIZABLES for track1, which may be proportional to the number of mentions of each type.

Track	Type	P	R	F1
Track1	NORMALIZABLES	0.9426	0.9291	0.9358
	NO_NORMALIZABLES	1.0000	0.2000	0.3333
	PROTEINAS	0.8787	0.8941	0.8863
	Overall	0.9123	0.9088	0.9105
Track2	Overall	0.8284	0.8502	0.8391

Table 3. The highest results of our system for PharmaCoNER. (P: micro-average precision; R: micro-average recall; F1: micro-average F1 score)

---

[3] https://en.wikipedia.org/wiki/Levenshtein_distance

[4] https://github.com/PlanTL-SANIDAD/PharmaCoNER-CODALAB-Evaluation-Script

[5] https://storage.googleapis.com/bert_models/2018_11_03/multilingual_L-12_H-768_A-12.zip

3.1 Ablation Study

Table 4 provided additional ablation study results analyzing the contribution of individual features on track 1 and reporting the performance of each standard terminology selection method (STS) on track 2. We found that both character-level embedding, POS tagging representation, and word shape representation contributed towards our system on track 1. They brought 1.69%, 0.51%, and 0.63% improvements on F1-score, respectively. On track 2, when removing the extended synonyms, the F1 score declined from 0.8048 to 0.7932.

Track	model	P	R	F1
Track1	BERT	0.8989	0.9087	0.9037
	- Character-embedding	0.8981	0.8757	0.8868
	- POS tagging	0.8986	0.8986	0.8986
	- Word shape	0.8874	0.9076	0.8974
Track2	STS 1	0.7722	0.8153	0.7932
	STS 2	0.7826	0.8284	0.8048

Table 4. Ablation study of track 1 and track 2 on the development set. (P: micro-average precision; R: micro-average recall; F1: micro-average F1 score)

4 Discussion

For task 1, our analysis found that data processing had a great influence on the NER offset results. Separating alphabets and digitals in a word , for example, "PaO2" was split into 'PaO' and '2' , caused some errors of entity boundary or entity type. Separating words by the hyphen '- ' also caused some errors. For example, "4-methyilumbelliferyl α-D-galactosidasa" is totally identified as 'PROTEINAS', but in "daclizumab-tacrolimus-MMF-esteroide", "daclizumab" is identified as "PROTEINAS", "tacrolimus", "MMF" and "esteroide" are identified as "NORMILIZED". Our experiments on the development set showed that the effect of tokenization on micro-average F1 score on NER was about 2%.

There were mainly the following three types of errors caused by our system. (1) abbreviation recognition errors: it is difficult to identify abbreviations in a record correctly; (2) long entity: entities consisting of four or more tokens are hard to identify correctly, such as 'anticuerpos antitransglutaminasa tisular IgA'. (3) drugs: model cannot recognize drugs such as 'dasatinib', 'nilotinib' and so on.

Since we experimented with a pipeline model, the mistakes of task 1 will be propagated to task 2 and there are about 8% errors caused by track1. In addition, about 10% errors are caused by the matching model. We summarized the modes of low recall rate by standard terminology selection methods when constructing <mention, standard terminology> pairs. The modes are: (1) about 40% entities are abbreviations, which is difficult to find the candidates from SNOMED-CT; (2) about 20% of entities have the same candidates in SNOMED-CT [6], which are not normalized entities in the shared task.

For further improvements, there may be two directions: (1) using joint learning methods for task 1 and task 2. (2) integrating knowledge graph into our system.

5 Conclusion

In this study, we developed a deep learning-based pipeline system for the PharmaCoNER shared task, a challenge specifically for clinical entity extraction from clinical text in Spanish.

Acknowledgments

This paper is supported in part by grants: NSFCs (National Natural Science Foundations of China) (U1813215, 61876052 and 61573118), Special Foundation for Technology Research Program of Guangdong Province (2015B010131010), Strategic Emerging Industry Development Special Funds of Shenzhen (JCYJ20170307150528934 and JCYJ20180306172232154), Innovation Fund of Harbin Institute of Technology (HIT.NSRIF.2017052).

References

Alexis Conneau, Douwe Kiela, Holger Schwenk, Loïc Barrault, and Antoine Bordes. 2017. Supervised Learning of Universal Sentence Representations from Natural Language Inference Data. In *Proceedings of the 2017 Conference on Empirical Methods in Natural Language Processing*, pages 670–680.

---

[6] https://browser.ihtsdotools.org/?perspective=full&conceptI d1=404684003&edition=MAIN/SNOMEDCT-ES&release=&languages=es,en

Jacob Devlin, Ming-Wei Chang, Kenton Lee, and Kristina Toutanova. 2019. BERT: Pre-training of Deep Bidirectional Transformers for Language Understanding. In *Proceedings of the 2019 Conference of the North American Chapter of the Association for Computational Linguistics: Human Language Technologies, Volume 1 (Long and Short Papers)*, pages 4171–4186.

Noémie Elhadad, Sameer Pradhan, Sharon Gorman, Suresh Manandhar, Wendy Chapman, and Guergana Savova. 2015. SemEval-2015 task 14: Analysis of clinical text. In *Proceedings of the 9th International Workshop on Semantic Evaluation (SemEval 2015)*, pages 303–310.

Gonzalez-Agirre, Aitor, Marimon, Montserrat, Marimon, Montserrat, Rabal, Obdulia, Villegas, Marta, and Krallinger, Martin. 2019. PharmaCoNER: Pharmacological Substances, Compounds and proteins Named Entity Recognition track. In *Proceedings of the BioNLP Open Shared Tasks (BioNLP-OST)*, pages 1--X. Association for Computational Linguistics, November.

Cyril Grouin, Asma Ben Abacha, Delphine Bernhard, Bruno Cartoni, Louise Deleger, Brigitte Grau, Anne-Laure Ligozat, Anne-Lyse Minard, Sophie Rosset, and Pierre Zweigenbaum. 2010. CARAMBA: concept, assertion, and relation annotation using machine-learning based approaches. In

Jianglu Hu, Xue Shi, Zengjian Liu, Xiaolong Wang, Qingcai Chen, and Buzhou Tang. 2017. HITSZ CNER: A hybrid system for entity recognition from chinese clinical text. In *CEUR Workshop Proceedings*, volume 1976, pages 25–30.

Zhiheng Huang, Wei Xu, and Kai Yu. 2015. Bidirectional LSTM-CRF models for sequence tagging. *arXiv preprint arXiv:1508.01991*.

Liadh Kelly, Lorraine Goeuriot, Hanna Suominen, Aurélie Névéol, Joao Palotti, and Guido Zuccon. 2016. Overview of the CLEF eHealth evaluation lab 2016. In *International Conference of the Cross-Language Evaluation Forum for European Languages*, pages 255–266. Springer.

John Lafferty, Andrew McCallum, and Fernando CN Pereira. 2001. Conditional random fields: Probabilistic models for segmenting and labeling sequence data.

Robert Leaman, Chih-Hsuan Wei, and Zhiyong Lu. 2013. NCBI at the BioCreative IV CHEMDNER Task: Recognizing chemical names in PubMed articles with tmChem. In *BioCreative Challenge Evaluation Workshop*, volume 2, page 34. Citeseer.

Haodi Li, Qingcai Chen, Buzhou Tang, Xiaolong Wang, Hua Xu, Baohua Wang, and Dong Huang. 2017. CNN-based ranking for biomedical entity normalization. *BMC bioinformatics*, 18(11):385.

Zengjian Liu, Yangxin Chen, Buzhou Tang, Xiaolong Wang, Qingcai Chen, Haodi Li, Jingfeng Wang, Qiwen Deng, and Suisong Zhu. 2015. Automatic de-identification of electronic medical records using token-level and character-level conditional random fields. *Journal of biomedical informatics*, 58:S47–S52.

Mizuki Morita, Yoshinobu Kano, Tomoko Ohkuma, Mai Miyabe, and Eiji Aramaki. 2013. Overview of the NTCIR-10 MedNLP Task. In *NTCIR*. Citeseer.

Ioannis Tsochantaridis, Thorsten Joachims, Thomas Hofmann, and Yasemin Altun. 2005. Large margin methods for structured and interdependent output variables. *Journal of machine learning research*, 6(Sep):1453–1484.

Özlem Uzuner, Brett R. South, Shuying Shen, and Scott L. DuVall. 2011. 2010 i2b2/VA challenge on concepts, assertions, and relations in clinical text. *Journal of the American Medical Informatics Association*, 18(5):552–556.

Deep neural model with enhanced embeddings for pharmaceutical and chemical entities recognition in Spanish clinical text

Renzo M. Rivera Zavala
Computer Science Department
Catholic University of Santa María
rriveraz@ucsm.edu.pe

Paloma Martínez
Computer Science Department
University Carlos III of Madrid
pmf@inf.uc3m.es

Abstract

In this work, we introduce a Deep Learning architecture for pharmaceutical and chemical Named Entity Recognition in Spanish clinical cases texts. We propose a hybrid model approach based on two Bidirectional Long Short-Term Memory (Bi-LSTM) network and Conditional Random Field (CRF) network using character, word, concept and sense embeddings to deal with the extraction of semantic, syntactic and morphological features. The approach was evaluated on the PharmaCoNER Corpus obtaining an F-measure of 85.24% for subtask 1 and 49.36% for subtask2. These results prove that deep learning methods with specific domain embedding representations can outperform the state-of-the-art approaches.

1 Introduction

Currently, the number of biomedical literature is growing at an exponential rate. Therefore, the efficient access to information on biological, chemical, and biomedical data described in scientific articles, patents, or e-health reports is a growing interest in biomedical research, industrial medicine manufacturing, and so forth. In this context, improved access to chemical and drug name mentions in biomedical texts is a crucial step downstream tasks such as drug and protein interactions, chemical compounds, adverse drug reactions, among others.

Named Entity Recognition (NER) is one of the fundamental tasks of biomedical text mining, intending to automatically extract and identify mentions of entities of interest in running text, typically through their mention offsets or by classifying individual tokens whether they belong to entity mentions or not. There are different approaches to address the NER task. Dictionary-based methods, which are limited by the size of the dictio-

nary, spelling errors, the use of synonyms, and the constant growth of vocabulary. Rule-based methods and Machine Learning methods usually require both syntactic and semantic features as well as specific language and domain features. One of the most effective methods is Conditional Random Fields (CRF) (Lafferty et al., 2001) since CRF is one of the most reliable sequence labeling methods. Recently, deep learning-based methods have also demonstrated state-of-the-art performance for English (Hemati and Mehler, 2019; Pérez-Pérez et al., 2017; Suárez-Paniagua et al., 2019) texts by automatically learning relevant patterns from corpora, which allows language and domain independence. However, until now, to the best of our knowledge, there is only one work that addresses the generation of Spanish biomedical word embeddings (Armengol-Estapé Jordi, 2019; Soares et al., 2019).

In this paper, we propose a hybrid model combining two Bi-LSTM layers with a CRF layer. To do this, we adapt the NeuroNER model proposed in (Dernoncourt et al., 2017) for track 1 (NER offset and entity classification) of the PharmaCoNER task (Gonzalez-Agirre et al., 2019). Specifically, we have extended NeuroNER by adding context information, Part-of-Speech (PoS) tags, and information about overlapping or nested entities. Moreover, in this work, we use existing pre-trained as well as our trained word embedding models: i) a word2vec/FastText Spanish Billion Word Embeddings models (Cardellino, 2016), which were trained on the 2014 dump of Wikipedia ii) our medical word embeddings for Spanish trained using the FastText model and iii) a sense-disambiguation embedding model (Trask et al., 2015). For track 2 (concept indexing) based on the output of the previous step, we use full-text search and fuzzy matching on the SNOMED-CT Spanish Edition dictionary to obtain the corre-

Proceedings of the 5th Workshop on BioNLP Open Shared Tasks, pages 38–46
Hong Kong, China, November 4, 2019. ©2019 Association for Computational Linguistics

sponding index.

Experiment results on PharmarCoNER tasks showed that our features representation improved each of separate representations, implying that LSTM-based compositions play different roles in capturing token-level features for NER tasks, thus making improvements in their combination. Moreover, the use of specific domain word vector representations (word embeddings) outperform general domain word vector and concept vector representations (concept embeddings).

2 Materials and Methods

In this section, we first describe the corpora, the training procedure and the word, concept, and sense embedding models used in our study. Then, we describe our system architecture for offset and entity classification.

2.1 Corpora

The corpus was gathered from Spanish biomedical texts from different multilingual biomedical sources:

1. The Spanish Bibliographical Index in Health Sciences (IBECS - `http://ibecs.isciii.es`) corpus that collects scientific journals covering multiple fields in health sciences,

2. Scientific Electronic Library Online (SciELO - `https://scielo.org/es/`) corpus gathers electronic publications of complete full-text articles from scientific journals of Latin America, South Africa, and Spain,

3. MedlineNLM corpus obtained from the PubMed free search engine (`https://www.ncbi.nlm.nih.gov/pubmed/`),

4. The MedlinePlus corpus (an online information service provided by the U.S. National Library of Medicine - `https://medlineplus.gov/`), consists of Health topics, Drugs and supplements, Medical Encyclopedia and Laboratory test information, and

5. The UFAL corpus (`https://ufal.mff.cuni.cz/ufal_medical_corpus`) is a collection of parallel corpora of medical and general domain texts.

Source corpus details are described in Table 1.

All the corpora are in XML (Dublin core format) and TXT format files. XML files were processed for extract only raw text from specific XML tags such as "title" and "description" from Spanish labels, based on the Dublin Core format as shown in Figure 1. TXT files were not processed. Raw texts from all files were compiled in a single TXT file. Texts were processed, setting all to lower, removing punctuations, trailing spaces and stop words and used as input to generate our word embeddings. Sentences pre-processing (split and tokenized) were made using Spacy [1], an open-source python library for advanced multi-language natural language processing.

2.2 Transfer Learning

Transfer learning aims to perform a task on a dataset using knowledge learned from a previous dataset (Giorgi and Bader, 2018). As shown in many works, such as speech recognition (Wang and Zheng, 2015), sentence classification (Mou et al., 2016) and Named Entity Recognition (Giorgi and Bader, 2018), transfer learning improves generalization of the model, reduces training times on the target dataset, and reduces the amount of labeled data needed to obtain high performance. In this work we used an existing generic word embedding (Word2Vec embedding trained on Spanish Wikipedia), a trained medical embedding model, and a medical/generic sense-disambiguation embedding.

Word embedding is an approach to represent words as vectors of real numbers. Word embedding models have gained much popularity among the NLP community because they are able to capture syntactic and semantic information among words. In this work, we used the Spanish Billion Words Corpora (SBWC) (Cardellino, 2016) (W2V-SBWC), which is a pre-trained model of word embeddings trained on different general domain text corpora written in Spanish (such Ancora Corpus (Martí et al., 2007) and Wikipedia) using the word2vec (Mikolov et al., 2013) implementation. The FastText-SBWC pre-trained word embeddings model was trained on the SBWC using the FastText implementation.

Furthermore, we used the sense2vec (Trask et al., 2015) model, which provides multiple dense vector representations for each word based on the

---

[1] `https://spacy.io/`

Collection\Corpus	IBECS	SciELO	MedlineNLM	MedlinePlus	UFAL
Documents	168,198	161,710	330,928	1,063	265,410
Words	23,648,768	26,169,655	4,710,191	217,515	41,604,517
Unique Words	184,936	159,997	20,942	5,099	198,424

Table 1: Biomedical Spanish corpus details.

```
<dc:description xml:lang="en">BACKGROUND Acinetobacter baumannii is an important nosocomial pathogen whose virulence
<dc:type>English Abstract</dc:type>
<dc:language>es</dc:language>
<dc:date>1998 Oct </dc:date>
<dc:title xml:lang="es">Adherencia de Acinetobacter baumannii al tejido de tráquea de la rata.</dc:title>
<dc:title xml:lang="en">[Adherence of Acinetobacter baumannii to rat tracheal tissue].</dc:title>
<dc:publisher>Revista medica de Chile</dc:publisher>
</metadata>
</record>
</pubmed-document>
```

Figure 1: Dublin core format for biomedical corpus.

sense of the word. This model is able to analyze the context of a word based on the lexical and grammatical properties of words and then assigns its more adequate vector. We used the Reddit Vector, a pre-trained model of sense-disambiguation representation vectors presented by (Trask et al., 2015). This model was trained on a collection of general domain comments published on Reddit (corresponding to the year 2015) written in Spanish and English.

2.3 Medical word and concept embeddings

We used the FastText (Bojanowski et al., 2016) implementation to train our word embeddings using the Spanish Biomedical Corpora (SBC) described in section 2.1 (FastText-SBC). Moreover, we trained a concept embedding model replacing biomedical concepts in the SBC with their unique SNOMED-CT Spanish Edition identifier (SNOMED-SBC). We used the PyMedTermino library (Lamy et al., 2015) for concept indexing. A full-text search with the Levenshtein distance algorithm (Miller et al., 2009) was applied in a first instance for concept indexing and fuzzy search with threshold using FuzzyDict implementation (Hemati and Mehler, 2019) as a second approach for concepts not found by partial matching. The FastText model uses a combination of various subcomponents to produce high-quality embeddings. It uses a standard CBOW or skip-gram models, with position-dependent weighting, phrase representations, and sub-word information in a combined manner. The training parameters for each model are shown in Table 2. Our pre-trained mod-

els can be found in Github [2] with the corpora sources, text preprocessing, and training information.

2.4 System Description

Our approach is based on a deep learning network with a preprocess step, learning transfer, two recurrent neural network layers and the last layer for CRF (see Figure 2) as proposed in (Dernoncourt et al., 2017). The input for the first Bi-LSTM layer are character embeddings. In the second layer, we concatenate character embeddings from the first layer with word, concept, and sense-disambiguate embeddings for the second Bi-LSTM layer. Finally, the last CRF layer obtains the most suitable labels for each token using a tag encoding format. For more details about NeuroNER, please refer to (Dernoncourt et al., 2017).

Our contribution consists of extending the NeuroNER system with additional features. In particular, Sense embeddigs (obtained using POS tags), concept embeddings (obtained using semantic features) and the extended BMEWO-V encoding format has been added to the network and were as a pre-preprocessing a step.

POS tags are concatenated to token in order to create dense vector representations containing word/POS information (sense embeddings) and include this in the token embedding layer of the network. Furthermore, concept features are dense vector representations generated replacing concepts with their unique SNOMED concept identi-

---

[2] https://github.com/rmriveraz/
PharmaCoNER

Parameter\Model	FastText-SBC	SNOMED-SBC
Number of negatives sampled	20	20
Sampling threshold	6e-5	6e-5
Minimum number of word occurrences	10	10
Minimum length of character n-gram	3	3
Maximum length of character n-gram	6	6
Size of word vectors	300	300
Epochs	10	10
Processor	4 Intel Xeon 2.00 Ghz, 8 Cores, 16 Logical Processors	4 Intel Xeon 2.00 Ghz, 8 Cores, 16 Logical Processors
RAM	32 Gb	32 Gb
Corpus Size	1Gb	1Gb
Training Time	4 hours	8 hours

Table 2: Training parameters for embeddings models built in this work.

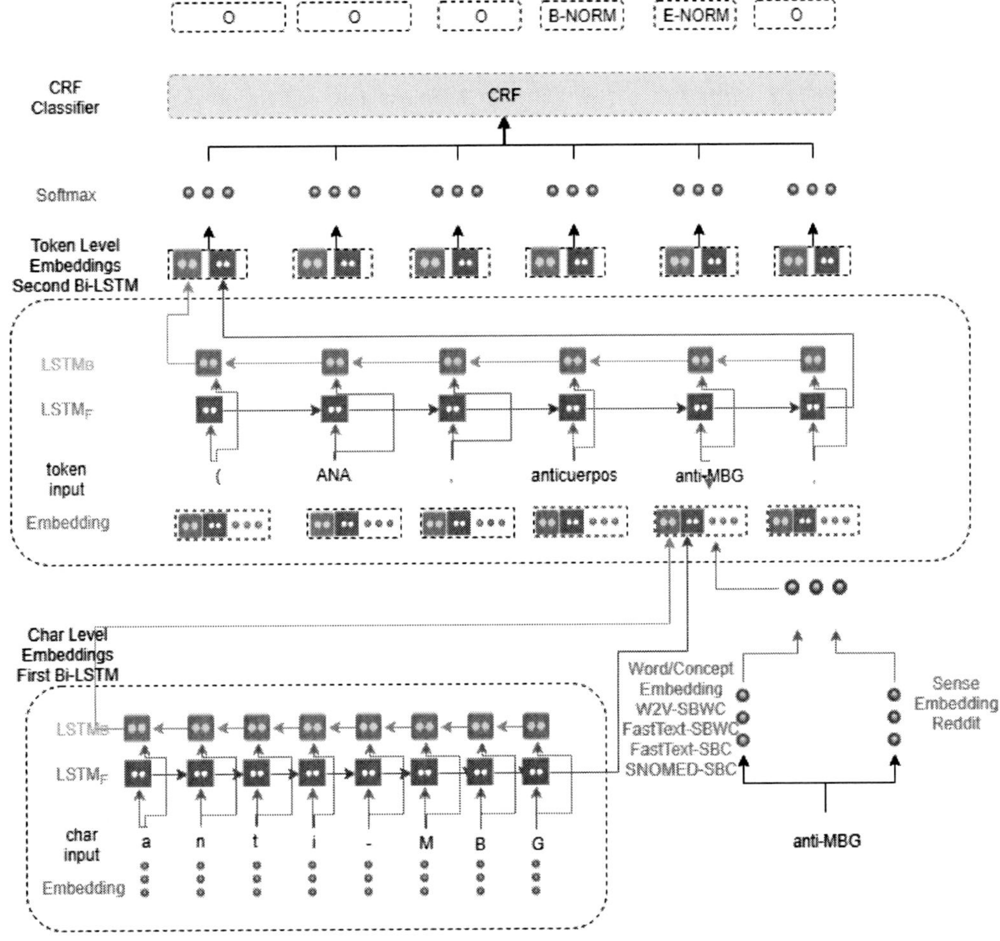

Figure 2: The architecture of the hybrid Bi-LSTM CRF model for drug and chemical compounds identifications.

fiers (concept embeddings) and include this in the token embedding layer of the network.

The BMEWO-V encoding format distinguishes the B tag for entity start, the M tag for entity continuity, the E tag for entity end, the W tag for a single entity, and the O tag for other tokens that do not belong to any entity. The V tag allows us to represent nested entities. BMEWO-V is similar to other previous encoding formarts (Borthwick et al., 1998); however, it allows the representation of nested and discontinuous entities. As a result, we obtain our sentences annotated in the CoNLL-2003 format (Tjong Kim Sang and De Meulder, 2003). An example of the BMEWO-V encoding format applied to the sentence "calcio iónico corregido 1,16 mmol/l y magnesio 1,9 mg/dl." ("ionic calcium corrected 1.16 mmol / l and magnesium 1.9 mg / dl.") can be seen in Figure 3 and Table 3.

2.4.1 First Bi-LSTM layer using character embeddings

Word embedding models are able to capture syntactic and semantic information. However, other linguistic information such as morphological information, orthographic transcription, or part-of-speech (POS) tags are not exploited. According to (Ling et al., 2015), the use of character embeddings improves learning for specific domains and is useful for morphologically rich languages. For this reason, we decided to include the character-level representations to obtain morphological and orthographic information from words. Each word is decomposed into its character n-grams and initialized with a random dense vector which is then learned. We used a 25- feature vector to represent each character. In this way, tokens in sentences are represented by their corresponding character embeddings, which are the input for our Bi-LSTM network.

2.4.2 Second Bi-LSTM layer using word and Sense embeddings

The input for the second Bi-LSTM layer is the concatenation of character embeddings from the first layer with the pre-trained word or concept embeddings and sense-disambiguation embeddings (described in sections 2.2 and 2.3) of the tokens in a given input sentence. The second layer goal is to obtain a sequence of probabilities for each tag in the BMEWO-V encoding format. In this way, for each input token, this layer returns six probabilities (one for each tag in BMEWO-V). The final

tag should be with the highest probability for each token.

2.5 Last layer based on Conditional Random Fields (CRF)

To improve the accuracy of predictions, a Conditional Random Field (CRF) (Lafferty et al., 2001) model is trained, which takes as input the label probability for each independent token from the previous layer and obtains the most probable sequence of predicted labels based on the correlations between labels and their context. Handling independent labels for each word shows sequence limitations. For example, considering the drug sequence labeling problem an "I-NORMALIZABLES" tag cannot be found before a "B- NORMALIZABLES" tag or a "B- NORMALIZABLES" tag cannot be found after an "I-NORMALIZABLES" tag. Finally, once tokens have been annotated with their corresponding labels in the BMEWO-V encoding format, the entity mentions must be transformed into the BRAT format. V tags, which identify nested or overlapping entities, are generated as new annotations within the scope of other mentions.

3 Evaluation

As it was described above, our system is based on a deep network with two Bi-LSTM layers and the last layer for CRF. We evaluate our NER system using the train, validation, and test datasets (SPACCC) provided by the PharmaCoNER task organizers (Gonzalez-Agirre et al., 2019). Detailed information for each datasets can be seen in Table 4. The PharmacoNER dataset is a manually annotated corpus of 1,000 clinical cases written in Spanish and annotated with mentions of chemical compounds, drugs, genes, and proteins. The dataset consists of Normalizables (4,398), No Normalizables (50), Proteins (3,009), and Unclear (167) labels. Further details can be found in (Gonzalez-Agirre et al., 2019).

The PharmaCoNER task considers two subtasks. Track 1 consider offset recognition and entity classification of pharmacological substances, compounds, and proteins. Track 2 considers concept indexing where for each entity, the list of unique SNOMED concept identifiers must be generated. Scope level F-measure is used as the main metric where true positives are entities which match with the gold standard clue words and scope

Figure 3: BRAT annotation example from PharmaCoNER corpus sentence.

token	start offset	end offset	tag	tag
calcio	0	6	V-NORMALIZABLES	W-NORMALIZABLES
iónico	8	14	V-NORMALIZABLES	O
corregido	16	25	V-NORMALIZABLES	O
1,16	27	31	O	O
mmol/l	33	39	O	O
y	41	42	O	O
magnesio	43	51	V-NORMALIZABLES	O
1,9	52	55	O	O
mg/dl	57	62	O	O
.	63	64	O	O

Table 3: Tokens annotated with BMEWO-V encoding in the ConLL-2003 format.

Dataset	Subset	Documents	Sentences	Entities
PharmaCoNER	Train	500	8036	3822
	Valid	250	3759	1926
	Test	3751	62000	

Table 4: PharmaCoNER subsets details.

boundaries assigned to the clue word. A detailed description of evaluation can be found in the PharmaCoNER web [3].

3.1 Track 1 - Offset detection and Entity Classification

The NER task is addressed as a sequence labeling task. For track 1 we tested different configurations with various pre-trained embeddings models. The embedding models and their parameters are summarized in Table 5. Table 6 describes our different experiments configurations.

In Table 8, we compare the different pre-trained models in Spanish on the validation dataset. As shown in Table 8 specific domain word embeddings outperform general domain models by almost 5 points. For the test dataset, we applied our best system configuration FastText-SBC + Reddit (see Table 8) obtaining an f-score of 85.24% for offset detection and entity classification. Furthermore, Table 7 shows the classification results obtained by our best system configuration for track 1 with a micro average of 88.10% for valid dataset.

Moreover, we compared our best system configuration (FastText-SBC + Reddit) with the baseline system (NeuroNER without POS and BMEWO-V format encoding) using the same pre-trained models and configuration. Table 9 shows that our extended system outperforms the baseline system, which has proven that POS and BMEWO-V format to be an additional source of information that can be leveraged by neural networks and keep our model domain agnostic. Furthermore, the use of specific domain word embeddings highly improve performance as shown in Table 8.

3.2 Track 2 - Concept Indexing

For track 2, we applied the same approach described for SNOMED-SBC model training in section 2.3 for entities obtained in the previous task. We used the PyMedTermino library employing a two-stage search using full-text search and fuzzy search for concepts not found by partial matching. Table 10 shows our result for valid and test dataset

---

[3] http://temu.bsc.es/pharmaconer/index.php/evaluations

Detail	W2V-SBWC	FastText-SBWC	FastText-SBC	SNOMED-SBC	Reddit
Type	Word	Word	Word	Concept	Sense
Corpus size	1.5 billion	1.5 billion	6 trillion	6 trillion	2 billion
Vocab size	1 million	1 million	2 million	2 million	1 million
Array size	300	300	300	300	128
Algorithm	Word2Vec Skip-gram BOW	FastText Skip-gram BOW	FastText Skip-gram BOW	FastText Skip-gram BOW	Sense2Vec

Table 5: Embedding models details.

Parameter	Run 1	Run 2	Run 3	Run 4
Sense-disambiguation embedding dimension	128	128	128	128
Pre-trained word embeddings	FastText-SBC + Reddit	W2V-SBWC + Reddit	FastText-SBWC + Reddit	SNOMED-SBC + Reddit
Word embeddings dimension	300	300	300	300
Character embedding dimension	50	50	50	50
Hidden layers dimension (for each LSTM)	100	100	100	100
Learning method	SGD	SGD	SGD	SGD
Dropout rate	0.5	0.5	0.5	0.5
Learning rate	0.005	0.005	0.005	0.005
Epochs	100	100	100	100

Table 6: System hyperparameters for each run.

Entity	Precision (%)	Recall (%)	F-score (%)
Normalizables	92.38	86.41	89.29
No_Normalizables	0.00	0.00	0.00
Proteins	93.29	85.35	89.14
Unclear	87.80	70.59	78.26
Micro-average	91.75	84.74	88.10

Table 7: Results for valid dataset entities.

Experiment	Embedding Model	Precision (%)	Recall (%)	F-score (%)
Run 4	SNOMED-SBC + Reddit	83.52	74.97	79.02
Run 2	W2V-SBWC + Reddit	83.85	75.75	79.60
Run 3	FastText-SBWC + Reddit	84.70	77.31	80.84
Run 1	FastText-SBC + Reddit	**89.13**	**82.61**	**85.75**

Table 8: Embeddings model results for track 1 on valid dataset.

for track 2.

Our results for track 2 are low due to a large number of misspellings that exceed the similarity threshold such as "diacepam" ("diazepam"), drug names where the identifier corresponds to the active substance as "durogesic" ("Duragesic") active ingredient "fentanyl" ("fentanyl"), identifiers not existing in SNOMED CT, such as CHEBI:135810

System	Precision (%)	Recall (%)	F-score (%)
NeuroNER	86.38	82.07	84.16
Extended NeuroNER	**89.13**	**82.61**	**85.75**

Table 9: Baseline comparison for track 1 on valid dataset.

and 373757009 and false positives, such as diseases identified as NORMALIZABLE entities and PROTEIN tokens not annotated in the corpus.

4 Conclusions

In this work, we propose a system for the detection of chemical compounds, drugs, genes, and proteins in clinical narrative written in Spanish. We address the named entity recognition task as a sequence labeling task. Our hybrid model based on machine and deep learning approaches only use dense vector representations features instead of hand-crafted word-based features. We proved that as in other tasks such as NER, the use of dense representation of words such as word-level embeddings, character-level embeddings, and sense embeddings are helpful for named entity recognition. The hybrid system achieves satisfactory performance with F-score over 85%. The extension of NeuroNER network is domain-independent and could be used in other fields, although generic prebuilt word embeddings are used, new medical Spanish word and concept embeddings have been generated for this work.

As future work, we plan to enhance the SNOMED-CT concept embeddings and analyze why its performance is lower than the medical word embeddings. We plan to test whether other supervised classifiers such as Markov Random Fields, Optimum-Path-Forest, or CRF as RNN would obtain more benefit from dense vector representation. That is to say, we would use the same continuous representations with the after-mentioned classifiers. Apart from that, we could train word embeddings obtained from multiple multilingual biomedical corpus to obtain multilingual word representations and test other word representation algorithms such as concept embeddings using UMLS or other biomedical unique concept identifier dictionary. The motivation would be to see whether word embeddings generated with multilingual biomedical domain texts can help to improve the results and provide a deep learning model language and domain-independent.

Funding

This work was supported by the Research Program of the Ministry of Economy and Competitiveness - Government of Spain, (DeepEMR project TIN2017-87548-C2-1-R).

References

Marimon Montserrat Krallinger Martin Armengol-Estapé Jordi, Soares Felipe. 2019. Pharmaconer tagger: a deep learning-based tool for automatically finding chemicals and drugs in spanish medical texts. *Genomics Inform*, 17(2):e15–.

Piotr Bojanowski, Edouard Grave, Armand Joulin, and Tomas Mikolov. 2016. Enriching word vectors with subword information. *CoRR*, abs/1607.04606.

Andrew Borthwick, John Sterling, Eugene Agichtein, and Ralph Grishman. 1998. Exploiting diverse knowledge sources via maximum entropy in named entity recognition. In *Sixth Workshop on Very Large Corpora*.

Cristian Cardellino. 2016. Spanish Billion Words Corpus and Embeddings.

Franck Dernoncourt, Ji Young Lee, and Peter Szolovits. 2017. NeuroNER: an easy-to-use program for named-entity recognition based on neural networks. In *Proceedings of the 2017 Conference on Empirical Methods in Natural Language Processing: System Demonstrations*, pages 97–102, Copenhagen, Denmark. Association for Computational Linguistics.

John M Giorgi and Gary D Bader. 2018. Transfer learning for biomedical named entity recognition with neural networks. *Bioinformatics*, 34(23):4087–4094.

Aitor Gonzalez-Agirre, Montserrat Marimon, Ander Intxaurrondo, Obdulia Rabal, Marta Villegas, and Martin Krallinger. 2019. Pharmaconer: Pharmacological substances, compounds and proteins named entity recognition track. In *Proceedings of the BioNLP Open Shared Tasks (BioNLP-OST)*, pages 1–X, Hong Kong, China. Association for Computational Linguistics.

Wahed Hemati and Alexander Mehler. 2019. Lstmvoter: chemical named entity recognition using a conglomerate of sequence labeling tools. *Journal of Cheminformatics*, 11(1):3.

Dataset	Precision (%)	Recall (%)	F-score (%)
valid	51.72	50.57	51.14
test	50.00	49.28	49.64

Table 10: Results for PharmaCoNER track 2 on valid and test dataset.

John D. Lafferty, Andrew McCallum, and Fernando C. N. Pereira. 2001. Conditional random fields: Probabilistic models for segmenting and labeling sequence data. In *Proceedings of the Eighteenth International Conference on Machine Learning*, ICML '01, pages 282–289, San Francisco, CA, USA. Morgan Kaufmann Publishers Inc.

Jean Baptiste Lamy, Alain Venot, and Catherine Duclos. 2015. PyMedTermino: An open-source generic API for advanced terminology services. *Studies in Health Technology and Informatics*, 210:924–928.

Wang Ling, Chris Dyer, Alan W. Black, Isabel Trancoso, Ramon Fermandez, Silvio Amir, Luis Marujo, and Tiago Luis. 2015. Finding function in form: Compositional character models for open vocabulary word representation. In *Proceedings of the 2015 Conference on Empirical Methods in Natural Language Processing*, pages 1520–1530, Lisbon, Portugal. Association for Computational Linguistics.

M.A. Martí, M. Taule, M. Bertran, and L. Márquez. 2007. AnCora: Multilingual and Multilevel Annotated Corpora.

Tomas Mikolov, Ilya Sutskever, Kai Chen, Greg Corrado, and Jeffrey Dean. 2013. Distributed representations of words and phrases and their compositionality. *CoRR*, abs/1310.4546.

Frederic P. Miller, Agnes F. Vandome, and John McBrewster. 2009. *Levenshtein Distance: Information Theory, Computer Science, String (Computer Science), String Metric, Damerau?Levenshtein Distance, Spell Checker, Hamming Distance*. Alpha Press.

Lili Mou, Zhao Meng, Rui Yan, Ge Li, Yan Xu, Lu Zhang, and Zhi Jin. 2016. How transferable are neural networks in NLP applications? *CoRR*, abs/1603.06111.

Martín Pérez-Pérez, Obdulia Rabal, Gael Pérez-Rodríguez, Miguel Vazquez, Florentino Fdez-Riverola, Julen Oyarzábal, Alfonso Valencia, Anália Lourenço, and Martin Krallinger. 2017. Evaluation of chemical and gene/protein entity recognition systems at biocreative v.5: the cemp and gpro patents tracks.

Felipe Soares, Marta Villegas, Aitor Gonzalez-Agirre, Martin Krallinger, and Jordi Armengol-Estapé. 2019. Medical word embeddings for Spanish: Development and evaluation. In *Proceedings of the 2nd Clinical Natural Language Processing Workshop*, pages 124–133, Minneapolis, Minnesota, USA. Association for Computational Linguistics.

Víctor Suárez-Paniagua, Renzo M. Rivera Zavala, Isabel Segura-Bedmar, and Paloma Martínez. 2019. A two-stage deep learning approach for extracting entities and relationships from medical texts. *Journal of Biomedical Informatics*, 99:103285.

Erik F. Tjong Kim Sang and Fien De Meulder. 2003. Introduction to the CoNLL-2003 shared task: Language-independent named entity recognition. In *Proceedings of the Seventh Conference on Natural Language Learning at HLT-NAACL 2003*, pages 142–147.

Andrew Trask, Phil Michalak, and John Liu. 2015. sense2vec - A fast and accurate method for word sense disambiguation in neural word embeddings. *CoRR*, abs/1511.06388.

Dong Wang and Thomas Fang Zheng. 2015. Transfer learning for speech and language processing. *CoRR*, abs/1511.06066.

A Neural Pipeline Approach for the PharmaCoNER Shared Task using Contextual Exhaustive Models

Mohammad Golam Sohrab[†], Pham Minh Thang[†],
Makoto Miwa[†,‡], and Hiroya Takamura[†]
[†]Artificial Intelligence Research Center (AIRC)
National Institute of Advanced Industrial Science and Technology (AIST),
2-4-7 Aomi, Koto-ku, Tokyo, 135-0064, Japan
[‡]Toyota Technological Institute, Japan
{sohrab.mohammad, pham.thang, takamura.hiroya}@aist.go.jp,
makoto-miwa@toyota-ti.ac.jp

Abstract

We present a neural pipeline approach that performs named entity recognition (NER) and concept indexing (CI), which links them to concept unique identifiers (CUIs) in a knowledge base, for the PharmaCoNER shared task on pharmaceutical drugs and chemical entities. We proposed a neural NER model that captures the surrounding semantic information of a given sequence by capturing the forward- and backward-context of bidirectional LSTM (Bi-LSTM) output of a target span using contextual span representation-based exhaustive approach. The NER model enumerates all possible spans as potential entity mentions and classify them into entity types or no entity with deep neural networks. For representing span, we compare several different neural network architectures and their ensembling for the NER model. We then perform dictionary matching for CI and, if there is no matching, we further compute similarity scores between a mention and CUIs using entity embeddings to assign the CUI with the highest score to the mention. We evaluate our approach on the two sub-tasks in the shared task. Among the five submitted runs, the best run for each sub-task achieved the F-score of 86.76% on Sub-task 1 (NER) and the F-score of 79.97% (strict) on Sub-task 2 (CI).

1 Introduction

The PharmaCoNER (Gonzalez-Agirre et al., 2019) shared task[1] is an open challenge that allows participants to use any methodology and knowledge sources for the clinical records with protected health information. The task aims at two sub-tasks in pharmaceuticals drug and clinical domain: named entity recognition (NER), which is officially called NER offset and entity classification, and concept indexing (CI). Among these sub-tasks, we focus on NER since NER has drawn considerable attentions as the first step towards many natural language processing (NLP) applications including relation extraction (Miwa and Bansal, 2016), event extraction (Feng et al., 2016), and co-reference resolution (Fragkou, 2017). Recently, deep neural networks have shown impressive performance on named entity recognition in several domains (e.g., Lample et al. (2016)). Such models achieved state-of-the-art results without requiring any hand-crafted features or external knowledge resources.

In this paper, we present a pipeline approach that addresses both NER and CI. We mainly focus on NER and employ a neural exhaustive model (Sohrab and Miwa, 2018; Sohrab et al., 2019) for NER. The model detects flat and nested entities by reasoning over all the spans within a specified maximum size. Unlike the existing models that rely on token-level labels, our model directly employs an entity type as the label of a span. Each span is represented as the combination of the boundary and inside representations by using the outputs of bidirectional long short-term memory (Bi-LSTM). We employ and compare different span representations following (Sohrab and Miwa, 2018; Sohrab et al., 2019) that leads to propose a new contextual exhaustive models. The original model (Sohrab and Miwa, 2018) simply treated all the tokens in a span equally by taking the average of LSTM outputs corresponding to tokens inside the span for inside representation and concatenated them with boundary representation where context of each span is totally ignored. Sohrab et al. (2019) proposed several extensions for the representation including contextual span representations and several different inside representations. In this approach, the contextual span representations are considered to capture only the previous and next time steps of LSTM output of

---

[1]https://2019.bionlp-ost.org/tasks

Proceedings of the 5th Workshop on BioNLP Open Shared Tasks, pages 47–55
Hong Kong, China, November 4, 2019. ©2019 Association for Computational Linguistics

a target span, where the surrounding context of a sequence from beginning to target span and end to target span as forward- and backward-context are ignored. Unlike the previous methods (Sohrab and Miwa, 2018; Sohrab et al., 2019), the proposed contextual exhaustive approach captures the surrounding context representation of a given sequence by capturing the forward- and backward-context of Bi-LSTM output of a target span; we describe the details in Section 3.1.3. Besides, the contextual exhaustive approach is extended to leverage the output of a morphological analyser. The spans with the representations are classified into their entity types or non-entity. With the mentions predicted by the NER module, we map them to a knowledge base (KB) (i.e., SNOMED-CT) by direct dictionary matching and similarity scores between mentions and the names of their candidate CUI terms. The best run for each sub-task achieved the F-score of 86.76% on sub-task 1 (NER) and the F-scores of 79.97% on sub-task 2 (CI).

2 Related Work

Most NER work focus on flat entities. Lample et al. (2016) proposed a LSTM-CRF (conditional random fields) model and this has been widely used and extended for the flat NER, e.g., Akbik et al. (2018). In recent studies of neural network based flat NER, Gungor et al. (2018, 2019) have shown that morphological analysis using additional word representations based on linguistic properties of the words, especially for morphologically rich languages such as Turkish and Finnish, improves the NER performances further compared with using only representations based on the surface forms of words.

Recently, nested NER has been widely interested in NLP. Zhou et al. (2004) detected nested entities in a bottom-up way. They detected the innermost flat entities and then found other NEs containing the flat entities as sub-strings using rules on the detected entities. The authors reported an improvement of around 3% in the F-score under certain conditions on the GENIA data set (Collier et al., 1999). Recent studies show that the conditional random fields (CRFs) can produce significantly higher tagging accuracy in flat or nested (stacking flat NER to nested representation) NERs (Son and Minh, 2017). Ju et al. (2018) proposed a novel neural model to address nested

entities by dynamically stacking flat NER layers until no outer entities are extracted. A cascaded CRF layer is used after the LSTM output in each flat layer. The authors reported that the model outperforms state-of-the-art results by achieving 74.5% in F-score on the GENIA data set.

Sohrab and Miwa (2018) proposed a neural model that detects nested entities using exhaustive approach and outperforms the state-of-the-art results by achieving 77.1% in terms of F-score on the GENIA data set. Sohrab et al. (2019) further extended the span representations for entity recognition and addressed sensitive span detection tasks in the MEDDOCAN (MEDical DOCument ANonymization) shared task[2], and the system achieved 93.12% and 93.52% in terms of F-score for NER and sensitive span detection, respectively.

3 Pipeline Approach for NER and Concept Indexing

The pipeline approach consists of two modules:

- Named entity recognition that uses a contextual neural exhaustive approach

- Concept indexing (CI) that generates the list of unique SNOMED concept identifiers of the mentions that are detected by the NER module for each document.

3.1 Neural Named Entity Recognition

We solve the NER task, first by employing a neural exhaustive model (Sohrab and Miwa, 2018; Sohrab et al., 2019) that leads to implement a new contextual exhaustive approach, exhaustively considers all possible contextual spans in a sentence using a single neural network. The model detects nested entities by enumerating all possible contextual spans. The model is built upon a shared bidirectional LSTM (Bi-LSTM) layer, and we consider several different representations for the contextual span using the outputs of Bi-LSTM. Figure 1 shows the contextual exhaustive model to detect the possible mentions. The proposed neural contextual exhaustive model consists of embedding, bidirectional LSTM and exhaustive layers. we will explain each layer in the following subsections.

---

[2]http://temu.bsc.es/meddocan/

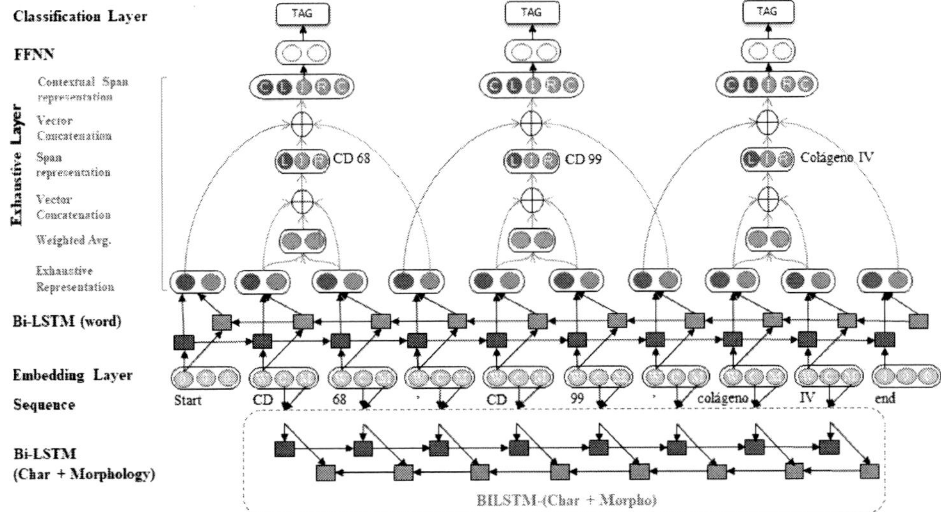

Figure 1: An overview of the exhaustive contextual span representations model. To compute the contextual span representations of 'CD 99', the model concatenates the left-, right-, and inside-representations of Bi-LSTM output vector and further concatenates the contextual information that are represented with the forward LSTM output vector of 'CD' in the previous time step and the backward LSTM output vector of '99' in the previous time step.

3.1.1 Embedding Layer

In the embedding layer, each word is represented by concatenating the pre-trained word embedding and character-based word representation, where we encode the character-level information of the word. The character-based word representation is obtained by feeding the sequence of character embeddings comprising a word to Bi-LSTM and concatenate the forward and backward output representations. Besides, we leverage the morphological analyzer[3] to generate morphological tags, where the tag for each input word is generated by merging the lemma and part-of-speech tag of the word. Then each tag produced by the morphological analyzer is treated as a sequence of characters and encoded using the character-level information using randomly initialized character embeddings. Specifically, we fed the sequence to a separate Bi-LSTM and concatenate the forward and backward outputs to obtain the morphological representation of a word.

3.1.2 Bidirectional LSTM Layer

Given an input sentence sequence $X = \{x_1, ..., x_n\}$ where x_i denotes the i-th word and n denotes the number of words in the sentence sequence, the distributed embeddings of the words

in the sequence from the embedding layer are fed into a Bi-LSTM layer. The Bi-LSTM layer computes the hidden vector sequence in forward $\overrightarrow{\mathbf{h}} = \left[\overrightarrow{\mathbf{h}_1}, \overrightarrow{\mathbf{h}_2}, \ldots, \overrightarrow{\mathbf{h}_n} \right]$ and backward $\overleftarrow{\mathbf{h}} = \left[\overleftarrow{\mathbf{h}_1}, \overleftarrow{\mathbf{h}_2}, \ldots, \overleftarrow{\mathbf{h}_n} \right]$ manners. We concatenate the forward and backward outputs as $\mathbf{h}_i = \left[\overrightarrow{\mathbf{h}_i}; \overleftarrow{\mathbf{h}_i} \right]$, where $[;]$ denotes concatenation.

3.1.3 Exhaustive Layer

The exhaustive layer enumerates all possible spans by exhaustive combination. We generate all possible spans with the sizes less than or equal to the maximum span size L, which is a predefined hyper-parameter. We use (i, k) to represent the span from i to k inclusive, where $1 \leq i < k \leq n$ and $k - i < L$. We represent each span using the outputs of the shared underlying LSTM layer and represent span with different ways as in explained later. We then feed the representation of each segmented span to a rectified linear unit (ReLU) as an activation function. Finally, the output of the activation layer is passed to a softmax output layer to classify the span into a specific entity type.

In the latter part of this section, we introduce the span representations and its several enhancements.

Contextual Span Representations with Averaging For contextual span representations (Sohrab

---

[3] https://github.com/PlanTL-SANIDAD/SPACCC_POS-TAGGER

et al., 2019), we represent the span with three separate representations: the surrounding context representation, the boundary representation for span detection and the inside representation for semantic type classification. We capture the context representation of a given sequence from Bi-LSTM output $\mathbf{h_i}$. Specifically, we obtain the contextual span representation by capturing the forward- and backward-context of Bi-LSTM output of a target span (i, k) by concatenating vector output of previous $\overrightarrow{\mathbf{h}}_{i-1}$ in forward manner, and output of previous $\overleftarrow{\mathbf{h}}_{i-1}$ in backward manner. The boundary representation is prepared to capture both ends of the span. For this, we rely on the outputs of the Bi-LSTM layer corresponding to the boundary words of a target span. The inside representation is prepared to capture its semantic type by encoding the whole semantic information of the span. We use the average of all the outputs corresponding to the words in the span for the inside representation. Following the above contextual, boundary, and inside representations, we represent the representation $\mathbf{R}(i, k)^{[F,L,A,R,B]}$ (Forward-context, Left-boundary, inside with Average, Right-boundary, and Backward-context) of the span (i, k) as follows:

$$\mathbf{R}(i, k)^{[F,L,A,R,B]} =$$

$$\left[\overrightarrow{\mathbf{h}}_{i-1}; \mathbf{h}_i; \frac{1}{k-i+1}\sum_{j=i}^{k}\mathbf{h}_j; \mathbf{h}_k; \overleftarrow{\mathbf{h}}_{i-1}\right]. \quad (1)$$

Contextual Span Representations using Attention We also try an attention mechanism (Bahdanau et al., 2015) instead of the average over words in each span. Specifically, we replace the inside representations using attention mechanism as follows:

$$\alpha_t = \mathbf{w}_\alpha FFNN_\alpha\left(\overleftrightarrow{\mathbf{x}}_t\right), \quad (2)$$

$$\alpha_{i,t} = \frac{\exp(\alpha_t)}{\sum_{k=start(i)}^{end(i)}\exp(\alpha_k)}, \quad (3)$$

$$\overline{\mathbf{x}}_i = \sum_{k=start(i)}^{end(i)} \alpha_{i,t}\overleftrightarrow{\mathbf{x}}_t, \quad (4)$$

where $\overleftrightarrow{\mathbf{x}}_t$ is the concatenated output of the Bi-LSTM layer over a span. $\overline{\mathbf{x}}_i$ is a weighted sum of word vectors in span (i, k). Instead of Equation (1), we obtain the representation

$\mathbf{R}(i, k)^{[F,L,\overline{A},R,B]}$ ($\overline{A}$ for inside with Attention-based representation) of the span (i, k) as follows:

$$\mathbf{R}(i, k)^{[F,L,\overline{A},R,B]} = \left[\overrightarrow{\mathbf{h}}_{i-1}; \mathbf{h}_i; \overline{\mathbf{x}}_i; \mathbf{h}_k; \overleftarrow{\mathbf{h}}_{i-1}\right]. \quad (5)$$

Contextual LSTM-Minus-based Span Representations We also try LSTM-Minus (Wang and Chang, 2016) for the boundary representation[4]. The left boundary is computed as the representation of the previous word of the span subtracted from the representation of the last word of the current span. Similarly, the right boundary is computed as the representation of the next word of the span subtracted from the representation of the first word of the current span. In contextual LSTM-Minus-based span representations of an input sequence, we compute the forward- and backward-context of a target span as the same manner that stated to represent the forward- and backward-context representations of $\mathbf{R}(i, k)^{[F,L,A,R,B]}$. We obtain the representation $\mathbf{R}(i, k)^{[F,\overline{L},A,\overline{R},B]}$ ($\overline{L}$ and $\overline{R}$ for Left- and Right-boundary based on LSTM-Minus, respectively) of the span (i, k) as follows:

$$\mathbf{R}(i, k)^{[F,\overline{L},A,\overline{R},B]} = [\overrightarrow{\mathbf{h}}_{i-1}; \mathbf{h}_k - \mathbf{h}_{i-1};$$

$$\frac{1}{k-i+1}\sum_{j=i}^{k}\mathbf{h}_j; \mathbf{h}_i - \mathbf{h}_{k+1}; \overleftarrow{\mathbf{h}}_{i-1}]. \quad (6)$$

Furthermore, the LSTM-Minus based representation using attention can be considered as:

$$\mathbf{R}(i, k)^{[F,\overline{L},\overline{A},\overline{R},B]} = [\overrightarrow{\mathbf{h}}_{i-1};$$

$$\mathbf{h}_k - \mathbf{h}_{i-1}; \overline{\mathbf{x}}_i; \mathbf{h}_i - \mathbf{h}_{k+1}; \overleftarrow{\mathbf{h}}_{i-1}]. \quad (7)$$

Base Span Representations We further consider representations without context representation (Sohrab and Miwa, 2018), which we denote base span representations. For the base span representations, we generate representations by eliminating forward- and backward-context from Equations (1), (5)–(7) and they can be rewritten respectively as:

$$\mathbf{R}(i, k)^{[L,A,R]} = \left[\mathbf{h}_i; \frac{1}{k-i+1}\sum_{j=i}^{k}\mathbf{h}_j; \mathbf{h}_k\right]. \quad (8)$$

---

[4]Note that we used the bi-directional representations to take the differences for LSTM-Minus unlike the original one (Wang and Chang, 2016). The investigation of different formulations is left for future work.

$$\mathbf{R}(i, k)^{[L,\overline{A},R]} = [\mathbf{h}_i; \overline{\mathbf{x}}_i; \mathbf{h}_k]. \qquad (9)$$

$$\mathbf{R}(i, k)^{[\overline{L},A,\overline{R}]} =$$

$$\left[\mathbf{h}_k - \mathbf{h}_{i-1}; \frac{1}{k-i+1}\sum_{j=i}^{k} \mathbf{h}_j; \mathbf{h}_i - \mathbf{h}_{k+1} \right]. \qquad (10)$$

$$\mathbf{R}(i, k)^{[\overline{L},\overline{A},\overline{R}]} = [\mathbf{h}_k - \mathbf{h}_{i-1}; \overline{\mathbf{x}}_i; \mathbf{h}_i - \mathbf{h}_{k+1}]. \qquad (11)$$

3.2 Concept Indexing

The concept indexing (CI) requires to identify a concept unique identifier (CUI) for every mention span of a concept in a document. SNOMED-CT knowledge-base is used to extract all candidates CUI and its term names. For CI, the input is all predicted mention span $M = \{m_1, m_2, \ldots, m_n\}$, where m_i denotes the i-th mention and n denotes the total number of predicted mentions. Each mention is represented as a word sequence $m_i = \{w_1, \ldots, w_k\}$. Each CUI c is an entry in a knowledge base (KB) (i.e., SNOMED-CT). For the CI task, the list of entity mention $\{m_i\}_{i=1,\ldots,T}$ needs to be mapped to a list of corresponding CUIs $\{c_i\}_{i=1,\ldots,T}$.

Using the SNOMED-CT database, we first conduct dictionary look-up matching for each mention m_i with CUIs' term names to retrieve an optimal CUI. If the CUI is not found for a mention, we then compute a similarity score using the dot-product with entity embeddings that supposedly should capture possible related CUIs and select the maximum score to predict the optimal CUI for a mention.

We use fixed, continuous, task-specific entity embeddings, namely the pre-trained entity embeddings of Spanish SNOMED-CT KB by extracting all CUIs term name using GloVe (Pennington et al., 2014). For the multi-token term name of a CUI, we simply compute the average embeddings.

4 Experimental Settings

We provide empirical evidence on the effectiveness of the pipeline architecture in both NER and concept indexing on the PharmaCoNER[5] task of the BioNLP-OST 2019[6]. The PharmaCoNER corpus with four entity types[7] is randomly split into three subsets: train, development and test sets, which contain 500, 250 and 250 clinical cases, respectively.

Our model is implemented in the Chainer[8] deep learning framework. We employed the official PharmCoNER evaluation script[9] to evaluate our system's performances on both tasks.

4.1 Data Pre-processing

Each text and the corresponding annotation file were processed by several simple rules only for tokenization. [10] After tokenization, each text with mapping annotation files were directly passed to the deep neural approach for mention detection and classification. Note that the offsets were restored to the original offsets in evaluation.

4.2 Hyper-parameters

Word representations We generated task specific word embeddings of Spanish PharmaCoNER corpus by merging the raw text of training, development, and test (including background set) sets using GloVe (Pennington et al., 2014). We set the dimension of word embeddings to 200, the dimension of character embeddings for character encoding to 25, and character embeddings for morphological analysis to 25.

Hidden dimensions The hidden states in the LSTMs had 200 dimensions. Each feed forward neural network consisted of two hidden layers with 150 dimensions.

Learning We chose Adam (Kingma and Ba., 2015) as the optimization algorithm with a mini-batch size of 10. We used the same hyper-parameters in all the experiments; we set the gra-

---

[5]http://temu.bsc.es/pharmaconer/

[6]https://2019.bionlp-ost.org/

[7](NORMALIZABLES: mentions of chemicals that can be manually normalized to a unique concept identifier, NO_NORMALIZABLES: mentions of chemicals that could not be normalized manually to a unique concept identifier, PROTEINAS: mentions of proteins, genes, peptides, peptide hormones and antibodies, and UNCLEAR: cases of general substance class mentions of clinical and biomedical relevance)

[8]https://chainer.org/

[9]https://github.com/PlanTL-SANIDAD/PharmaCoNER-Evaluation-Script

[10]Unlike the traditional NER models, our model is independent from traditional 'BIO' tagging scheme, where 'B', 'I', and 'O' stand for 'Begin', 'Inside', and 'Outside' of named entities respectively, so we do not need to assign such tags to the tokens.

	Sub-task 1: NER			Sub-task 2: CI		
Span representation	P(%)	R(%)	F(%)	P(%)	R(%)	F(%)
Ensemble	86.88	86.65	**86.76**	87.53	73.61	**79.97**
CSR-Attn	87.08	85.61	86.34	88.01	73.25	79.95
CLM-Attn	85.32	83.93	84.62	87.98	72.54	79.52
CSR-Avg	84.53	83.67	84.09	86.81	71.83	78.61
BLM-Attn	77.58	88.48	82.67	88.41	68.54	77.22

Table 1: Performance of NER and CI on the test set

Label	P(%)	R(%)	F(%)	Prediction	Annotation	Correct
NORMALIZABLES	89.64	88.08	88.85	956	973	857
UNCLEAR	92.00	67.65	77.97	25	34	23
PROTEINAS	84.19	83.70	83.95	854	859	719
NO_NORMALIZABLES	99.99	10.01	18.18	1	10	1
Overall (micro)	86.88	86.65	86.76	1,836	1,876	1,600

Table 2: Sub-task 1: Categorical performance on the test set

dient clipping to 5, the dropout rate to 0.0 and the Adam hyper-parameters to the default values (Kingma and Ba., 2015). The model was trained for up to 10 epochs, with early stopping based on the performance on the development set.

5   Results and Discussions

In order to evaluate the performance of NER and concept indexing, we conducted experiments on different sets of span representations, including contextual span representation (CSR) with averaging (CSR-Avg), CSR using attention (CSR-Attn), contextual LSTM-Minus-based span representations (CLM) with averaging (CLM-Avg), CLM using attention (CLM-Attn). Besides for base span representations (BSR), BSR with averaging (BSR-Avg), BSR using attentions (BSR-Attn), base LSTM-Minus-based span representation (BLM) with averaging (BLM-Avg), BLM using attention (BLM-Attn) are also considered. We also report the result of ensemble learning that combines the predictions using different span representations to reduce the variance of predictions and reduce the generalization error.

Table 1 shows the five submitted results of NER and CI in terms of F-score on the test set. The top five span representations are chosen based on development score to submit the results. In this table, it is shown that the ensemble approach using maximum voting of all the approaches is effective to improve the system performance both in NER and CI tasks with achieving 86.67% in terms of F-score on NER. In contrast, the CSR-Attn shows the best performance as an individual span representation on NER with achieving 86.34% in terms of F-score.

In the CI task, the ensemble approach shows the best performance by achieving 79.97% in terms of F-score. CSR-Attn achieved 79.95% in terms of F-score as the best individual span representation. The pipeline approach may not be a perfect solution to solve the concept indexing task, where wrong predictions from the NER module will affect the results in the second step.

Table 2 shows the categorical performances using ensemble learning of NER on the test set. In this table, we also break down the number of predicted and correct mentions among the gold annotations. In this table, it can be observed that for the classes of NORMALIZABLES and PROTEINAS, the model shows high performance because there are a reasonable number of training instances for the classes and the mentions in these two classes appeared in the same documents. In contrast, for the rare classes UNCLEAR and NO_NORMALIZABLES, the performances are low. This may be partly due to their low frequency in the training set, making it hard to learn their representation in the network.

5.1   Ablation Study

We show the performances of different NER models for Sub-tasks 1 and 2 on the development set in Table 3 to compare the possible scenarios of the

	Sub-task 1: NER			Sub-task 2: CI		
Span representation	P(%)	R(%)	F(%)	P(%)	R(%)	F(%)
Ensemble	90.82	87.80	**89.28**	88.91	68.46	**77.36**
CSR-Attn	91.54	86.50	88.95	88.33	68.19	76.96
CSR-Avg	87.89	85.93	86.90	86.96	67.25	75.84
CLM-Attn	89.87	84.27	86.98	88.73	68.33	77.20
CLM-Avg	86.80	85.36	86.07	87.51	67.85	76.44
BSM-Attn	86.43	85.31	85.86	87.22	68.39	76.67
BSM-Avg	84.99	86.45	85.71	87.33	68.12	76.54
BLM-Attn	87.15	85.20	86.16	88.91	67.38	76.66
BLM-Avg	87.32	84.37	85.82	87.57	67.25	76.07

Table 3: Performance of NER and CI on the development set

Label	P(%)	R(%)	F(%)	Prediction	Annotation	Correct
NORMALIZABLES	91.61	88.58	90.07	1,084	1,121	993
UNCLEAR	91.11	93.18	92.13	45	44	41
PROTEINAS	89.63	86.98	88.28	723	745	648
NO_NORMALIZABLES	90.00	56.25	69.23	10	16	9
Overall (micro)	90.82	87.80	89.28	1,862	1,926	1,691

Table 4: Sub-task 1: Categorical performances on the development set

given solutions and to report the best system submissions for NER and CI. The Sub-tasks 1 and 2 results in Table 3 shows that almost all the results in different approaches are close to each other to solve the Sub-tasks 1 and 2. The top four models (i.e., CSR-Attn, CLM-Attn, CSR-Avg, and BLM-Attn) and the ensemble of eight models are considered for test evaluation. As for the single NER model, the results on Sub-tasks 1 and 2 in Table 3 show that attention performs better than averaging when the other settings are same. LSTM-Minus helps when there is no contextual information, but it does not help when there is contextual information.

In the CI task on development set, the ensemble approach shows the best performance by achieving 77.36% in terms of F-score. CLM-Attn achieved 77.20% in terms of F-score as the best individual span representation.

Table 4 shows the categorical performances using ensemble learning of NER on the development set. In this table, it seems that the model is well generalized to detect the mentions of each classes including rare classes such as UN-CLEAR and NO_NORMALIZABLES on development set. The categorical performances of NORMALIZABLES and PROTEINAS in terms of F-score are dropped marginally from devel-

opment to test scores by 1.22% and 4.33%, respectively. But it is surprising that the categorical performances of the rare classes UN-CLEAR and NO_NORMALIZABLES, where the performances in terms of F-score are significantly dropped by 14.16% and 51.05% respectively, that affect the overall F-score of test set. We remain this analysis for our future work.

6 Conclusion

This paper presented a pipeline approach that integrates the contextual that captures the surrounding context of a target span and non-contextual neural exhaustive models, which consider all possible spans exhaustively, for named entity recognition (NER) and dictionary and similarity score-based matching for concept indexing (CI), without depending on any external NLP tools. The proposed contextual exhaustive model is capable to detect flat and nested entities from the generated mention candidates of all possible spans. The model obtains the representation of each span using the outputs of the underlying shared bidirectional LSTM layer, and it represents the different spans by concatenating forward- and backward-context, boundary and inside representations of the span. Several enhancements, namely contextual span representation, average representation, attention mechanism, LSTM-Minus, and ensembling are investigated for the representations. It then classifies the span into an entity type or nonentity. To predict the concept unique identifier (CUI) of a mention, the system performs dictionary matching and then computes a similarity score for a mention with no matching using entity embeddings. Among the five submitted runs, the best run for each Sub-task achieved the F-score of 86.76% on Sub-task 1 (NER) and the F-scores of 79.97% on Sub-task 2 (CI).

In the future direction, we will implement a joint modeling that directly recognize entity mentions and link them to a concept unique identifier in an end-to-end manner.

Acknowledgments

We thank the anonymous reviewers for their valuable comments. This paper is based on results obtained from a project commissioned by the New Energy and Industrial Technology Development Organization (NEDO).

References

Alan Akbik, Duncan Blythe, and Roland Vollgraf. 2018. Contextual string embeddings for sequence labeling. In *COLING 2018, 27th International Conference on Computational Linguistics*, pages 1638–1649.

Dzmitry Bahdanau, Kyunghyun Cho, and Yoshua Bengio. 2015. Neural machine translation by jointly learning to align and translate. In *ICLR 2015*.

N. Collier, H. S. Park, N. Ogata, Y. Tateisi, C. Nobata, T. Ohta, T. Sekimizu, H. Imai, K. Ibushi, and Jun'ichi Tsujii. 1999. The GENIA Project: Corpus-based Knowledge Acquisition and Information Extraction from Genome Research Papers. In *Proceedings of EACL*, pages 171–172. ACL.

Xiaocheng Feng, Lifu Huang, Duyu Tang, Heng Ji, Bing Qin, and Ting Liu. 2016. A Language-Independent Neural Network for Event Detection. In *Proceedings of the 54th Annual Meeting of the ACL (Volume 2: Short Papers)*, pages 66–71, Berlin, Germany.

Pavlina Fragkou. 2017. Applying named entity recognition and co-reference resolution for segmenting english texts. *Progress in Artificial Intelligence*, 6(4):325–346.

Aitor Gonzalez-Agirre, Montserrat Marimon, Ander Intxaurrondo, Obdulia Rabal, Marta Villegas, and Martin Krallinger. 2019. Pharmaconer: Pharmacological substances, compounds and proteins named entity recognition track. In *Proceedings of the BioNLP Open Shared Tasks (BioNLP-OST)*, pages 1–X, Hong Kong, China. Association for Computational Linguistics.

Onur Gungor, Tunga Gungor, and Suzan Uskudarli. 2019. The effect of morphology in named entity recognition with sequence tagging. *Natural Language Engineering*, 25(1):147–169.

Onur Gungor, Suzan Uskudarli, and Tunga Gungor. 2018. Improving named entity recognition by jointly learning to disambiguate morphological tags. In *COLING 2018, 27th International Conference on Computational Linguistics*, pages 2082–2092.

Meizhi Ju, Makoto Miwa, and Sophia Ananiadou. 2018. A Neural Layered Model for Nested Named Entity Recognition. In *Proceedings of the 2018 Conference of the North American Chapter of the Association for Computational Linguistics: Human Language Technologies, Volume 1 (Long Papers)*, pages 1446–1459, New Orleans, Louisiana. ACL.

Diederik Kingma and Jimmy Ba. 2015. Adam: A Method for Stochastic Optimization. In *ICLR*.

Guillaume Lample, Miguel Ballesteros, Sandeep Subramanian, Kazuya Kawakami, and Chris Dyer. 2016. Neural Architectures for Named Entity Recognition. In *Proceedings of the 2016 Conference of the North American Chapter of the ACL: Human Language Technologies. ACL*, volume 1, pages 260–270, San Diego, California. ACL.

Makoto Miwa and Mohit Bansal. 2016. End-to-End Relation Extraction using LSTMs on Sequences and Tree Structures. In *Proceedings of the 54th Annual Meeting of the ACL*, pages 1105–1116, Berlin, Germany. ACL.

Jeffrey Pennington, Richard Socher, and Christopher D Manning. 2014. Glove: Global vectors for word representation. In *Proceedings of EMNLP 2014*.

Mohammad Golam Sohrab and Makoto Miwa. 2018. Deep exhaustive model for nested named entity recognition. In *Proceedings of the 2018 Conference on Empirical Methods in Natural Language Processing*, pages 2843–2849, Brussels, Belgium. Association for Computational Linguistics.

Mohammad Golam Sohrab, Pham Minh Thang, and Makoto Miwa. 2019. A generic neural exhaustive approach for entity recognition and sensitive span detect. In *Proceedings of the Iberian Languages Evaluation Forum (IberLEF 2019)*, pages 735–743, Span. IberLEF 2019.

Nguyen Truong Son and Nguyen Le Minh. 2017. Nested Named Entity Recognition Using Multilayer Recurrent Neural Networks. In *Proceedings of PACLING 2017*, pages 16–18, Sedona Hotel, Yangon, Myanmar.

Wenhui Wang and Baobao Chang. 2016. Graph-based dependency parsing with bidirectional LSTM. In *Proceedings of the 54th Annual Meeting of the Association for Computational Linguistics (Volume 1: Long Papers)*, pages 2306–2315, Berlin, Germany. Association for Computational Linguistics.

Guodong Zhou, Jie Zhang, Jian Su, Dan Shen, and Chewlim Tan. 2004. Recognizing Names in Biomedical Texts: a Machine Learning Approach. *Bioinformatics*, 20(7):1178–1190.

Biomedical Named Entity Recognition with Multilingual BERT

Kai Hakala, Sampo Pyysalo
Turku NLP Group, University of Turku, Finland
{first.last}@utu.fi

Abstract

We present the approach of the Turku NLP group to the PharmaCoNER task on Spanish biomedical named entity recognition. We apply a CRF-based baseline approach and multilingual BERT to the task, achieving an F-score of 88% on the development data and 87% on the test set with BERT. Our approach reflects a straightforward application of a state-of-the-art multilingual model that is not specifically tailored to either the language nor the application domain. The source code is available at: https://github.com/chaanim/pharmaconer

1 Introduction

Named entity recognition (NER) is a fundamental task in information extraction, and the ability to detect mentions of domain-relevant entities such as chemicals and proteins is required for the analysis of texts in specialized domains such as biomedicine. Although a wealth of manually annotated corpora and dedicated NER methods have been introduced for the analysis of English biomedical and clinical texts (e.g. (Leaman and Lu, 2016; Crichton et al., 2017; Weber et al., 2019)), there has been comparatively little work on these basic resources for other languages, including Spanish.

The PharmaCoNER task focuses on pharmacological compound mentions in Spanish clinical texts, promoting the development of biomedical text mining tools for non-English data (Gonzalez-Agirre et al., 2019). Track 1 involves the recognition and classification of entity mentions into upper-level ontological categories (chemical, protein, etc.), and Track 2 the normalization (grounding) of these mentions to identifiers in external resources. We participate in Track 1.

We participate in the PharmaCoNER task using a collection of tools developed for English as well

Item	Train	Devel
Documents	500	250
Tokens	177 022	85 148
Annotations	3 822	1 926
Protein	1 405	745
Chemical(+)	2 304	1 121
Chemical(-)	24	16
Other	89	44

Table 1: Data statistics.

as out-of-domain multilingual models. In particular, we use a freely available NER toolkit, NER-suite, tailored for English biomedical literature and a multilingual neural model, BERT, pretrained on general domain Wikipedia articles. Thus, the emphasis of this work is on analyzing how well such tools can be adapted to new languages and domains with minimal effort. We cast the task as sequence labeling using a conventional in-out-begin (IOB) representation of the data for learning and prediction. The used tools are described in detail in Section 3.

2 Data

The annotation involves four types of entities, labeled in the data as PROTEINAS (proteins, genes, and related entities), NORMALIZABLES (chemicals that can be normalized to external resources), NO_NORMALIZABLES (chemicals that cannot), and UNCLEAR (miscellaneous related entities). In the following, we refer to these respectively as Protein, Chemical(+), Chemical(-) and Other. Table 1 briefly summarizes data statistics. We note that compared to English language biomedical NER resources, the number of annotations is somewhat limited; for example, the JNLPBA shared task (Kim et al., 2004) data contains over 50,000 training examples of similar

Proceedings of the 5th Workshop on BioNLP Open Shared Tasks, pages 56–61
Hong Kong, China, November 4, 2019. ©2019 Association for Computational Linguistics

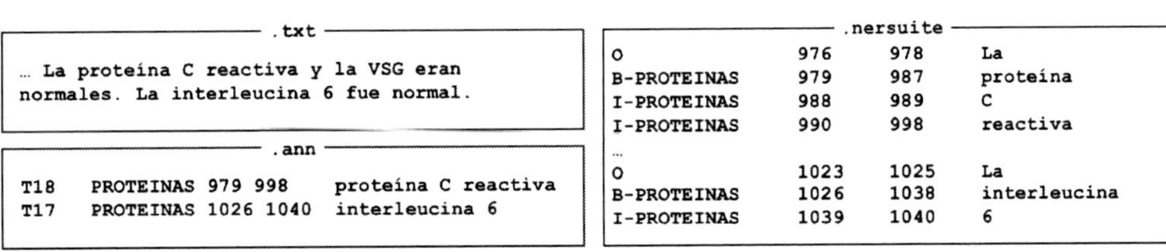

Figure 1: Illustration of data formats. Left: task data in separate `.txt` and `.ann` files. Right: NERsuite format.

types, the BioCreative II GM data (Smith et al., 2008) over 18,000 gene mentions, and the BioCreative CHEMDNER (Krallinger et al., 2015) data over 80,000 chemical mentions. We thus expect that methods addressing the PharmaCoNER task to benefit from pretraining or other similar methods of incorporating information from outside of just the task data.

The task data is distributed in the simple standoff format first introduced for the BioNLP Shared Task 2009 (Kim et al., 2009). To convert this data into a version the column-based IOB format popularized by the CoNLL NER tasks and used by many NER tools, we apply a simple conversion script provided with the BRAT annotation tool[1] (Stenetorp et al., 2012). We note that conversion between the standoff and the token-based IOB representations is lossless if and only if there are no overlapping annotations in the source data and the boundaries of the annotations match token boundaries. Based on an experiment on the training data, we estimate that the conversion preserves the original annotations exactly over 99% of the time. Figure 1 illustrates the two formats.

We note that one training file[2] failed conversion due to an off-by-one offset error. We excluded this file in all of our experiments.

3 Methods

3.1 NERsuite

Conditional Random Fields (CRF) (Lafferty et al., 2001) are a popular and effective model for sequence labeling and thus a relevant baseline in NER work. We perform experiments with NERsuite[3], an NER toolkit that is based on the CRFsuite (Okazaki, 2007) CRF implementation and includes rich features optimized for English biomedical text. In particular, NERsuite incorporates fea-

tures derived from analysis by the GENIA tagger (Tsuruoka et al., 2005), which performs part-of-speech tagging, chunking and lemmatization and has been trained on English text. When applied on Spanish input, the tags and lemmas will necessarily very frequently be incorrect. We nevertheless opted to apply the system as an off-the-shelf baseline as its rich feature set also includes many language-independent features. We leave the NERsuite parameters at their defaults.

3.2 BERT

In our second experiment we utilize BERT (Devlin et al., 2018), a transformer (Vaswani et al., 2017) based attentive neural architecture. Whereas pretrained BERT models have shown strong performance for English NER tasks (Peng et al., 2019), to our knowledge no pretrained Spanish BERT models are readily available[4]. Thus we conduct our experiments with the multilingual BERT model (Pires et al., 2019) trained on a Wikipedia corpus, covering 104 languages. Whereas Spanish is one of the pretraining languages used for the model, the used Wikipedia corpus is not specific to clinical or biomedical content. We use the cased variant of the model, which preserves the case and accents of the characters. BERT relies on subword units, shared between all the used languages, leading to subword embeddings which can benefit from the commonalities of similar languages, yet are a compromise across different uses in different languages and domains. For fine-tuning the model, we use the Keras BERT Python library [5].

When fine-tuning the model for the NER task at hand, we replace the original pretraining output layers with a CRF layer and allow the optimizer to adjust all layers of the network. The model is optimized with Adam (Kingma and Ba, 2014) with a

---

[1] `http://brat.nlplab.org`
[2] `S0211-69952015000200015-1`
[3] `http://nersuite.nlplab.org/`

[4] Shallow word embeddings for Spanish are studied e.g. by Soares et al. (2019)
[5] `https://github.com/CyberZHG/keras-bert`

batch size of 16 and a learning rate of 2e-5 warmed up from 2e-7 over the first training epoch.

We train the model for 50 epochs and evaluate the model after every epoch on the development set using entity-level F-scores on subword tokens. The best performing checkpoint is used as the final prediction model, i.e. we use early stopping with a decreasing patience. In addition to early stopping, the model is regularized with dropout (Srivastava et al., 2014) within each transformer block and weight decay (Loshchilov and Hutter, 2017). The dropout is set to 0.1, but the weight decay is selected in a grid search, being the only hyperparameter optimized in our experiments.

As the input for the BERT model we use the CoNLL formatted data identical to the CRF experiments (see Section 2), which is split into sentences and tokenized on word-level. As the BERT model utilizes subword units, we further retokenize every word independently. Due to computational reasons we use a maximum length of 128 subword units for the input. This limit permits running the model on low memory consumer-grade GPUs instead of requiring data center hardware. Sentences longer than the limit are split into separate input sequences for the network. Note that this may occasionally split entities into separate example sequences leading to sequences starting with I tags. When converting the predictions back to the word-level CoNLL format, we assign the predicted entity label of the first subword unit for the entire token.

4   Results and discussion

The official PharmaCoNER evaluation criteria measure performance on the level of entity mentions (rather than e.g. tokens) and require exact identification of the offset where each mention occurs and the type of the mentioned entity. We note that this common but fairly stringent criterion penalizes many small divergences from the reference annotation twice: if a predicted entity is otherwise correct but e.g. differs in its ending offset from a gold standard entity, the predicted mention is considered a false positive, and the corresponding gold standard entity a false negative. Performance is evaluated in terms of precision, recall, and balanced F-score over all entity types (microaverage). To provide a more fine-grained look into the performance of our approach, we perform additional analyses breaking down performance by type as

well as considering approximate matching criteria, namely *left* boundary matching where only the start offsets of mentions is required to match, *right* boundary matching where only end is required, and *overlap* matching, where any overlapping spans are considered a match. We require entity types to match for all criteria.

The detailed evaluation for the NERsuite and BERT models on the development set are listed in Tables 2 and 3, where *exact* matching criterion corresponds to the official evaluation. The NERsuite model achieves and overall F-score of 82% showing surprisingly strong performance considering the fact that it relies on English part-of-speech tagging, chunking and lemmatization models. The BERT model surpasses this baseline by +6.5pp with an overall F-score of 88%. We used the BERT model as our official submission to the shared task resulting in an F-score of 87.38% on the test set according to the organizers. For both of these models the *overlap* evaluation shows an improvement of 3–4pp, suggesting that the models are in effect better at detecting the entities, but suffer from slightly inaccurate boundary detection. For the BERT model the difference between *exact* and *overlap* results is slightly larger, which might be caused by the additional retokenization to subword units and detokenization back to the original CoNLL format. As the overall performance of the BERT model is notably better than NERsuite's, we focus on the former in all further analyses.

To measure the BERT model's ability to generalize to unseen entity mentions, we analyze how many of the development data entity spans are not present in the training data and how well the model performs on these entities in comparison to entity spans which the model has seen during training. We observe that 55% of the unique entity spans, covering 36% of all occurrences, in the development set are not present in the training data. This suggests that strong generalization abilities are required from the model to perform well in the task.

To obtain a rough understanding of how well the model performs on the entities unseen during training, we measure the recall of the model separately for entity spans seen and not seen during training (Table 4). As can be expected the model has an extremely high recall of 96% for spans present in the training data, but also relatively strong performance with recall of 72% for previously unseen spans. This suggests that the

Criterion	Protein	Chemical(+)	Chemical(-)	Other	Overall
Exact	88.89/74.09/80.82	93.41/75.76/83.66	0.00/0.00/0.00	79.31/52.27/63.01	91.35/73.95/81.73
Left	92.43/77.05/84.04	95.16/77.18/85.24	0.00/0.00/0.00	82.76/54.55/65.75	93.85/75.97/83.97
Right	91.63/76.38/83.31	94.73/76.83/84.84	0.00/0.00/0.00	79.31/52.27/63.01	93.21/75.45/83.40
Overlap	95.01/79.19/86.38	96.15/78.16/86.23	0.00/0.00/0.00	82.76/54.55/65.75	95.45/77.37/85.47

Table 2: NERsuite development set results for various boundary matching criteria (precision/recall/F-score).

Criterion	Protein	Chemical(+)	Chemical(-)	Other	Overall
Exact	84.87/88.86/86.82	92.99/87.51/90.17	40.00/12.50/19.05	76.47/88.64/82.11	89.05/87.44/88.24
Left	89.36/93.56/91.41	95.73/90.10/92.83	40.00/12.50/19.05	78.43/90.91/84.21	92.49/90.81/91.64
Right	87.44/91.54/89.44	93.65/88.14/90.81	40.00/12.50/19.05	76.47/88.64/82.11	90.48/88.84/89.65
Overlap	91.92/95.30/93.58	96.02/90.72/93.30	40.00/12.50/19.05	78.43/90.91/84.21	93.71/91.85/92.77

Table 3: BERT development set results for various boundary matching criteria (precision/recall/F-score).

Entities	Pretraining	No pretraining
All	87.44	54.00
Seen	96.13	70.16
Unseen	71.72	24.78

Table 4: Recall of the BERT model on development set with and without pretraining on all entities, entity spans which are also present in the training data (seen) and entity spans which do not appear in the training data (unseen).

Pretraining	Precision	Recall	F-score
Yes	89.05	87.44	88.24
No	57.62	54.00	55.75

Table 5: Development set results for BERT model with and without pretraining.

model has either learned suitable subword representations during the pretraining for detecting pharmacological entities or is able to effectively utilize the context in which they appear.

As the model is pretrained on multilingual out-of-domain data, we are also interested in the benefits of such pretraining. To this end we train an identical model with randomly initialized weights as the starting point. The same subword unit vocabulary is used. This model results in far inferior performance with an F-score of 56% (see Table 5). Moreover the recall of unseen entity spans is mere 25%, whereas for previously seen spans the recall is 70%. Thus the pretraining, even with multilingual Wikipedia data, seems to offer drastic improvements to the model, particularly for detecting entity spans not seen during training. However, using the same vocabulary makes this comparison slightly unfair as subword embeddings are left in their random initial state if not present in the training data. In the development set this impacts around 12% of the unique subword units, which however constitute only 2% of all subword occurrences.

We also note that although we have used a CRF layer as the output of the BERT model, in our preliminary experiments we observed similar results with a fully connected output layer. This suggests that the transformer architecture has the capability of implicitly modelling sequential dependencies of the output labels, unlike earlier neural models such as bidirectional LSTM networks, which still substantially benefit from the added CRF output layer (Ma and Hovy, 2016; Lample et al., 2016).

5 Conclusions and future work

In this study we have demonstrated that strong results for Spanish clinical NER can be achieved with straightforward adaptation of multilingual or English text mining tools. In particular the multilingual BERT model pretrained on general domain Wikipedia articles shows competitive performance with an F-score of 87% in the official PharmaCoNER evaluation.

As prior studies have shown that the multilingual BERT model can also be utilized in zero-shot settings (Pires et al., 2019), as a future work, we will look into optimal ways of incorporating English NER datasets in this task. This can be either achieved in zero-shot setting, training the model purely on English NER datasets and applying on Spanish texts or by combining both English and Spanish training data in a multitask setting.

In addition to studying the BERT model, we have demonstrated that a strong baseline system for this task can also be achieved with the NERsuite toolkit, even though it relies on feature representations built upon POS tagging and chunking models trained on English data, warranting the use of such freely available tools even in cross-lingual settings.

Acknowledgments

We are grateful to CSC – IT Center for Science for computational resources used to train our models.

References

Gamal Crichton, Sampo Pyysalo, Billy Chiu, and Anna Korhonen. 2017. A neural network multi-task learning approach to biomedical named entity recognition. *BMC bioinformatics*, 18(1):368.

Jacob Devlin, Ming-Wei Chang, Kenton Lee, and Kristina Toutanova. 2018. BERT: Pre-training of deep bidirectional transformers for language understanding. *arXiv preprint arXiv:1810.04805*.

Aitor Gonzalez-Agirre, Montserrat Marimon, Ander Intxaurrondo, Obdulia Rabal, Marta Villegas, and Martin Krallinger. 2019. PharmaCoNER: Pharmacological substances, compounds and proteins named entity recognition track. In *Proceedings of the BioNLP Open Shared Tasks (BioNLP-OST)*, Hong Kong, China. Association for Computational Linguistics.

Jin-Dong Kim, Tomoko Ohta, Sampo Pyysalo, Yoshinobu Kano, and Jun'ichi Tsujii. 2009. Overview of BioNLP'09 shared task on event extraction. In *Proceedings of BioNLP Shared Task*, pages 1–9.

Jin-Dong Kim, Tomoko Ohta, Yoshimasa Tsuruoka, Yuka Tateisi, and Nigel Collier. 2004. Introduction to the bio-entity recognition task at JNLPBA. In *Proceedings of JNLPBA*, pages 70–75.

Diederik P Kingma and Jimmy Ba. 2014. Adam: A method for stochastic optimization. *arXiv preprint arXiv:1412.6980*.

Martin Krallinger, Obdulia Rabal, Florian Leitner, Miguel Vazquez, David Salgado, Zhiyong Lu, Robert Leaman, Yanan Lu, Donghong Ji, Daniel M Lowe, et al. 2015. The CHEMDNER corpus of chemicals and drugs and its annotation principles. *Journal of cheminformatics*, 7(1):S2.

John Lafferty, Andrew McCallum, and Fernando CN Pereira. 2001. Conditional random fields: Probabilistic models for segmenting and labeling sequence data. In *Proceedings of the ICML*.

Guillaume Lample, Miguel Ballesteros, Sandeep Subramanian, Kazuya Kawakami, and Chris Dyer. 2016. Neural architectures for named entity recognition. In *Proceedings of the 2016 Conference of the North American Chapter of the Association for Computational Linguistics: Human Language Technologies*, pages 260–270, San Diego, California. Association for Computational Linguistics.

Robert Leaman and Zhiyong Lu. 2016. TaggerOne: joint named entity recognition and normalization with semi-markov models. *Bioinformatics*, 32(18):2839–2846.

Ilya Loshchilov and Frank Hutter. 2017. Fixing weight decay regularization in Adam. *arXiv preprint arXiv:1711.05101*.

Xuezhe Ma and Eduard Hovy. 2016. End-to-end sequence labeling via bi-directional LSTM-CNNs-CRF. In *Proceedings of the 54th Annual Meeting of the Association for Computational Linguistics (Volume 1: Long Papers)*, pages 1064–1074, Berlin, Germany. Association for Computational Linguistics.

Naoaki Okazaki. 2007. CRFsuite: a fast implementation of Conditional Random Fields (CRFs).

Yifan Peng, Shankai Yan, and Zhiyong Lu. 2019. Transfer learning in biomedical natural language processing: An evaluation of BERT and ELMo on ten benchmarking datasets. *arXiv preprint arXiv:1906.05474*.

Telmo Pires, Eva Schlinger, and Dan Garrette. 2019. How multilingual is multilingual BERT? *arXiv preprint arXiv:1906.01502*.

Larry Smith, Lorraine K Tanabe, Rie Johnson nee Ando, Cheng-Ju Kuo, I-Fang Chung, Chun-Nan Hsu, Yu-Shi Lin, Roman Klinger, Christoph M Friedrich, Kuzman Ganchev, et al. 2008. Overview of BioCreative II gene mention recognition. *Genome biology*, 9(2):S2.

Felipe Soares, Marta Villegas, Aitor Gonzalez-Agirre, Martin Krallinger, and Jordi Armengol-Estapé. 2019. Medical word embeddings for Spanish: Development and evaluation. In *Proceedings of the 2nd Clinical Natural Language Processing Workshop*, pages 124–133, Minneapolis, Minnesota, USA. Association for Computational Linguistics.

Nitish Srivastava, Geoffrey Hinton, Alex Krizhevsky, Ilya Sutskever, and Ruslan Salakhutdinov. 2014. Dropout: a simple way to prevent neural networks from overfitting. *The journal of machine learning research*, 15(1):1929–1958.

Pontus Stenetorp, Sampo Pyysalo, Goran Topić, Tomoko Ohta, Sophia Ananiadou, and Jun'ichi Tsujii. 2012. BRAT: a web-based tool for NLP-assisted text annotation. In *Proceedings of EACL demonstrations*, pages 102–107.

Yoshimasa Tsuruoka, Yuka Tateishi, Jin-Dong Kim, Tomoko Ohta, John McNaught, Sophia Ananiadou, and Junichi Tsujii. 2005. Developing a robust part-of-speech tagger for biomedical text. In *Panhellenic Conference on Informatics*, pages 382–392. Springer.

Ashish Vaswani, Noam Shazeer, Niki Parmar, Jakob Uszkoreit, Llion Jones, Aidan N Gomez, Łukasz Kaiser, and Illia Polosukhin. 2017. Attention is all you need. In *Advances in neural information processing systems*, pages 5998–6008.

Leon Weber, Jannes Münchmeyer, Tim Rocktäschel, Maryam Habibi, and Ulf Leser. 2019. HUNER: Improving biomedical NER with pretraining. *Bioinformatics*.

Gap in pagination due to unavailable paper.

Pages 62-71

DX-HITSZ at BioNLP-OST 2019: Trigger Word Detection and Thematic Role Identification via BERT and Multitask Learning

Dongfang Li[1], Ying Xiong[1], Baotian Hu[1], Hanyang Du[1]
Buzhou Tang[1,2], Qingcai Chen[1,2]
[1]Harbin Institute of Technology (Shenzhen), Shenzhen, China
[2]Peng Cheng Laboratory, Shenzhen, China
{crazyofapple, xiongying0929, dhy1996525}@gmail.com
{hubaotian, tangbuzhou, qingcai.chen}@hit.edu.cn

Abstract

The prediction of the relationship between the disease with genes and its mutations is a very important knowledge extraction task that can potentially help drug discovery. In this paper, we present our approaches for trigger word detection (task 1) and the identification of its thematic role (task 2) in AGAC track of BioNLP Open Shared Task 2019. Task 1 can be regarded as the traditional name entity recognition (NER), which cultivates molecular phenomena related to gene mutation. Task 2 can be regarded as relation extraction which captures the thematic roles between entities. For two tasks, we exploit the pre-trained biomedical language representation model (i.e., BERT) in the pipe of information extraction for the collection of mutation-disease knowledge from PubMed. And also, we design a fine-tuning technique and extra features by using multi-task learning. The experiment results show that our proposed approaches achieve 0.60 (ranks 1) and 0.25 (ranks 2) on task 1 and task 2 respectively in terms of F_1 metric.

1 Introduction

Using the natural language processing methods to discover and mine drug-related knowledge from text has been a hot topic in recent years. For the goal of drug repurposing, an active gene annotation corpus (AGAC) was developed as a benchmark dataset (Wang et al., 2018b). The AGAC track is part of the BioNLP Open Shared Task 2019, aims to gather text mining approaches among the BioNLP community to propel drug-oriented knowledge discovery. It consists of three tasks for the extraction of mutation-disease knowledge from PubMed abstracts: trigger words NER, thematic roles identification, and mutation-disease knowledge discovery. We participated in the trigger words NER and thematic roles identification tasks.

Recently, pre-trained models have been the dominant paradigm in natural language processing. They achieved remarkable state-of-the-art performance across a wide range of related tasks, such as textual entailment, natural language inference, question answering, etc. BERT, proposed by Devlin et al. (2019), has achieved a better-marked result in GLUE leaderboard with a deep transformer architecture (Wang et al., 2018a). BERT first trains a language model on an unsupervised large-scale corpus, and then the pre-trained model is fine-tuned to adapt to downstream tasks. This fine-tuning process can be seen as a form of transfer learning, where BERT learns knowledge from the large-scale corpus and transfer it to downstream tasks. While BERT was built for general-purpose language understanding, there are also some pre-trained models following BERT architecture that effectively leverage domain-specific knowledge from a large set of unannotated biomedical texts (e.g. PubMed abstracts, clinical notes), such as SciBERT (Beltagy et al., 2019), BioBERT (Lee et al., 2019), NCBI BERT (Peng et al., 2019), etc. These models can effectively transfer knowledge from a large amount of unlabeled texts to biomedical text mining models with minimal task-specific architecture modifications.

In this paper, we investigate different methods to combine and transfer the knowledge from the three different sources and illustrate our results on the AGAC corpus. Our method is based on fine-tuning $BERT_{base}$, NCBI BERT and BioBERT using multi-task learning, which has demonstrated the efficiency of knowledge transformation (Liu et al., 2019) and integrating models for both tasks with ensembles. The proposed methods are proved effective for natural language understanding in the biomedical domain, and we rank first place on task 1 (Trigger words NER) and second place on task 2 (Thematic roles identification).

Proceedings of the 5th Workshop on BioNLP Open Shared Tasks, pages 72–76
Hong Kong, China, November 4, 2019. ©2019 Association for Computational Linguistics

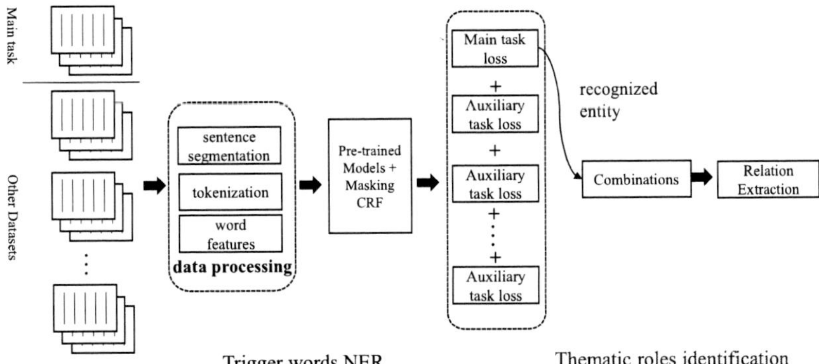

Figure 1: The pipeline of our approach. We first split PubMed abstracts into sentences, tokenize them into words and extract some features like POS tags, then a BERT-based method for NER offset and entity recognition, and finally predict relations for each potential entity pair.

2 Background

The model architecture of BERT (Devlin et al., 2019) is a multi-layer bidirectional Transformer encoder based on the original Transformer model (Vaswani et al., 2017). The input representation is a concatenation of WordPiece embeddings (Wu et al., 2016), positional embeddings, and the segment embedding. A special classification embedding ([CLS]) is inserted as the first token and a special token ([SEP]) is added as the final token. It is firstly pre-trained with two strategies on large-scale unlabeled text, i.e., masked language model and next sentence prediction. The pre-trained BERT model provides a powerful context-dependent sentence representation and can be used for various target tasks, i.e., text classification and machine comprehension, through the fine-tuning procedure.

Hence, the BERT model can be easily extended to the medical domain information extraction pipeline, first extracting the trigger words and then determining the relationship between them, as illustrated in Figure 1.

3 Our Approach

3.1 Task 1: Trigger Words NER

Task 1 aims to identify trigger words in the PubMed digest and annotating them as correct trigger markers or entities (Var, MPA, Interaction, Pathway, CPA, Reg, PosReg, NegReg, Disease, Gene, Protein, Enzyme). It can be seen as an NER task involving the identification of many domain-specific proper nouns in the biomedical corpus.

We first split each PubMed abstracts into sentences using '\n' or '.', and convert each sentence into words by NLTK[1] tokenizer. After that, words are further tokenized into its word pieces $x = (x_1, \ldots, x_T)$. Then we use a representation based on the BERT from the last layer $\mathbf{H} = (h_1, \ldots, h_T)$. In order to make better use of the word-level information, POS tagging labels and word shape embedding representation (Liu et al., 2015) of each word [2] are also concatenated into the output of BERT, passing through a single projection layer, followed by a conditional random fields (CRF) layer with a masking constraint [3] to calculate the token-level label probability $\mathbf{p} = (p_1, \ldots, p_T)$. When fine-tuning the BERT, we found that the performance of the model performed better in the case of BIO for the selection of the tagging schemes compared to BIOES. We further extend our model to multi-task learning joint trained by sharing the architecture and parameters. Although the differences in different datasets, multi-task means joint learning with other biomedical corpora. The assumption is to make more efficient use of the data and to encourage the models to learn more generalized representations. More specially, the same token-level information and BERT encoder are shared and each data set has a specific output layer, e.g., CRF layer. Our final loss function is obtained as follows:

$$-\sum \lambda_{c_i} log P(y_{c_i}|x_{c_i}) + \lambda_r \|W\|_2 \quad (1)$$

---

[1] https://www.nltk.org/

[2] If a word is tokenized into several tokens, each token will be given the same tagging labels.

[3] Transition mask with invalid moves as 0 and valid as 1.

where y_{c_i} denote true tag sequence and x_{c_i} denote the input tokens for corpora c_i, λ_{c_i} and λ_r are weighted parameters.

3.2 Task 2: Thematic Roles Identification

Task 2 is to identify the thematic roles (ThemeOf, CauseOf) between trigger words.

We treat it as a multi-label classification problem by introducing "no relation (NA)" label. When constructing the training data of task 2, we use the relationship of two entities with a distance of no more than one sentence. For NA label, random sampling is performed. In the testing process, relation label will be assigned to the corresponding thematic role when its probability is maximum and larger than the threshold. Otherwise, it will be predicted as no relation. We also anonymously use a predefined tag (such as $\%Disease$) to represent a target named entity. And we additionally append two concrete predicted entity words separated by the [SEP] tag after each sentence. Following Shi and Lin (2019), we also add the token-level relative distance to the subject entity information for each token, i.e. 0 for the position t between two entities, $t - s$ for tokens before first entity and $t - e$ for tokens after second entity, where s, e are the starting and ending positions of first and second entity after tokenization, respectively. The relation logits of two entities are performed using a single output layer from the BERT, as

$$y = \text{softmax}(\mathbf{W}h_{cls} + b) \qquad (2)$$

where h_{cls} denotes the hidden state of the first special token ([CLS]).

4 Experiments

In this section, we provide the leaderboard performance and conduct an analysis of the effect of models from different settings.

4.1 Experimental Setup

The AGAC track organizers develop an active gene annotation corpus (AGAC) (Wang et al., 2018b; Gachloo et al., 2019), for the sake of knowledge discovery in drug repurposing. The track corpus consists of 1250 PubMed abstracts: 250 for public, 1000 for final evaluation. We randomly split the public texts into train and development data sets with the radio of 8:2. The training set is used to learn model parameters, the development set to select optimal hyper-parameters. For

Dataset	#Train	#Dev	#Test
BC5CDR	4,559	4,580	4,796
NCBI disease	5,423	922	939
BC2GM	12,573	2,518	5,037
2010 i2b2/VA	16,315	-	27,626

Table 1: Datasets for joint learning in recognizing the trigger words.

evaluation results, we measure the trigger words recognition and thematic roles extraction performance with F_1 score. Table 1 shows the external data sets used under the joint learning method. The BIO form of these data sets is different from that of task 1, hence we use different projection and CRF layers. But not the more data sets, the better. We found that the NCBI disease (Doğan et al., 2014) and BC5CDR (Li et al., 2016) datasets are helpful for the final results, and the performance is reduced when using BC2GM (Smith et al., 2008) and 2010 i2b2VA dataset (Uzuner et al., 2011).

4.2 Implementation and Hyperparameters

We tried the original BERT[4], BioBERT[5] and NCBI BERT[6] pre-trained models. Each training example is pruned to at most 384 and 512 tokens for named entity recognition (NER) and relation extraction (RE). We use a batch size of 5 for NER, and 32 for RE. We also use the hierarchical learning rate in the training process so that the pre-trained parameters and the newly added parameters converge at different optimization processes. For fine-tuning, we train the models for 20 epochs using a learning rate of 2×10^{-5} for pre-trained weights and 3×10^{-5} for others. The learning parameters were selected based on the best performance on the dev set. For NER, we ensemble 5 models from 5-fold cross-validation and 2 models using the normal training-validation approach. For RE, we ensemble 3 models that used all the construction data in training.

4.3 Main Results

Table 2 compares the results of the two tasks of the pre-trained model in trigger words NER and thematic roles identification. We report the impact of using different pre-training models on the

---

[4]https://github.com/google-research/bert

[5]https://github.com/dmis-lab/biobert

[6]https://github.com/ncbi-nlp/NCBI_BERT

Task	Model	P	R	F1
Trigger Words Recognition	BiLSTM+CRF	0.478	0.408	0.440
	$BERT_{base}$	0.497	0.448	0.471
	NCBI BERT	0.553	0.453	0.498
	BioBERT	0.511	0.529	**0.519**
Thematic Roles Identification	$BERT_{base}$	0.758	0.890	0.818
	NCBI BERT	0.778	0.879	0.826
	BioBERT	0.807	0.891	**0.847**

Table 2: Model comparision in development set with different pre-trained models

development set results. We found that even pre-trained models in the general field are superior to the classic BiLSTM+CRF tagging method (Lample et al., 2016). From the last three lines of each task, we can see that different pre-trained models have different results under the same experimental settings. It proves the effectiveness of pre-training tasks in specific domain. During the fine-tuning process of task 1, we found that the joint extraction of entities with other datasets improved our final results.

Label	P	R	F1
CPA	0.39	0.27	0.32
Disease	0.57	0.57	0.57
Enzyme	0.75	0.16	0.26
Gene	0.71	0.64	0.68
Interaction	0.50	0.29	0.36
MPA	0.46	0.47	0.47
NegReg	0.71	0.62	0.66
Pathway	0.83	0.36	0.50
PosReg	0.64	0.61	0.63
Protein	0.32	0.17	0.22
Reg	0.75	0.50	0.60
Var	0.64	0.63	0.64
ALL	0.63	0.56	0.60

Table 3: Precision (P), Recall (R) and F1 scores in test set of Task 1.

The results for task 1 is summarized in Table 3. The difference in the performance in the different labels is partly sourced by the imbalance distribution of trigger labels in the corpus. Our method ends up first place on the leaderboard and substantially improving upon previous state-of-the-art methods. The results for task 2 is summarized in Table 4. Our method ends up second place on the leaderboard. Our method has a large discrepancy between the development set performance and test set performance. It may be the test set is quite different from our constructed data set. This is

also related to how we use recognized entities, sentence- or document-level combinations.

Label	P	R	F1
CauseOf	0.60	0.26	0.36
ThemeOf	0.63	0.11	0.19
ALL	0.61	0.16	0.25

Table 4: Precision (P), Recall (R) and F1 scores in test set of Task 2.

4.4 Ablation Study

As shown in Table 5, we found that adding a layer of BiLSTM behind the BERT encoder did not improve the performance of the model, resulting in a 0.04 loss of F_1. For NER tasks, external features are effective for the model's performance. So we verified the efficacy of word shape and POS tags on task 1, and we found that adding this information can increase the F_1 value of our model by more than 0.01.

Model	P	R	F1
BioBERT	0.511	0.529	0.519
+ BiLSTM	0.502	0.448	0.473
- Word shape	0.539	0.453	0.492
- POS tags	0.518	0.482	0.499

Table 5: Ablation study of Task 1 in development set.

5 Conclusion

In this paper, we have explored the value of integrating pre-trained biomedical language representation models into a pipe of information extraction methods for collection of mutation-disease knowledge from PubMed. In particular, we investigate the use of three pre-trained models, $BERT_{base}$, NCBI BERT and BioBERT, for fine-tuning on the new task and reducing the risk of overfitting. By considering the relationship between different data sets, we achieve better results. Experimental results on a benchmark annotation of genes with active mutation-centric function changes corpus show that pre-trained representations help improve baseline to attain state-of-the-art performance. In future work, we would like to train the entity recognition and relation extraction tasks simultaneously, reducing the cascading error caused by the pipeline model in biomedical information extraction.

Acknowledgment: This work was supported by Natural Science Foundation of China (Grant No.61872113), and the joint project with Baidu Inc.

References

Iz Beltagy, Arman Cohan, and Kyle Lo. 2019. Scibert: Pretrained contextualized embeddings for scientific text. *CoRR*, abs/1903.10676.

Jacob Devlin, Ming-Wei Chang, Kenton Lee, and Kristina Toutanova. 2019. BERT: pre-training of deep bidirectional transformers for language understanding. In *Proceedings of the 2019 Conference of the North American Chapter of the Association for Computational Linguistics: Human Language Technologies, NAACL-HLT 2019, Minneapolis, MN, USA, June 2-7, 2019, Volume 1 (Long and Short Papers)*, pages 4171–4186.

Rezarta Islamaj Doğan, Robert Leaman, and Zhiyong Lu. 2014. Ncbi disease corpus: a resource for disease name recognition and concept normalization. *Journal of biomedical informatics*, 47:1–10.

Mina Gachloo, Yuxing Wang, and Jingbo Xia. 2019. A review of drug knowledge discovery using bionlp and tensor or matrix decomposition. *Genomics & Informatics*, 17(2).

Guillaume Lample, Miguel Ballesteros, Sandeep Subramanian, Kazuya Kawakami, and Chris Dyer. 2016. Neural architectures for named entity recognition. *arXiv preprint arXiv:1603.01360*.

Jinhyuk Lee, Wonjin Yoon, Sungdong Kim, Donghyeon Kim, Sunkyu Kim, Chan Ho So, and Jaewoo Kang. 2019. Biobert: a pre-trained biomedical language representation model for biomedical text mining. *CoRR*, abs/1901.08746.

Jiao Li, Yueping Sun, Robin J Johnson, Daniela Sciaky, Chih-Hsuan Wei, Robert Leaman, Allan Peter Davis, Carolyn J Mattingly, Thomas C Wiegers, and Zhiyong Lu. 2016. Biocreative v cdr task corpus: a resource for chemical disease relation extraction. *Database*, 2016.

Xiaodong Liu, Pengcheng He, Weizhu Chen, and Jianfeng Gao. 2019. Multi-task deep neural networks for natural language understanding. *arXiv preprint arXiv:1901.11504*.

Zengjian Liu, Yangxin Chen, Buzhou Tang, Xiaolong Wang, Qingcai Chen, Haodi Li, Jingfeng Wang, Qiwen Deng, and Suisong Zhu. 2015. Automatic de-identification of electronic medical records using token-level and character-level conditional random fields. *Journal of biomedical informatics*, 58:S47–S52.

Yifan Peng, Shankai Yan, and Zhiyong Lu. 2019. Transfer learning in biomedical natural language processing: An evaluation of BERT and elmo on ten benchmarking datasets. In *Proceedings of the 18th BioNLP Workshop and Shared Task, BioNLP@ACL 2019, Florence, Italy, August 1, 2019.*, pages 58–65.

Peng Shi and Jimmy Lin. 2019. Simple bert models for relation extraction and semantic role labeling. *arXiv preprint arXiv:1904.05255*.

Larry Smith, Lorraine K Tanabe, Rie Johnson nee Ando, Cheng-Ju Kuo, I-Fang Chung, Chun-Nan Hsu, Yu-Shi Lin, Roman Klinger, Christoph M Friedrich, Kuzman Ganchev, et al. 2008. Overview of biocreative ii gene mention recognition. *Genome biology*, 9(2):S2.

Özlem Uzuner, Brett R South, Shuying Shen, and Scott L DuVall. 2011. 2010 i2b2/va challenge on concepts, assertions, and relations in clinical text. *Journal of the American Medical Informatics Association*, 18(5):552–556.

Ashish Vaswani, Noam Shazeer, Niki Parmar, Jakob Uszkoreit, Llion Jones, Aidan N. Gomez, Lukasz Kaiser, and Illia Polosukhin. 2017. Attention is all you need. In *Advances in Neural Information Processing Systems 30: Annual Conference on Neural Information Processing Systems 2017, 4-9 December 2017, Long Beach, CA, USA*, pages 5998–6008.

Alex Wang, Amanpreet Singh, Julian Michael, Felix Hill, Omer Levy, and Samuel R Bowman. 2018a. Glue: A multi-task benchmark and analysis platform for natural language understanding. *arXiv preprint arXiv:1804.07461*.

Yuxing Wang, Xinzhi Yao, Kaiyin Zhou, Xuan Qin, Jin-Dong Kim, Kevin Bretonnel Cohen, and Jingbo Xia. 2018b. Guideline design of an active gene annotation corpus for the purpose of drug repurposing. In *2018 11th International Congress on Image and Signal Processing, BioMedical Engineering and Informatics (CISP-BMEI)*, pages 1–5. IEEE.

Yonghui Wu, Mike Schuster, Zhifeng Chen, Quoc V. Le, Mohammad Norouzi, Wolfgang Macherey, Maxim Krikun, Yuan Cao, Qin Gao, Klaus Macherey, Jeff Klingner, Apurva Shah, Melvin Johnson, Xiaobing Liu, Lukasz Kaiser, Stephan Gouws, Yoshikiyo Kato, Taku Kudo, Hideto Kazawa, Keith Stevens, George Kurian, Nishant Patil, Wei Wang, Cliff Young, Jason Smith, Jason Riesa, Alex Rudnick, Oriol Vinyals, Greg Corrado, Macduff Hughes, and Jeffrey Dean. 2016. Google's neural machine translation system: Bridging the gap between human and machine translation. *CoRR*, abs/1609.08144.

Gap in pagination due to unavailable paper.

Pages 77-83

Biomedical relation extraction with pre-trained language representations and minimal task-specific architecture

Ashok Thillaisundaram
BenevolentAI
4-6 Maple Street
Bloomsbury, London
W1T 5HD
ashok@benevolent.ai

Theodosia Togia
BenevolentAI
4-6 Maple Street
Bloomsbury, London
W1T 5HD
sia@benevolent.ai

Abstract

This paper presents our participation in the AGAC Track from the 2019 BioNLP Open Shared Tasks. We provide a solution for Task 3, which aims to extract "gene – function change – disease" triples, where "gene" and "disease" are mentions of particular genes and diseases respectively and "function change" is one of four pre-defined relationship types. Our system extends BERT (Devlin et al., 2018), a state-of-the-art language model, which learns contextual language representations from a large unlabelled corpus and whose parameters can be fine-tuned to solve specific tasks with minimal additional architecture. We encode the pair of mentions and their textual context as two consecutive sequences in BERT, separated by a special symbol. We then use a single linear layer to classify their relationship into five classes (four pre-defined, as well as 'no relation'). Despite considerable class imbalance, our system significantly outperforms a random baseline while relying on an extremely simple setup with no specially engineered features.

1 Introduction

Bidirectional Encoder Representations from Transformers (BERT) is a language representation model that has recently advanced the state of the art in a wide range of NLP tasks (e.g. natural language inference, question answering, sentence classification etc.) (Devlin et al., 2018). This is due to its capacity for learning lexical and syntactic aspects of language (Clark et al., 2019; Goldberg, 2019) using large unlabelled corpora. BERT achieves much of its expressive power using a bi-directional Transformer encoder (Vaswani et al., 2017) and a 'predict the missing word" training objective based on Cloze tasks (Taylor, 1953). In the biomedical domain, BioBERT (Lee et al., 2019) and SciBERT (Beltagy et al., 2019)

learn more domain-specific language representations. The former uses the pre-trained BERT-Base model and further trains it with biomedical text (Pubmed[1] abstracts and Pubmed Central[2] full-text articles). The latter trains a BERT model from scratch on a large corpus of scientific text (over 80% biomedical) and learns a domain-specific vocabulary using WordPiece tokenisation (Wu et al., 2016).

BERT has been adapted for use in relation extraction as a basis for supervised, unsupervised and few-shot learning models (Soares et al., 2019). A recent model, Transformer for Relation Extraction (TRE) (Alt et al., 2019) uses an architecture similar to that of BERT by extending the OpenAI Generative Pre-trained Transformer (Radford et al., 2018), in order to perform relation classification for entity mention pairs. In contrast to BERT, TRE uses a next word prediction objective. The model encodes the pairs and their context in a sequence separated by a special symbol. In our model, we use a similar way of encoding gene-disease pairs and their textual context in order to predict their 'function change' relationship, but in contrast to TRE, we leverage SciBERT's domain-specific vocabulary and representations learnt from scientific text.

2 Task and data

Task description Task 3 of the AGAC track of BioNLP-OST 2019 involves Pubmed abstract-level relation extraction of gene-disease relations. The relations of interest concern the function change of the gene which affects the disease. The four relation types are:

- Loss of Function (LOF): a gene undergoes a mutation leading to a loss of function which

---

[1] https://www.ncbi.nlm.nih.gov/pubmed/
[2] https://www.ncbi.nlm.nih.gov/pmc/

Proceedings of the 5th Workshop on BioNLP Open Shared Tasks, pages 84–89
Hong Kong, China, November 4, 2019. ©2019 Association for Computational Linguistics

then affects the disease.

- Gain of Function (GOF): a gene mutation causes a gain of function.

- Regulation (REG): the function change is either neutral or unknown.

- Complex (COM): the function change is too complex to be described as any of the former relations.

To illustrate these relation types more concretely, we repeat the examples given on the task webpage. The following sentence depicts the Gain of Function relation between SHP-2 and juvenile myelomonocytic leukemia: *'Mutations in SHP-2 phosphatase that cause hyperactivation of its catalytic activity have been identified in human leukemias, particularly juvenile myelomonocytic leukemia.'* In this case, 'hyperactivation of catalytic activity' indicates Gain of Function.

An example of the Regulation relation, on the other hand, would be the following sentence: *'Lynch syndrome (LS) caused by mutations in DNA mismatch repair genes MLH1.'*. The phrase 'caused by' demonstrates an association between MLH1 and Lynch syndrome but no information is given on the specific nature of the mechanism relating them.

Annotated corpus The training data provided consist of 250 PubMed abstracts with annotations of the form 'Gene; Function change; Disease' for each abstract. For test data, a further 1000 PubMed abstracts have been provided (without annotations).

Train/dev split Given that no development set had been explicitly provided, we divided the PubMed ids of the original training set into a smaller training and a development set using an 80/20 split, in order to be able to prevent overfitting and perform early stopping. We assigned Pubmed ids to each one of the two sets in the prespecified proportions randomly but choosing a random seed that ensures a small KL divergence between the train and dev class distributions. In the rest of the paper, we use the terms 'train set' or 'training data' to refer to 80% of the original annotated data that we use to train our model.

Generation of negative labels The training data contain some Pubmed ids that have no

relation annotations whatsoever (either from the four pre-defined classes or explicitly negative). However, negative examples are crucial for training a model for such a task given that the majority of gene-disease pair mentions that are found in a randomly selected abstract are not expected to be related with a function change relationship. To generate pairs of negative mentions, we used a widely available Named Entity Recognition (NER) and Entity Linking system (see Section 3) to find mentions of genes and diseases in the abstracts. An entity mention predicted by NER was aligned to a labelled entity mention in the training data if they are both grounded to the same identifier. A pair was aligned if both its entity mentions (gene and disease) could be aligned. In less than 20% of the pairs we performed manual alignment in order to improve the training signal. The dev set, however, was kept intact to ensure strict evaluation. The resulting distribution of relations is highly skewed towards the negative labels ('No relation'); the training set has the following distribution (`No relation: 0.939, GOF: 0.017, LOF: 0.03, REG: 0.012, COM: 0.0007`) while for the dev set, it is (`No relation: 0.935, LOF: 0.027, GOF: 0.019, REG: 0.016, COM: 0.003`). 'COM' is the least represented relationship with only two examples in the train set and two in the dev set.

3 Method

This task can be decomposed into an NER step to obtain all gene-disease mention pairs in an abstract followed by a relation extraction (RE) step to predict the relation type for each mention pair found.

For NER, we use Pubtator (Wei et al., 2013) to recognise spans tagged as genes or diseases. The main focus of our paper is performing relation extraction given NER labels. The reported results, however, don't assume gold NER labels.

Relation Extraction Model Our model is a simple extension of SciBERT (Beltagy et al., 2019) for use in relation extraction, inspired by the encoding of mention pairs and textual context used in (Alt et al., 2019). SciBERT, which utilises the same model architecture as BERT-base, consists of 12 stacked transformer encoders each with 12 attention heads. It is pre-trained using two objectives: Masked language modelling (Cloze task (Taylor, 1953)) and next sentence prediction. When trained, it is provided with sentence pairs represented as follows: `[CLS] This is sentence 1 [SEP] This is sentence 2 [SEP]`. The `[SEP]` token indicates when each sequence ends. The final hidden state of the `[CLS]` token is fed into a classifier to predict whether the two sentences appear sequentially in the corpus. As a result, the final hidden state of the `[CLS]` token learns a representation of the relationship between the two input sentences.

We adapt SciBERT for relation extraction by fine-tuning the representation learnt by the `[CLS]` hidden state. We encode each pair of gene-disease mentions along with the corresponding PubMed abstract in the following format: `[CLS] gene-mention disease-mention [SEP] This is the PubMed abstract [SEP]`. This input data is fed into SciBERT and the final hidden state of its `[CLS]` token is passed into a single linear layer to predict the relation type expressed in that abstract for that gene-disease mention pair. The `[CLS]` hidden state which was pre-trained to learn a representation of the relationship between two sentences is now fine-tuned to learn which relationship class exists between a gene-disease pair (first 'sentence') and a PubMed abstract (second 'sentence'). Our encoding is similar to the approach proposed in (Alt et al., 2019). This adaptation, while simple, is powerful because it is completely agnostic to domain-specific idiosyncrasies; for example, it can be used for any entity types and relation labels. Further, as it has already been pre-trained on a large unstructured corpus, it can be fine-tuned using a considerably smaller dataset.

Model training We use negative log likelihood of the true labels as a loss function. We train for at most 40 epochs with early stopping based on the dev set performance. We used two early stopping criteria alternatives: the macro-averaged F1-score over all labels and over just the positive labels. Training stops if the score used as stopping criterion does not increase on the dev set for 10 consecutive epochs or the maximum number of epochs has been reached. The batch size is chosen to be 32 and the maximum sequence length of each input sequence is set to be 350 Wordpiece (subword) tokens. This is due to memory constraints.

	P	R	F1	Supp.
No rel	0.934	0.372	0.532	627
REG	0.174	0.087	0.116	11
COM	0	0	0	2
LOF	0.076	0.307	0.122	19
GOF	0.022	0.577	0.042	12
Micro-all	0.368	0.368	0.368	
Macro-all	0.241	0.268	0.162	
Micro-pos	0.033	0.322	0.060	
Macro-pos	0.068	0.243	0.070	

Table 1: Model results on the four pre-defined classes, as well as 'No rel' (the negative class) when the macro-averaged F1-score (over the positive labels only) is used as our early stopping criterion. P, R and F1 stand for Precision, Recall and F1-score respectively; support = true positives + false negatives. Micro-all and Macro-all are the micro- and macro-averaged metrics for all classes while Micro-pos and Macro-pos are the micro- and macro-averaged metrics for only the positive classes (i.e. four classes excluding 'No rel').

For each batch, we used down-sampling to ensure that each class was represented equally on average. When training, we observed that our results were very sensitive to the classifier layer weight initialisations. This same behaviour was reported in the original BERT paper (Devlin et al., 2018). To address this, we performed 20 random restarts and selected the model that performs the best on the dev set (for each of the two stopping criteria).

4   Experiments and results

We report the standard classification metrics on the dev set: precision (**P**), recall (**R**), and F1-score (**F1**). For each one of these metrics, we include the macro-averaged values, the micro-averaged values **i)** over all relation labels and **ii)** restricted to just the positive ones. We also report the per-class values (in a one-vs-all fashion). The best results are shown for both of the early stopping criteria used (see Tables 1 and 2).

Random sampling-based baseline We compare our model performance against a simple baseline that predicts the class label by sampling from the categorical distribution of labels as calculated from the training set. Given the strongly skewed class distribution (which has low entropy of 0.46 bits, compared to 2.32 bits for a 5-class uniform distribution, and is therefore highly predictable), this is a strong baseline, especially for metrics re-

	P	R	F1	Supp.
No rel	0.937	0.761	0.840	627
REG	0	0	0	11
COM	0	0	0	2
LOF	0.214	0.040	0.067	19
GOF	0.038	0.429	0.070	12
Micro-all	0.722	0.722	0.722	
Macro-all	0.238	0.246	0.196	
Micro-pos	0.037	0.141	0.059	
Macro-pos	0.063	0.117	0.034	

Table 2: Model results on the four pre-defined classes, as well as 'No rel' (the negative class) when the macro-averaged F1-score (over all labels) is used as our early stopping criterion. All terms used here as defined in Table 1.

	P	R	F1	Supp.
No rel	0.934	0.92	0.927	627
REG	0	0	0	11
COM	0	0	0	2
LOF	0.043	0.053	0.048	19
GOF	0	0	0	12
Micro-all	0.862	0.863	0.863	
Macro-all	0.195	0.195	0.195	
Micro-pos	0.019	0.023	0.021	
Macro-pos	0.011	0.013	0.012	

Table 3: Baseline results on the four pre-defined classes, as well as 'No rel' (the negative class). All terms used here as defined in Table 1.

ported on frequent classes. Table 3 summarises the results, which have been averaged over 1,000 random sampling experiments. As expected, all metrics can achieve high scores on the negative (and by far the largest) class, illustrating how misleading micro-averaging with large classes can be as an indicator of model performance. Some classes have zero scores, which is unsurprising given their very low support in the dev set.

Discussion For both early stopping criteria mentioned above, our model significantly outperformed the random baseline on macro-averaged metrics and per-class metrics. The model obtained relatively good performance on the positive labels especially when taking into account the considerable class imbalance. When optimised to the macro-averaged F1-score over just the positive labels, the model performance was unsurprisingly slightly superior over the positive labels compared

to when optimised using the macro-averaged F1-score over all labels. However, this came at the expense of a loss in recall on the negative labels. To generate predictions on the test set, we chose the model optimised using the macro-averaged F1-score over just the positive labels.

Pubtator NER performance The performance of our relation extraction model is dependent on the results of the named entity recognition tool. Here we briefly summarise the performance of the Pubtator NER tool on the dev set. There are 44 entity pairs with positive labels in the dev set. Of these 44, Pubtator correctly identified 24 of them with an exact string match. For the remaining 20, 14 were identified but it was not an exact string match, and for the other 6, at least one of the entities was not found. We were fairly strict for our dev set evaluation, and so unless there was a perfect string match, the entities were not considered aligned to the labelled data. This would have degraded our performance metrics.

5 Related work

Many biomedical relation extraction systems have often relied hand-crafted linguistic features (Gu et al., 2016; Peng et al., 2016) but recently also convolutional neural networks (Nguyen and Verspoor, 2018; Choi, 2018), LSTM (Li et al., 2017; Sahu and Anand, 2018) or a combination of machine learning models and neural-network-based encoders (Zhang et al., 2018; Peng et al., 2018). A recent paper (Verga et al., 2018) achieves state-of-the-art results on biomedical relation classification for chemically-induced diseases (CDR (Li et al., 2016)) and ChemProt (CPR (Krallinger M., 2017)), by using a Transformer encoder (Vaswani et al., 2017) and end-to-end Named Entity Recognition and relation extraction, without, however, leveraging transformer-based language model pre-training. In the general domain, (Pawar et al., 2017) and (Smirnova and Cudr-Mauroux, 2019) provide a comprehensive review of different relation extraction paradigms and methods that have been developed to date.

6 Conclusions and further work

We have presented a system that extracts mentions of biomedical entities and classifies them into one of four function change relations (or absence of a relation). Our system leverages widely available language representations pre-trained on biomedical data and utilises minimal task-specific architecture, while not relying on specially engineered linguistic features. Despite the model simplicity and the class imbalance in the data (even within the four non-negative classes), our model is able to significantly outperform the random baseline.

Our model can be improved by using more recent language modeling methods, such as XLNet (Yang et al., 2019), and different ways of encoding the mention pairs and textual context (e.g. by using not only the hidden state of the [CLS] token but also the hidden states of the entity mentions as input to the relationship classifier). Different methods can be explored for addressing class imbalance (e.g. a cost-sensitive classifier, data augmentation etc). Further, an end-to-end Named Entity Recognition and Relation Extraction architecture can be devised. It would also be interesting to compare our model against more competitive baselines.

Acknowledgments

We would like to thank Nathan Patel for his engineering support as well as Angus Brayne, Julien Fauqueur and other colleagues working on NLP for insightful discussions.

References

Christoph Alt, Marc Hbner, and Leonhard Hennig. 2019. Improving relation extraction by pre-trained language representations. *Proceedings of Automated Knowledge Base Construction (AKBC'19)*.

Iz Beltagy, Arman Cohan, and Kyle Lo. 2019. Scibert: Pretrained contextualized embeddings for scientific text. https://arxiv.org/pdf/1903.10676.pdf.

Sung-Pil Choi. 2018. Extraction of protein-protein interactions (ppis) from the literature by deep convolutional neural networks with various feature embeddings. *Journal of Information Science*, 44(1):60–73.

Kevin Clark, Urvashi Khandelwal, Omer Levy, and Christopher D. Manning. 2019. What Does BERT Look At? An Analysis of BERT's Attention. https://arxiv.org/pdf/1906.04341.pdf.

Jacob Devlin, Ming-Wei Chang, Kenton Lee, and Kristina Toutanova. 2018. Bert: Pre-training of deep bidirectional transformers for language understanding. *Proceedings of Empirical Methods in Natural Language Processing (EMNLP'18)*.

Yoav Goldberg. 2019. Assessing bert's syntactic abilities. http://arxiv.org/abs/1901.05287.

Jinghang Gu, Longhua Qian, and Guodong Zhou. 2016. Chemical-induced disease relation extraction with various linguistic features. *Database*, 2016.

Akhondi S.A. et al. (eds). Krallinger M., Rabal O. 2017. Overview of the biocreative vi chemical-protein interaction track. *Proceedings of the BioCreative VI Workshop , Bethesda, MD. pp. 141146.*

Jinhyuk Lee, Wonjin Yoon, Sungdong Kim, Donghyeon Kim, Sunkyu Kim, Chan Ho So, and Jaewoo Kang. 2019. Biobert: a pre-trained biomedical language representation model for biomedical text mining. https://arxiv.org/pdf/1901.08746.pdf.

Fei Li, Meishan Zhang, Guohong Fu, and Donghong Ji. 2017. A neural joint model for entity and relation extraction from biomedical text. *BMC Bioinformatics*, 18(1):198:1–198:11.

Jiao Li, Yueping Sun, Robin J. Johnson, Daniela Sciaky, Chih-Hsuan Wei, Robert Leaman, Allan Peter Davis, Carolyn J. Mattingly, Thomas C. Wiegers, and Zhiyong Lu. 2016. Biocreative v cdr task corpus: a resource for chemical disease relation extraction. *Database*, 2016.

Dat Quoc Nguyen and Karin Verspoor. 2018. Convolutional neural networks for chemical-disease relation extraction are improved with character-based word embeddings. *Proceedings of the Association for Computational Linguistics (ACLi18), BioNLP*.

Sachin Pawar, Girish K. Palshikar, and Pushpak Bhattacharyya. 2017. Relation extraction : A survey. *CoRR*. https://arxiv.org/pdf/1712.05191.pdf.

Yifan Peng, Anthony Rios, Ramakanth Kavuluru, and Zhiyong Lu. 2018. Chemical-protein relation extraction with ensembles of svm, cnn, and rnn models. Http://arxiv.org/abs/1802.01255.

Yifan Peng, Chih-Hsuan Wei, and Zhiyong Lu. 2016. Improving chemical disease relation extraction with rich features and weakly labeled data. *Journal of Cheminformatics*, 8(1):53:1–53:12.

Alec Radford, Karthik Narasimhan, Tim Salimans, and Ilya Sutskever. 2018. Improving language understanding by generative pre-training. https://s3-us-west-2.amazonaws.com/openai-assets/research-covers/language-unsupervised/language$_u$nderstanding$_p$aper.pdf.

Sunil Kumar Sahu and Ashish Anand. 2018. Drug-drug interaction extraction from biomedical texts using long short-term memory network. *Journal of Biomedical Informatics*, 86:15–24.

Alisa Smirnova and Philippe Cudr-Mauroux. 2019. Relation extraction using distant supervision: A survey. *ACM Computing Surveys*, 51(5):106:1–106:35.

Livio Baldini Soares, Nicholas FitzGerald, Jeffrey Ling, and Tom Kwiatkowski. 2019. Matching the blanks: Distributional similarity for relation learning. *Proceedings of the Association for Computational Linguistics (ACL'19)*.

Wilson Taylor. 1953. Cloze procedure: A new tool for measuring readability. *Journalism Bulletin, 30(4):415433*.

Ashish Vaswani, Noam Shazeer, Niki Parmar, Jakob Uszkoreit, Llion Jones, Aidan N. Gomez, Lukasz Kaiser, and Illia Polosukhin. 2017. Attention is all you need. In *Proceedings on Neural Information Processing Systems (NIPS'17)*, pages 5998–6008.

Patrick Verga, Emma Strubell, and Andrew McCallum. 2018. Simultaneously self-attending to all mentions for full-abstract biological relation extraction. *Proceedings of the North Americal Association for Computational Linguistics (NAACL'18)*, pages 872–884.

Chih-Hsuan Wei, Hung-Yu Kao, and Zhiyong Lu. 2013. Pubtator: a web-based text mining tool for assisting biocuration. *Nucleic Acids Research*, 41(Webserver-Issue):518–522.

Yonghui Wu, Mike Schuster, Zhifeng Chen, Quoc V. Le, Mohammad Norouzi, Wolfgang Macherey, Maxim Krikun, Yuan Cao, Qin Gao, Klaus Macherey, Jeff Klingner, Apurva Shah, Melvin Johnson, Xiaobing Liu, ukasz Kaiser, Stephan Gouws, Yoshikiyo Kato, Taku Kudo, Hideto Kazawa, Keith Stevens, George Kurian, Nishant Patil, Wei Wang, Cliff Young, Jason Smith, Jason Riesa, Alex Rudnick, Oriol Vinyals, Greg Corrado, Macduff Hughes, and Jeffrey Dean. 2016. Google's neural machine translation system: Bridging the gap between human and machine translation.

Zhilin Yang, Zihang Dai, Yiming Yang, Jaime Carbonell, Ruslan Salakhutdinov, and Quoc V. Le. 2019. Xlnet: Generalized autoregressive pretraining for language understanding. https://arxiv.org/pdf/1906.08237.pdf.

Yijia Zhang, Hongfei Lin, Zhihao Yang, Jian Wang, Shaowu Zhang, Yuanyuan Sun, and Liang Yang. 2018. A hybrid model based on neural networks for biomedical relation extraction. *Journal of Biomedical Informatics*, 81:83–92.

RACAI's System at PharmaCoNER 2019

Radu Ion
Institute for AI
Romanian Academy
13 "Calea 13 Septembrie"
Bucharest 050711, Romania
radu@racai.ro

Vasile Florian Păiș
Institute for AI
Romanian Academy
13 "Calea 13 Septembrie"
Bucharest 050711, Romania
vasile@racai.ro

Maria Mitrofan
Institute for AI
Romanian Academy
13 "Calea 13 Septembrie"
Bucharest 050711, Romania
maria@racai.ro

Abstract

This paper describes the Named Entity Recognition system of the Institute for Artificial Intelligence "Mihai Drăgănescu" of the Romanian Academy (RACAI for short). Our best F1 score of **0.84984** was achieved using an ensemble of two systems: a gazetteer-based baseline and a RNN-based NER system, developed specially for PharmaCoNER 2019. We will describe the individual systems and the ensemble algorithm, compare the final system to the current state of the art, as well as discuss our results with respect to the quality of the training data and its annotation strategy. The resulting NER system is language independent, provided that language-dependent resources and preprocessing tools exist, such as tokenizers and POS taggers.

1 Introduction

Named entity recognition (NER) efforts present two challenges: entity detection, identifying the portion of text associated with an entity, and disambiguation, assigning the identified text to a specific entity class. At the Institute for Artificial Intelligence "Mihai Drăgănescu" of the Romanian Academy, one of the research goals focuses on constructing an improved named entity recognition system for Romanian language, including biomedical entities. In this context, the current PharmaCoNER 2019 competition (Gonzalez-Agirre et al., 2019) offered the opportunity to reconsider the existing Romanian NER system which provided the grounds for developing new approaches that are language-independent and more accurate. With respect to our current Romanian biomedical NER system, Mitrofan (2017) presents a neural network based NER system that is able to detect the beginnings and insides of entities with four labels: anatomical parts, disorders, medical procedures and chemical compounds. Although we did not have the time to train this NER system on PharmaCoNER 2019 data, our F1 score on Spanish, when compared to the reported F1 score for the Romanian chemical compounds (the label that best overlaps with the labels of PharmaCoNER 2019), is a strong indicator that we can greatly improve the Romanian biomedical NER system (by how much is the subject for a future paper).

We begin by looking at state of the art approaches for NER systems, presented in Section 2 "Related work", then we continue with the resources used for this specific task, in Section 3 "Resources", followed by a presentation of our implemented algorithms and methods, in Section 4 "RACAI Systems". Finally, system evaluation results are presented in Section 5 "System evaluation", followed by conclusions.

2 Related work

To tackle the challenges posed by BioNER, different NER approaches were proposed. Even though high performances have been obtained by applying classical NER approaches such as dictionary-based methods (Sekine and Nobata, 2004), rule-based methods (Rau, 1991), Hidden Markov Models (Zhou and Su, 2002), Conditional Random Fields (Dingare et al., 2005), the current dominant techniques are based on neural methods, which will also be our focus in this paper, mainly because we think that this is the current state of the art approach to NER.

Deep learning methods have shown impressive results when applied to NLP and, since (Hochreiter and Schmidhuber, 1997) proposed Long-Short Term Memory neural networks and Bidirectional Long-Short Term Memory (BiLSTM) networks (Graves, 2012), a wide variety of NER systems have been created based on these methods.

Proceedings of the 5th Workshop on BioNLP Open Shared Tasks, pages 90–99
Hong Kong, China, November 4, 2019. ©2019 Association for Computational Linguistics

Santos and Guimaraes (2015) presented a language-independent approach for NER based on a deep neural architecture that uses word and character-level embeddings to perform sequential classification. In order to demonstrate the language-independence of the system, two annotated corpora in two different languages were used: a Portuguese corpus - HAREM I (Milidiú et al., 2008) and a Spanish corpus - SPA CoNLL-2002 (Sang and F., 2002). The system obtained an F1 score of 79% when trained on HAREM I corpus and an F1 score of 82.2% for the SPA CoNLL-2002 corpus.

Chiu and Nichols (2016) presented a NER system based on stacked BiLSTM architecture trained to detect four types of entities such as: "PERSON", "ORGANIZATION", "LOCATION" and "MISC", each of the entity being annotated in BIOES format (Beginning, Inside, Outside, Ending and Single). Using two lexicons extracted from publicly-available resources the system obtained an F1-score of 91.62% on CoNLL-2003 (Sang and De Meulder, 2003) corpus and 86.28% on OntoNotes (Pradhan et al., 2013) corpus.

Shao et al. (2016) evaluated the performances of three types of neural networks based systems for multilingual NER. They compared a windows-based feed-forward network, a standard BiLSTM and a window-based BiLSTM. Word embeddings combined with word-level features were used and the annotation format was also BIOES. Based on the experiments the authors concluded that: the feed-forward neural network was outperformed in accuracy by the standard BiLSTM and when less information is available, the window-based BiLSTM is more robust than the standard BiLSTM.

Soares et al. (2019) used NeuroNER (Dernoncourt et al., 2017) framework in order to perform NER for medical domain. The Spanish Clinical Cases Corpus (SPACCC) was used to train the system, which is based on a LSTM neural network. The biomedical corpus was previously annotated with four entity types, a subset of the types PharmaCoNER 2019 uses. Using medical word-embeddings, the system achieved an F1 score of 88.18%, outperforming the baseline system which scored 87.76%.

3 Resources

In order to develop, train and test a NER system several resources are needed. In this section we review the main types of linguistic resources used in our work:

3.1 Corpora

When applied to general domain, most of the state of the art systems make use of the CoNLL-2002 corpus (Sang and F., 2002), which contains six files that cover two languages: Dutch and Spanish. The set of entity labels used for this corpus contains four types of entities: PER (persons), ORG (organizations), LOC (locations) and MISC (miscellaneous).

In order to perform named entity recognition on biomedical textual data several annotated corpora were developed. For English there are several annotated corpora used for biomedical NER such as: **NCBI** (Doğan et al., 2014) a gold-standard corpus for disease mentions and concepts that contains 793 abstracts extracted from PubMed; **CHEMD-NER** (Krallinger et al., 2015) a corpus of 10,000 abstracts collected from PubMed annotated with two types of NEs: chemicals and drugs.

Lately a slightly increasing number of resources specific to this field have been created for languages other than English. For example for French there is the **Quaero** corpus (Névéol et al., 2014) which contains 103,056 words annotated with ten types of NEs defined using UMLS: anatomy, chemical and drugs, devices, disorders, geographic areas, living beings, objects, phenomena, physiology, procedures. For Romanian there is the **MoNERo** (Maria Mitrofan, 2019) corpus which is a biomedical gold standard corpus and contains 154,825 words annotated with four types of entities: anatomy, chemicals and drugs, disorders and procedures. For Spanish **IxaMedGS** (Oronoz et al., 2015) is a corpus that contains 142,154 discharge records out of which 75 were annotated with two types of NEs: diseases and drugs; **DrugSemantics** corpus (Moreno et al., 2017) has 226,729 tokens annotated with ten types of NEs: chemical composition, disease, drug, excipient, food, medicament, pharmaceutical form, route, therapeutic action and unit of measurement.

3.2 Word embeddings

Continuous word representations, trained on large corpora have been proven to be useful for many NLP tasks, including NER. It is known that neural word representations have the ability to capture useful semantic properties and linguistic relationships between words (Bakarov, 2018). Therefore pre-trained word embeddings are available for different languages, including Romanian and Spanish. For example in Romanian we have a set of word embeddings (Păiș and Tufiș, 2018) computed on the Reference Corpus for Contemporary Romanian Language (CoRoLa) (Barbu Mititelu et al., 2018) corpus.

Grave et al. (2018) released a set of pre-trained embeddings for 157 languages calculated on texts extracted from Wikipedia. Also for Spanish there is a different set of pre-trained embeddings made available by the Chile NLP group[1] and calculated using the Spanish Billion Word Corpus (SB-WCE)[2].

Chiu and Nichols (2016) showed that word embeddings vectors calculated on a specific domain produce better results than those obtained from general-domain texts. Therefore (Soares et al., 2019) calculated a set of medical word embeddings for Spanish. They used text from two sources: full medical articles from SciELo database[3] (100 million tokens) and biomedical texts from Wikipedia (82 million tokens). The experiments performed using this resource generated more accurate results than those calculated based on general-domain texts.

3.3 SNOMED CT

SNOMED CT (Systematized Nomenclature of Medicine - Clinical Terms)[4] is a multilingual healthcare terminology built around a concept-based ontology. It contains more than 1 million distinct medical terms, 326,734 concepts and 19 hierarchies. Concepts are classified under hierarchies, of which most of them corresponding to the types of entities instances of which are encountered by clinicians during their work (body parts, diseases, substances, procedures, etc.). A concept in SNOMED CT has a unique name, unique numeric code, and more descriptions (one main definition, several secondary and more synonyms). This resource is available in both English and Spanish. To use it for scientific purposes, a license is required after completing a form. We used this resource to extract all the available proteins and genes. Using the SNOMED browser for Spanish[5]

we extracted 9,556 proteins names.

4 RACAI Systems

4.1 RACAI Baseline

Our baseline system is an enhanced gazetteer-based annotation tool. It takes as input multiple files, each containing an entity list of the same type. For example, in the `PROTEINAS.txt` file there will be a list of proteins. On each line, there will be a string containing a word or an expression denoting a protein.

Various gazetteer annotation systems already exist. We recall here Stanford TokensRegex (Chang and Manning, 2014) and Stanford RegexNER part of Stanford Core NLP (Manning et al., 2014). However, these and similar other systems, impose the format to correspond to some specific regular expression syntax (or at least to a certain fixed form textual representation). In our case, the gazetteer resources are partially generated directly from the training annotations provided for the task. Therefore, the format used is not directly checked and validated by a human operator.

Therefore, our system does not look for expressions exactly as they are provided. Instead it implements additional rules to improve matching such as:

- ignore special characters (example: '-' , '¡', '(' etc.) in both provided expressions and the searched text;

- recognize words followed by numbers regardless of the way they are written (for example: "CAP-57", "CAP 57", "CAP57").

Finally, in the case of overlapping entities being found, the longer one is kept. The software program allows for such overlapping entities to be saved for manual examination, but this particular feature did not seem useful for this task. The resulting annotation file is in the ".ann" format.

4.2 RPCN

RPCN stands for the "RACAI PharmaCoNER neural network" and is, as its name suggests, a neural network that we specifically designed for this competition and that, ultimately, will also be run for Romanian for which we have BioNER training data (Maria Mitrofan, 2019).

---

[1] https://github.com/dccuchile/spanish-word-embeddings
[2] http://crscardellino.github.io/SBWCE/
[3] https://www.scielo.org/en/
[4] https://www.snomed.org/snomed-ct
[5] https://browser.ihtsdotools.org

4.2.1 Comparison with the state of the art and design choices

As already discussed in Section 2, NER systems based on BiLSTMs and using convolutional neural networks (CNNs) to encode character-based features of the input (Chiu and Nichols, 2016) represent the current state of the art for NER task. Other approaches used stacked BiLSTM layers in an attempt to increase the generalization power of the network or decoders which chose the most probable label output given the LSTM encoding of the featurized input (Dernoncourt et al., 2017).

Our research goal was to test an approach based on BiLSTMs, given the abundance of papers using this type of artificial cell and reporting very good results. At this point, we have to mention that *all design choices of RPCN presented below were driven by intense experimentation* with the provided training data, aiming at *short training and evaluation* loops. Because the training data is rather small in size (a bit more than 3800 training examples), we quickly realized that running with more complex architectures (which have more parameters) leads to overfitting. Thus, all architectures with two BiLSTM layers and/or CNNs encoding character features were dropped early on from our experiments.

The RPCN network differs by mainstream BiLSTM NER networks by attempting to use an attention mechanism, like the one in (Anh Nguyen et al., 2019) (of which we did not know at the time of our experiments), whose main function is to model how much words surrounding labeled entities contribute to the label prediction. Also, RPCN tries to combine (by a simple addition) independently trained word embeddings from the medical domain with the embeddings extracted directly from the training corpus. We found that this approach gives a significant boost of performance (more than 10% in the F1 score) when compared to the usage of either word embedding sources in isolation or with general-purpose embeddings extracted from Wikipedia. We are thus able to confirm and supplement the findings of Soares et al. (2019).

In relation to the featurized input that we designed for RPCN, we were guided by the following assumptions and intuitions:

- all NEs are mostly noun phrases and in Spanish, as in Romanian, noun phrases have a well-defined syntactic structure which prompted the usage of POS tags as features;

- all NEs are medical substances obeying some naming patterns, so a feature regarding words affixes was needed;

- some proteins have specific character patterns, so a "word shape" feature was also thought to be useful (see the next subsection for the "shape features" details);

- with an eye to the rank of our system in the PharmaCoNER 2019 competition, we also thought that including the gazetteer feature (if available) directly into RPCN would increase the performance of the system.

4.2.2 Architecture

RPCN is a RNN which uses LSTM cells to encode the feature descriptions of the words coming in, remembering the information from both left and right contexts of the target word, which makes it BiLSTM RNN. The network was trained to label each word in the sequence with one of the PharmaCoNER target labels or with the "nothing interesting here" label which we called NONE.

The RPCN architecture is presented in Figure 1. We have tried the vanilla variant and the variant enhanced with an attention mechanism, as described by Bahdanau et al. and retained the latter for further development, as the better approach. RPCN is written in Java 1.8, using the DeepLearning4J deep neural network Java library, version 1.0.0-beta3.

Figure 1 shows the input vectors and the BiLSTM cell for a single input word, for example `cadenas`, but we consider sequences of words, each with its own BiLSTM cell (but shared parameters among words). The input vectors that go into the BiLSTM cell are as follows:

- the `WE Layer` is the word embedding layer for the input word; its output size was chosen by our hyperparameter grid search procedure to be 64 (see the Training subsection 4.2.3). The word is one-hot encoded and fed to this layer which compresses it to a 64 dimensional vector;

- the `External WEs` resource refers to our pretrained Spanish medical word embeddings (Soares et al., 2019). Because the size of these embeddings is larger than 64, one such embedding is fed to a fully connected

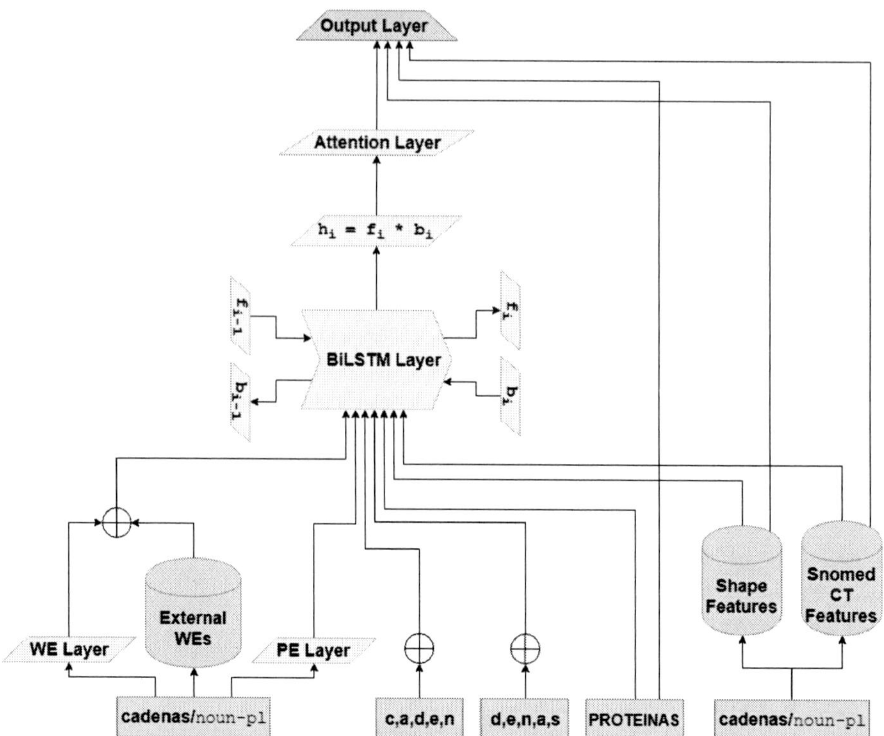

Figure 1: The RPCN neural network

layer with an output size of 64 so that we can add (element-wise) the output of the `WE Layer` with the output of this fully connected layer to obtain a "unified" word embedding representation for the input word;

- the `PE Layer` is the POS tag embedding layer for the POS tag of the input word (a plural noun for our example word); its output size was chosen to be 16 by the hyperparameter grid searching procedure. Each POS tag is encoded as a one-hot vector and fed to this layer which compresses it to a 16 dimensional vector;

- `c,a,d,e,n` and `d,e,n,a,s` are the "relative-index-hot" representations of the 5 character prefix and suffix of the input word; the vectors for each character are added to form a single output vector. The "relative-index-hot" stands for the use of the $1/(i+1)$ quantity instead of a 1 on the corresponding vector position, where i is the index of the character in the input word (0-based numbering), and this trick allows us to encode in a single vector both the prefix and suffix vectors which are sensitive to the character ordering in the word;

- `PROTEINAS` is the one-hot representation of the gazetteer label that is (optionally) available for the input word (if it is not available, we use the the one-hot representation of the "default" label NONE);

- the `Shape Features` resource refers to our word shape extraction algorithm that does the following:

  - using regular expressions, sets one bit in the feature binary vector if the input word looks like a substance, e.g. "CD34", "CAM5.2", "Tc99m-MDP", etc.

  - sets one bit in the the feature binary vector if the word is a "dash prefix word", e.g. "alfa", "Beta", "β", etc. and it is "glued" (no spaces) to the next word; the list of dashed prefix words has been automatically generated from the train set.

- the `SNOMED CT Features` resource refers to our Spanish SNOMED CT "word as a feature" algorithm. Based on our tokenized SNOMED CT gazetteer list, in which we labeled each "(sustancia)" concept description with either the `PROTEINAS` or the `NORMALIZABLES` labels, we counted each word and label pair and then computed the probability distribution $P(\text{PROTEINAS}|word)$ and $P(\text{NORMALIZABLES}|word)$. Each word is represented by a 2 float vector on how probable it is to point to either of these two labels. If the input word is not found in this resource, $P(\text{PROTEINAS}|word) = P(\text{NORMALIZABLES}|word) = 0.5$. To the 2 float vector we also append the relative frequency of the word in the tokenized SNOMED CT gazetteer list; naturally, we skip functional words from this computation. Some example vectors, computed as described above, are presented in Table 1.

The BiLSTM cell will combine the forward f_i and backward b_i states by multiplying the state vectors, element-wise: $h_i = f_i \cdot b_i$. This method proved to increase the precision of the system as the signal will be strong only if left and right evidence is strong (i.e. close to 1.0). We also found out that if we average the forward and backward states as in $h_i = (f_i + b_i)/2$, we can increase the recall at the cost of a lower precision. The same effect (recall increase) is obtained when the forward and backward states are concatenated.

Besides the weighted sum of the combined BiLSTM outputs h_i given by the attention layer, the output layer (a `softmax` layer with the output size equal to the number of target labels) also receives the raw inputs from the gazetteer feature, the shape features and the SNOMED CT features, in an effort to boost the precision of the system.

4.2.3 Training

The input text is tokenized first, using an in-house built tokenizer for Spanish, specifically designed for this task. The tokenizer will split words at the dash ('-') boundary because we observed that some entities contained the dash while others did not. The tokenizer will recognize (and thus generalize) the following types of tokens: numbers (integers, reals, Roman numerals), amounts (e.g. "305mg"), units of measure (e.g. "mg/g"), temperatures (e.g. "30°C") and area/volume expressions (e.g. "3x2cm2"). After tokenization, the text is POS tagged using the Stanford Core NLP suite with the Spanish POS tagging model and the sentence boundaries are detected using a simple regular expression: end of sentence punctuation followed by whitespace and then by an uppercase letter. No named entity is allowed to cross a sentence boundary.

We used a grid searching procedure, together with the supplied train and development data, to optimize the hyperparameters of RPCN. The hyperparameters are as follows:

- the *number of time steps* in the sequence: how many words are in a window of consecutive words that the RPCN can consider as a training example. Tried values were in the set $\{7, 11, 15, 19, 21, 25\}$ and the best value was set to 21;

- the *size of the LSTM state* vector; tried values were in the set $\{64, 128, 256\}$ and the best value was set to 128;

- the *size of the trained word embedding* vector, i.e. the size of the `WE Layer`. Tried values were in the set $\{32, 64, 128\}$ and the best value was set to 64;

- the *size of the POS tag embedding* vector, i.e. the size of the `PE Layer`. Tried values were in the set $\{8, 16\}$ and the best value was set to 16.

The train and development sets that were made available by the task organizers were distributed as follows: 3822 training annotations (T entries in the ".ann" files) and 1926 development annotations. We have randomly reshuffled the whole data set (training plus development) into 90% training set and 10% development set.

As far as the configuration of the computation graph goes, we used the Xavier weight initialization method together with the Stochastic Gradient Descent optimization algorithm and the Adam updater with the default parameters. The reader can refer to the documentation of the DeepLearning4J library for a description of these methods.

4.2.4 Running

The incoming text is tokenized, POS tagged and sentence split. Then, RPCN is run on consecutive sequences of adjacent words of length 21, each

Word	$P(\texttt{PROTEINAS}|word)$	$P(\texttt{NORMALIZABLES}|word)$	$P(word)$
lormetazepam	0.0	1.0	8.379841E-6
antinuclear	0.5	0.5	1.005581E-4
oxigenasa	0.625	0.375	6.703873E-5
carveol	1.0	0.0	8.379841E-6

Table 1: SNOMED CT word features for labels `PROTEINAS` and `NORMALIZABLES`

word receiving the best label by the `softmax` output, accumulating labels as the window passes by. The label with the highest accumulated score wins for each word. Spans of consecutive tokens having the same non-`NONE` labels are the new detected named entities.

The raw label assignments are post-processed to enforce the following:

- a recognized named entity will not start or end with a functional word;

- if there is a gazetteer annotation for a RPCN detected span then the labels must agree and the gazetteer span boundaries will be preferred. If the labels do not agree, both spans are deleted.

Finally, we also apply some regular expression based rules to catch some expressions which RPCN was not able to learn, e.g. `CD[0-9]+` (a protein) or the pattern `W1` "de" `W2` in which "de" `W2` receive the same label as `W1`.

4.3 Ensemble methods

Given the different annotator systems described above, an ensemble system was needed. Its aim was to take the resulting annotations from two or more runs, with the same or different system, and combine them using different rules in order to improve the overall results. The idea behind it is that each system could be better at detecting certain types of entities and the combined annotation would be better overall.

Our combining system takes as input two ".ann" files and produces another ".ann" file by applying rules. The rules are especially useful in the case of overlapping entities. If there are no overlapping entities, then the input annotations are simply merged. Currently there are 5 rules available:

- "PRIO1": gives priority to the first input file, retaining the corresponding entity annotation;

- "PRIO2": gives priority to the second input file;

- "SMALLER": keeps the smaller annotation, discarding the longer one in case of entity overlap;

- "LARGER": keeps the longer annotation;

5 System Evaluation

5.1 Working methodology

We mentioned that the initial distribution of annotations in the training and development sets was not satisfactory and thus, we have proceeded to the random reshuffling of the whole data set followed by a 90%/10% split. We have selected our best ensemble method on such a random reshuffling and training/development split.

The RACAI baseline system worked with the annotations from the training set plus the gazetteer list based on the Spanish SNOMED CT "(sustancia)" concept descriptions which we automatically extracted and labeled as either `PROTEINAS` or `NORMALIZABLES` and then manually validated.

RPCN was trained on the training set and evaluated, along with the RACAI baseline system, on the development set. For the *official evaluation run*, we used all annotations from the provided data set and the SNOMED CT entries as the gazetteer list.

5.2 Results

Table 2 presents the runs of the RACAI baseline system, RPCN and of four ensemble methods applied to the baseline (first input) and RPCN (second input).

The highest scores are bold-faced for the Precision (P), Recall (R) and F1 columns. According to our evaluations, the best ensemble method (by the F1 score which was the optimization target) is the "LARGER" (or C4 to match the name of the submitted zip file) ensemble method. Knowing that we are allowed to submit five different runs, based

System	P	R	F1
Baseline	0.8986	0.6915	0.7816
RPCN	**0.9025**	0.7539	0.8215
PRIO1 (C1)	0.8733	0.7764	0.8220
PRIO2 (C2)	0.8871	0.7764	0.8281
SMALLER (C3)	0.8694	0.7730	0.8183
LARGER (C4)	0.8911	**0.7799**	**0.8318**

Table 2: Development results of RACAI's NER systems

System	P	R	F1
Baseline	**0.92530**	0.71281	0.80527
RPCN	0.89327	0.76330	0.82319
PRIO1 (C1)	0.90189	0.80347	**0.84984**
LARGER (C4)	0.90043	0.79533	0.84462
C4M	0.78281	**0.84528**	0.81284

Table 3: Official PharmaCoNER 2019 results of RACAI's NER systems

on these evaluations, we decided to submit the output of the following systems: RPCN (best precision), LARGER (C4, best F1 score) and Baseline (official reference system). Before the submission deadline, we also sent the PRIO1 (C1, the Baseline priority) and an ensemble between the Baseline and one other system that we developed for PharmaCoNER 2019 (C4M). Table 3 presents the official results that were communicated to us by the task organizers.

6 Discussion and conclusions

The official evaluation results confirmed the results we obtained during development: the PRIO1 and LARGER ensembles between the Baseline and the RPCN systems are better than each of them, individually. RPCN definitely learned to recognize new entities, as its recall is larger with more than 5% than the recall of the Baseline system.

We can also see that the precision of RPCN dropped, as compared to the precision of the Baseline system, with more than 3% in the official evaluation. This discrepancy appeared during development as well and the main reason we found for it was that the training data *was not consistently annotated*. That is, the same expression (same words, same casing) was annotated in a document and was not annotated in another document. We do not think that at this specialization level we can justify this at a semantic level (i.e. the expression does not mean the same thing in the two documents). Thus, during development, we automatically re-annotated the whole supplied data, making sure the same expression is annotated everywhere with the same label (if there was an ambiguity, the re-annotation was cancelled for the expression). By doing this, we were able to close the precision gap between the Baseline and the RPCN systems.

While we do not know the rank of our system yet, our best system was scored with an F1 score of **0.84984**, which, we feel, is good performance. We will put this system to the tests of scalability and language-independence by using it unchanged (but with the specialized computational resources) in two Romanian-related tasks: as already stated, in the identification of Romanian biomedical NEs and in the rather different task of legal terminology identification (e.g. EuroVoc[6]) in Romanian legal texts, to be performed in the MARCELL project[7]. For the latter task, we will have the chance to determine if our system is able to reliably detect new terms which are missing from the legal terminology dictionaries.

Acknowledgments

The reported research was supported by the EC grant MARCELL (Multilingual Resources for CEF.AT in the Legal Domain), TENtec no. 27798023.

References

Kim Anh Nguyen, Ngan Dong, and Cam-Tu Nguyen. 2019. Attentive Neural Network forNamed Entity Recognition in Vietnamese. *2019 IEEE-RIVF International Conference on Computing and Communication Technologies (RIVF)*.

Dzmitry Bahdanau, Kyunghyun Cho, and Yoshua Bengio. Neural machine translation by jointly learning to align and translate. In *Proceedings of ICLR 2015*.

Amir Bakarov. 2018. A Survey of Word Embeddings Evaluation Methods.

Verginica Barbu Mititelu, Dan Tufiș, and Elena Irimia. 2018. The reference corpus of contemporary Romanian language (CoRoLa). In *Proceedings of the 11th Language Resources and Evaluation Conference – LREC 2018*, pages 1178–1185, Miyazaki,

[6]`https://data.europa.eu/euodp/en/data/dataset/eurovoc`
[7]`http://marcell-project.eu/`

Japan. European Language Resources Association (ELRA).

Angel X. Chang and Christopher D. Manning. 2014. Tokensregex: Defining cascaded regular expressions over tokens.

Jason PC Chiu and Eric Nichols. 2016. Named entity recognition with bidirectional lstm-cnns. *Transactions of the Association for Computational Linguistics*, 4:357–370.

Franck Dernoncourt, Ji Young Lee, and Peter Szolovits. 2017. Neuroner: an easy-to-use program for named-entity recognition based on neural networks. *arXiv preprint arXiv:1705.05487*.

Shipra Dingare, Malvina Nissim, Jenny Finkel, Christopher Manning, and Claire Grover. 2005. A system for identifying named entities in biomedical text: how results from two evaluations reflect on both the system and the evaluations. *International Journal of Genomics*, 6(1-2):77–85.

Rezarta Islamaj Doğan, Robert Leaman, and Zhiyong Lu. 2014. Ncbi disease corpus: a resource for disease name recognition and concept normalization. *Journal of biomedical informatics*, 47:1–10.

Aitor Gonzalez-Agirre, Montserrat Marimon, Ander Intxaurrondo, Obdulia Rabal, Marta Villegas, and Martin Krallinger. 2019. Pharmaconer: Pharmacological substances, compounds and proteins named entity recognition track. In *Proceedings of the BioNLP Open Shared Tasks (BioNLP-OST)*, pages 1–X, Hong Kong, China. Association for Computational Linguistics.

Edouard Grave, Piotr Bojanowski, Prakhar Gupta, Armand Joulin, and Tomas Mikolov. 2018. Learning word vectors for 157 languages. In *Proceedings of the International Conference on Language Resources and Evaluation (LREC 2018)*.

Alex Graves. 2012. Supervised sequence labelling. In *Supervised sequence labelling with recurrent neural networks*, pages 5–13. Springer.

Sepp Hochreiter and Jürgen Schmidhuber. 1997. Long short-term memory. *Neural computation*, 9(8):1735–1780.

Martin Krallinger, Obdulia Rabal, Florian Leitner, Miguel Vazquez, David Salgado, Zhiyong Lu, Robert Leaman, Yanan Lu, Donghong Ji, Daniel M Lowe, et al. 2015. The chemdner corpus of chemicals and drugs and its annotation principles. *Journal of cheminformatics*, 7(1):S2.

Christopher D. Manning, Mihai Surdeanu, John Bauer, Jenny Finkel, Prismatic Inc, Steven J. Bethard, and David Mcclosky. 2014. The Stanford CoreNLP Natural Language Processing toolkit. In *Proceedings of the 52nd Annual Meeting of the Association for Computational Linguistics: System Demonstrations*, pages 55–60.

Grigorina Mitrofan Maria Mitrofan, Verginica Barbu Mititelu. 2019. Monero: a biomedical gold standard corpus for the romanian language. In *Proceedings of 18th ACL Workshop on Biomedical Natural Language Processing*, volume (In press).

Ruy Luiz Milidiú, Cícero Nogueira dos Santos, and Julio Cesar Duarte. 2008. Portuguese corpus-based learning using etl. *Journal of the Brazilian Computer Society*, 14(4):17–27.

Maria Mitrofan. 2017. Bootstrapping a Romanian Corpus for Medical Named Entity Recognition. In *Proceedings of Recent Advances in Natural Language Processing*, pages 501–509, Varna, Bulgaria.

Isabel Moreno, Ester Boldrini, Paloma Moreda, and M Teresa Romá-Ferri. 2017. Drugsemantics: a corpus for named entity recognition in spanish summaries of product characteristics. *Journal of biomedical informatics*, 72:8–22.

Aurélie Névéol, Cyril Grouin, Jeremy Leixa, Sophie Rosset, and Pierre Zweigenbaum. 2014. The quaero french medical corpus: A ressource for medical entity recognition and normalization. In *In Proc BioTextM, Reykjavik*. Citeseer.

Maite Oronoz, Koldo Gojenola, Alicia Pérez, Arantza Díaz de Ilarraza, and Arantza Casillas. 2015. On the creation of a clinical gold standard corpus in spanish: Mining adverse drug reactions. *Journal of biomedical informatics*, 56:318–332.

Sameer Pradhan, Alessandro Moschitti, Nianwen Xue, Hwee Tou Ng, Anders Björkelund, Olga Uryupina, Yuchen Zhang, and Zhi Zhong. 2013. Towards robust linguistic analysis using ontonotes. In *Proceedings of the Seventeenth Conference on Computational Natural Language Learning*, pages 143–152.

Vasile Păiș and Dan Tufiș. 2018. Computing distributed representations of words using the COROLA corpus. *Proceedings of the Romanian Academy, Series A*, 19(2):403–409.

Lisa F Rau. 1991. Extracting company names from text. In *[1991] Proceedings. The Seventh IEEE Conference on Artificial Intelligence Application*, volume 1, pages 29–32. IEEE.

Erik F Tjong Kim Sang and Fien De Meulder. 2003. Introduction to the conll-2003 shared task: Language-independent named entity recognition. corr. *arXiv preprint cs.CL/0306050*.

Tjong Kim Sang and Erik F. 2002. Introduction to the CoNLL-2002 shared task. *Proceedings of the 6th Conference on Natural Language Learning - COLING-02*.

Cicero Nogueira dos Santos and Victor Guimaraes. 2015. Boosting named entity recognition with neural character embeddings. *arXiv preprint arXiv:1505.05008*.

Satoshi Sekine and Chikashi Nobata. 2004. Definition, dictionaries and tagger for extended named entity hierarchy. In *LREC*, pages 1977–1980. Lisbon, Portugal.

Yan Shao, Christian Hardmeier, and Joakim Nivre. 2016. Multilingual named entity recognition using hybrid neural networks. In *The Sixth Swedish Language Technology Conference (SLTC)*.

Felipe Soares, Marta Villegas, Aitor Gonzalez-Agirre, Martin Krallinger, and Jordi Armengol-Estapé. 2019. Medical word embeddings for spanish: Development and evaluation. In *Proceedings of the 2nd Clinical Natural Language Processing Workshop*, pages 124–133.

GuoDong Zhou and Jian Su. 2002. Named entity recognition using an hmm-based chunk tagger. In *proceedings of the 40th Annual Meeting on Association for Computational Linguistics*, pages 473–480. Association for Computational Linguistics.

Transfer Learning in Biomedical Named Entity Recognition: An Evaluation of BERT in the PharmaCoNER task

Cong Sun
Dalian University of Technology
suncong132@mail.dlut.edu.cn

Zhihao Yang *
Dalian University of Technology
yangzh@dlut.edu.cn

Abstract

To date, a large amount of biomedical content has been published in non-English texts, especially for clinical documents. Therefore, it is of considerable significance to conduct Natural Language Processing (NLP) research in non-English literature. PharmaCoNER is the first Named Entity Recognition (NER) task to recognize chemical and protein entities from Spanish biomedical texts. Since there have been abundant resources in the NLP field, how to exploit these existing resources to a new task to obtain competitive performance is a meaningful study. Inspired by the success of transfer learning with language models, we introduce the BERT benchmark to facilitate the research of PharmaCoNER task. In this paper, we evaluate two baselines based on Multilingual BERT and BioBERT on the PharmaCoNER corpus. Experimental results show that transferring the knowledge learned from source large-scale datasets to the target domain offers an effective solution for the PharmaCoNER task.

1 Introduction

Currently, most biomedical Natural Language Processing (NLP) tasks focus on English documents, while only few research has been carried out on non-English texts. However, it is essential to note that there is also a considerable amount of biomedical literature published in other languages than English, especially for clinical documents. Therefore, it is of considerable significance to conduct NLP research in non-English literature. PharmaCoNER(Gonzalez-Agirre et al., 2019) is the first Named Entity Recognition (NER) task to recognize chemical and protein entities from Spanish biomedical texts. Biomedical NER task is the foundation of biomedical NLP research, which is

---

*Corresponding author

often utilized as the first step in relation extraction, information retrieval, question answering, etc.

The existing biomedical NER methods can be roughly classified into two categories: traditional machine learning-based methods and deep learning-based methods. Traditional machine learning-based methods (Settles, 2005; Campos et al., 2013; Wei et al., 2015; Leaman et al., 2015, 2016) mainly depend on feature engineering, i.e., the design of useful features using various NLP tools. Overall, this is a labor-intensive and skill-dependent process. In contrast, deep learning-based methods are more promising in biomedical NER tasks. Since deep learning-based methods can automatically learn features, these methods no longer need to construct feature engineering and exhibit more encouraging performance. For examples, (Luo et al., 2017) proposed an attention-based BiLSTM-CRF approach to document-level chemical NER. (Dang et al., 2018) proposed a D3NER model, using CRF and BiLSTM improved with fine-tuned embeddings of various linguistic information to recognize disease and protein/gene entities. Recently, the language model pre-training (Peters et al., 2018; Radford et al., 2018; Devlin et al., 2019) has proven to be effective for improving many NLP tasks. The fine-tuning language model (Radford et al., 2018; Devlin et al., 2019) can transfer the knowledge learned from large-scale datasets to domain-specific tasks by simply fine-tuning the pre-trained parameters.

Inspired by the success of transfer learning with language models, we would like to make full use of the existing language model resources to implement the PharmaCoNER task. In this paper, we introduce the BERT (Devlin et al., 2019) benchmark to facilitate the research of PharmaCoNER task. We regard the large-scale dataset used to train the BERT model as the source do-

Proceedings of the 5th Workshop on BioNLP Open Shared Tasks, pages 100–104
Hong Kong, China, November 4, 2019. ©2019 Association for Computational Linguistics

main, and the PharmaCoNER dataset as the target domain, thus considering the PharmaCoNER task as a transfer learning problem. We evaluate two baselines based on Multilingual BERT and BioBERT. Experimental results show that transferring the knowledge learned from source large-scale datasets to the target domain offers an effective solution for the PharmaCoNER task.

2 Related Work

2.1 Language Model

Learning widely used representations of words has been an active area of research for decades. To date, pre-trained word embeddings are considered to be an integral part of modern NLP systems, offering significant improvements over embeddings learned from scratch (Turian et al., 2010). Recently, ELMo (Peters et al., 2018) has been proposed to generalize traditional word embedding research (Mikolov et al., 2013; Pennington et al., 2014; Bojanowski et al., 2017) to extract context-sensitive features. When integrating contextual word embeddings with existing task-specific architectures, ELMo achieves competitive performance for many major NLP benchmarks. More recent studies (Radford et al., 2018; Devlin et al., 2019) tend to exploit language models to pre-train some model architecture on a language model objective before fine-tuning that the same model for downstream tasks. BERT (Devlin et al., 2019), which stands for Bidirectional Encoder Representations from Transformers (Vaswani et al., 2017), is designed to pre-train deep bidirectional representations by jointly conditioning on both left and right context in all layers. The pre-trained BERT can be fine-tuned to create competitive models for a wide range of tasks.

2.2 Transfer Learning

Many machine learning methods work well only under a common assumption: the training and test data are drawn from the same feature space and distribution (Pan and Yang, 2009). When the distribution changes, most models need to be rebuilt from scratch using newly annotated training data. However, it is an expensive and challenging process. Therefore, it would be meaningful to reduce the need and effort to recollect the annotated training data. In such scenarios, transfer learning between task domains would be useful. For example, (Cui et al., 2018) demonstrate the effects of transfer learning in the computer vision domain. They explore transfer learning via fine-tuning the knowledge learned from large-scale datasets to small-scale domain-specific fine-grained visual categorization datasets. For NLP tasks, (Conneau et al., 2017) and (McCann et al., 2017) also demonstrate the effects of transfer learning on the natural language inference and machine translation tasks, respectively. These methods demonstrate the significance of transfer learning in machine learning methods.

3 Methods

3.1 Problem Definition

The PharmaCoNER task is structured into two sub-tracks: 'NER offset and entity classification' and 'concept indexing'. Since we only participate in the first track, we will explain the fist track in detail. There are three entity types for evaluation in the PharmaCoNER corpus, namely 'normalizables', 'notnormalizables' and 'proteins'. Specifically, 'normalizables' is the mentions of chemicals that can be manually normalized to a unique concept identifier. 'notnormalizables' is the mentions of chemicals that could not be normalized manually to a unique concept identifier. 'proteins' is the mentions of proteins and genes. We used the extended BIO (Begin, Inside, Other) tagging scheme in our experiments. Formally, we formulate the PharmaCoNER task as a multiclass classification problem. Given an input sequence $S = \{w_1, \cdots, w_i, \cdots, w_n\}$ which has processed by WordPiece, the goal of PharmaCoNER is to classify the tag t of token w_i. Essentially, the model estimates the probability $P(t|w_i)$, where $T = \{$B-normalizables, I-normalizables, B-notnormalizables, I-notnormalizables, B-proteins, I-proteins, O, X, CLS, SEP$\}, t \in T, 1 \leq i \leq n$.

3.2 Model Architecture

BERT (Devlin et al., 2019), which stands for bidirectional encoder representations from Transformers, is designed to learn deep bidirectional representations by jointly conditioning on both left and right context in all layers. The architecture of BERT is illustrated in Figure 1. The pre-trained BERT can be fine-tuned to create competitive models for a wide range of downstream tasks, such as named entity recognition, relation extraction, and question answering.

Here, we explain the architecture of BERT for

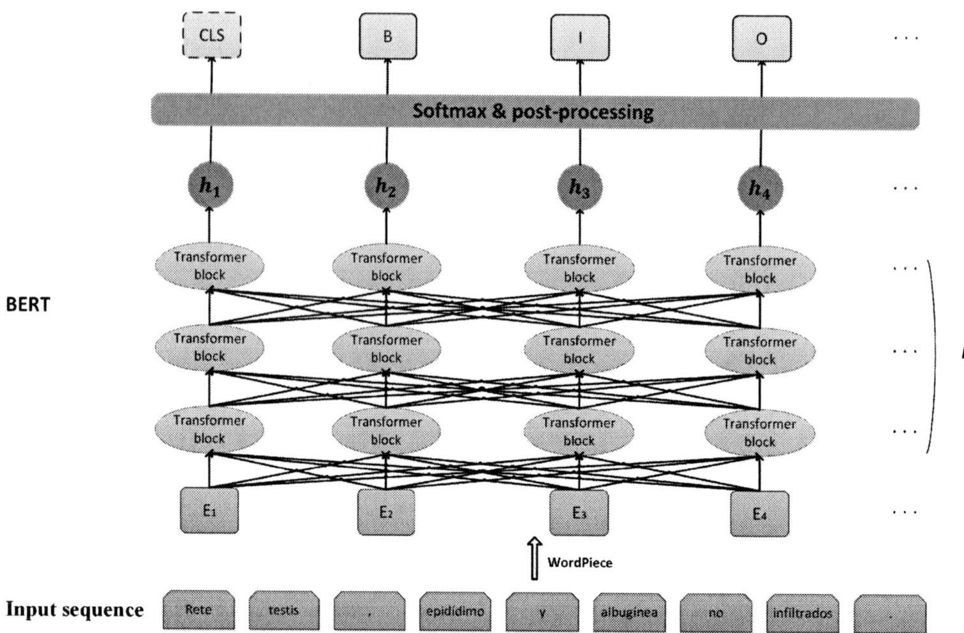

Figure 1: The Architecture of BERT.

NER tasks. The input of BERT can represent both a single text sentence or a pair of text sentences in one sequence. BERT differentiates the text sentences as follows: first, they separate them with a special token ([SEP]); second, they add the sentence A embedding to every token of the first text sentence and the sentence B embedding to every token of the second text sentence. Furthermore, every sequence starts with a special token ([CLS]). For a given token, the input representation is constructed by integrating the corresponding token, segment, and position embeddings. BERT provides two model sizes: $BERT_{BASE}$ and $BERT_{LARGE}$. For the BERT model, the number of layers L, the hidden size H and the number of self-attention heads A are listed as follows:

- $BERT_{BASE}$: L=12, H=768, A=12, Total Parameters=110M.

- $BERT_{LARGE}$: L=24, H=1024, A=16, Total Parameters=340M.

During the shared task, we exploit Multilingual BERT (Devlin et al., 2019) and BioBERT (Lee et al., 2019) to implement the PharmaCoNER task. Both the multilingual BERT and BioBERT models are pre-trained based on the $BERT_{BASE}$ size. The multilingual BERT model is pre-trained on Wikipedia in multiple languages. The BioBERT model is pre-trained on Wikipedia, BooksCorpus, PubMed (PubMed abstracts) and PMC (PubMed Central full-text articles). The pre-training process of Multilingual BERT and BioBERT is similar to the pre-training process of $BERT_{base}$. More details about Multilingual BERT and BioBERT can be found in the studies (Devlin et al., 2019; Lee et al., 2019).

For the output layer, we feed the final hidden representation h_i of each token i into the softmax function. The probability P is calculated as follows:

$$P(t|h_i) = softmax(W_o h_i + b_o) \qquad (1)$$

where $T = \{$B-normalizables, I-normalizables, B-notnormalizables, I-notnormalizables, B-proteins, I-proteins, O, X, CLS, SEP$\}$, $t \in T$, W_o and b_o are weight parameters. Furthermore, during the training, we use the categorical cross-entropy as the loss function. Finally, as shown in Figure 1, we removed the special tokens (labeled by 'X', 'CLS' and 'SEP') and obtained the final BIO labels at the post-processing step.

4 Results and Discussion

4.1 Experimental Settings

In this section, we introduce the dataset, evaluation metrics and details of the training process of our model.

Dataset. The PharmaCoNER corpus has been randomly sampled into three subsets: the training set, the development set and the test set. The training set contains 500 clinical cases, and the development set and the test set include 250 clinical cases, respectively.

Evaluation Metrics. We apply the standard measures precision, recall and micro-averaged F1-score to evaluate the effectiveness of our model. These metrics are also adopted as the evaluation metrics during the PharmaCoNER task.

Training Details. During the PharmaCoNER task, we utilized the training set for training the model and exploited the development set to choose the hyper-parameters of our model. In the prediction stage, we combined the training and development sets for training our model, and the organizers used the gold-standard test set to evaluate the final results. The detailed hyper-parameter settings are illustrated in Table 1. 'Opt.' denotes optimal.

Parameters	Tuned range	Opt.
Sequence length	128	128
Train batch size	[8, 16, 32]	32
Dev batch size	8	8
Test batch size	8	8
Learning rate	[1e-5, 2e-5, 3e-5]	2e-5
Epoch number	[10, 50, 100, 200]	100
Warmup	0.1	0.1
Dropout	0.1	0.1

Table 1: Detailed Hyper-parameter Settings in the PharmaCoNER task.

4.2 Experimental Results

We applied Multilingual BERT and BioBERT on the PharmaCoNER corpus, respectively. The experimental results are shown in Table 2. 'P', 'R', 'F' denote precision, recall, and micro-averaged F1-score, respectively. It is encouraging to see that the performance of both models is quite competitive. For the multilingual BERT model, since the model learned the Spanish language information during the pre-training process, its F1-score is higher, reaching 89.24%. For the BioBERT model, it also achieves an F1-score of 89.02%. While BioBERT was only pre-trained on the English biomedical texts, applying it to the Spanish PharmaCoNER task still yields competitive performance. The primary reason may be that there are a large number of chemical and protein mentions sharing the same name in English and Spanish in biomedical literature. Therefore, it is feasible to use the existing model pre-trained on English biomedical corpora to fine-tune the PharmaCoNER task. These results indicate that transferring the knowledge learned from source large-scale datasets via fine-tuning to the target-specific domain is an effective solution to the PharmaCoNER task.

Models	P(%)	R(%)	F(%)
Multilingual BERT	90.46	88.06	89.24
BioBERT	90.70	87.41	89.02

Table 2: The Experimental Results of Multilingual-BERT and BioBERT.

Furthermore, we manually analyzed the errors generated by our models on the corpus test set after the PharmaCoNER task. The main errors can be classified into three categories: (1) incorrect boundaries, (2) missing the chemical/protein mention, (3) and incorrectly distinguishing the chemical and protein mentions. By analyzing these error examples, we infer that document-level information or biomedical knowledge may be helpful for the PharmaCoNER task.

5 Conclusion

In this paper, we introduce the BERT benchmark to facilitate the research of PharmaCoNER task. We evaluate two baselines based on Multilingual BERT and BioBERT on the PharmaCoNER corpus. It is encouraging to see that the performance of both models is quite competitive, reaching F1-scores of 89.24% and 89.02%, respectively. Experimental results demonstrate that transferring the knowledge learned from source large-scale datasets to the target domain offers an effective solution for the PharmaCoNER task.

In future work, we would like to explore an appropriate way to integrate document-level information or biomedical knowledge to improve the performance of the model.

Acknowledgments

This work was supported by the grants from the National Key Research and Development Program of China (No.2016YFC0901902), Natural Science Foundation of China (No.61272373, 61572102 and 61572098), and Trans-Century Training Program Foundation for the Talents by the Ministry of Education of China (NCET-13-0084).

References

Piotr Bojanowski, Edouard Grave, Armand Joulin, and Tomas Mikolov. 2017. Enriching word vectors with subword information. *Transactions of the Association for Computational Linguistics*, 5:135–146.

David Campos, Sérgio Matos, and José Luís Oliveira. 2013. Gimli: open source and high-performance biomedical name recognition. *BMC bioinformatics*, 14(1):54.

Alexis Conneau, Douwe Kiela, Holger Schwenk, Loïc Barrault, and Antoine Bordes. 2017. Supervised learning of universal sentence representations from natural language inference data. In *Proceedings of the 2017 Conference on Empirical Methods in Natural Language Processing*, pages 670–680.

Yin Cui, Yang Song, Chen Sun, Andrew Howard, and Serge Belongie. 2018. Large scale fine-grained categorization and domain-specific transfer learning. In *Proceedings of the IEEE conference on computer vision and pattern recognition*, pages 4109–4118.

Thanh Hai Dang, Hoang-Quynh Le, Trang M Nguyen, and Sinh T Vu. 2018. D3ner: biomedical named entity recognition using crf-bilstm improved with fine-tuned embeddings of various linguistic information. *Bioinformatics*, 34(20):3539–3546.

Jacob Devlin, Ming-Wei Chang, Kenton Lee, and Kristina Toutanova. 2019. BERT: Pre-training of deep bidirectional transformers for language understanding. In *Proceedings of the Conference of the North American Chapter of the Association for Computational Linguistics*, pages 4171–4186.

Aitor Gonzalez-Agirre, Montserrat Marimon, Ander Intxaurrondo, Obdulia Rabal, Marta Villegas, and Martin Krallinger. 2019. Pharmaconer: Pharmacological substances, compounds and proteins named entity recognition track. In *Proceedings of the BioNLP Open Shared Tasks (BioNLP-OST)*.

Robert Leaman, Chih-Hsuan Wei, and Zhiyong Lu. 2015. tmchem: a high performance approach for chemical named entity recognition and normalization. *Journal of cheminformatics*, 7(1):S3.

Robert Leaman, Chih-Hsuan Wei, Cherry Zou, and Zhiyong Lu. 2016. Mining chemical patents with an ensemble of open systems. *Database*, 2016.

Jinhyuk Lee, Wonjin Yoon, Sungdong Kim, Donghyeon Kim, Sunkyu Kim, Chan Ho So, and Jaewoo Kang. 2019. BioBERT: pre-trained biomedical language representation model for biomedical text mining. *arXiv preprint arXiv:1901.08746*.

Ling Luo, Zhihao Yang, Pei Yang, Yin Zhang, Lei Wang, Hongfei Lin, and Jian Wang. 2017. An attention-based bilstm-crf approach to document-level chemical named entity recognition. *Bioinformatics*, 34(8):1381–1388.

Bryan McCann, James Bradbury, Caiming Xiong, and Richard Socher. 2017. Learned in translation: Contextualized word vectors. In *Advances in Neural Information Processing Systems*, pages 6294–6305.

Tomas Mikolov, Kai Chen, Greg Corrado, and Jeffrey Dean. 2013. Efficient estimation of word representations in vector space. *Computer Science*.

Sinno Jialin Pan and Qiang Yang. 2009. A survey on transfer learning. *IEEE Transactions on knowledge and data engineering*, 22(10):1345–1359.

Jeffrey Pennington, Richard Socher, and Christopher Manning. 2014. Glove: Global vectors for word representation. In *Proceedings of the Conference on Empirical Methods in Natural Language Processing*, pages 1532–1543.

Matthew Peters, Mark Neumann, Mohit Iyyer, Matt Gardner, Christopher Clark, Kenton Lee, and Luke Zettlemoyer. 2018. Deep contextualized word representations. In *Proceedings of the Conference of the North American Chapter of the Association for Computational Linguistics*, pages 2227–2237.

Alec Radford, Karthik Narasimhan, Time Salimans, and Ilya Sutskever. 2018. Improving language understanding with unsupervised learning. Technical report, OpenAI.

Burr Settles. 2005. Abner: an open source tool for automatically tagging genes, proteins and other entity names in text. *Bioinformatics*, 21(14):3191–3192.

Joseph Turian, Lev Ratinov, and Yoshua Bengio. 2010. Word representations: a simple and general method for semi-supervised learning. In *Proceedings of the 48th annual meeting of the association for computational linguistics*, pages 384–394.

Ashish Vaswani, Noam Shazeer, Niki Parmar, Jakob Uszkoreit, Llion Jones, Aidan N Gomez, Łukasz Kaiser, and Illia Polosukhin. 2017. Attention is all you need. In *Advances in neural information processing systems*, pages 5998–6008.

Chih-Hsuan Wei, Hung-Yu Kao, and Zhiyong Lu. 2015. Gnormplus: an integrative approach for tagging genes, gene families, and protein domains. *BioMed research international*, 2015.

A Multi-Task Learning Framework for Extracting Bacteria Biotope Information

Qi Zhang[1], Chao Liu[2,*], Ying Chi[2], Xuansong Xie[2] & Xiansheng Hua[2]
[1]Zhejiang University, [2]Alibaba DAMO Academy
zhangqihit@zju.edu.com
{ maogong.lc, xinyi.cy, xiansheng.hxs}@alibaba-inc.com
xingtong.xss@taobao.com

Abstract

This paper presents a novel transfer multi-task learning method for Bacteria Biotope rel+ner task at BioNLP-OST 2019. To alleviate the data deficiency problem in domain-specific information extraction, we use BERT(Devlin et al., 2018) (Bidirectional Encoder Representations from Transformers) and pre-train it using mask language models and next sentence prediction (Devlin et al., 2018) on both general corpus and medical corpus like PubMed. In fine-tuning stage, we fine-tune the relation extraction layer and mention recognition layer designed by us on the top of BERT to extract mentions and relations simultaneously. The evaluation results show that our method achieves the best performance on all metrics (including slot error rate, precision and recall) in the Bacteria Biotope rel+ner subtask.

1 Introduction

Information extraction aims to recognize the entities and classify the relations between them in given unstructured text. It provides cornerstone for many downstream applications such as information extraction, knowledge base population, and question-answering. It is a challenging task partly because it requires elaborative human annotations (Riedel et al., 2010), which could be slow or expensive to get.

Bacteria Biotope (BB) task is an interesting information extraction task aiming at extracting knowledge about bacteria biotope from bioinfomatics literature related to microorganism. Rel+ner subtask focuses on extracting entity mentions of following types: Microorganism (MI), Habitat (HA), Phenotype (PH), Geographical (GE) and identification of the *Lives_In* relation between a Habitat/Geographical mention and a Microorganism mention as well as the *Exhibits*

relation between a Phenotype mention and a Microorganism mention. This task intends to extract structured triple of microorganism from unstructured biomedical text.

Some previous work has been done in handling such an information extraction problem, including some joint entity and relation extraction methodology and pipeline method which firstly do named entity recognition (NER) and then do relation extraction on the results of NER. (Zheng et al., 2017) proposes a novel tagging schema (NTS) that encodes relation type in the NER tag to recognize the named entity and extract the relation between them jointly. This methodology has a fatal flaw that it can not handle relation facts that share the same entity and this phenomenon is common in BB task. (Bekoulis et al., 2018) proposes a multi-head selection layer (MHS) to model the relation of each entity pair which is similar to our method. (Zeng et al., 2018) proposes a sequence to sequence model with copy mechanism (Copy RE). However, all above the previous work has been done on a large-scale general dataset. While the Bacteria Biotope rel+ner task only bases on a domain-specific and comparatively small dataset. Under this background, we adapt a recently widely used transfer learning framework, BERT(Devlin et al., 2018), and pre-train it on large-scale corpus using two novel unsupervised prediction tasks to mitigate the problem of insufficient data.

2 Model Architecture

The overall framework of the model is shown in Figure 1. Bottom parts of the model (including input representation, transformer encoder) are shared by both named entity recongnition task and relation extraction task.

Corresponding author

Proceedings of the 5th Workshop on BioNLP Open Shared Tasks, pages 105–109
Hong Kong, China, November 4, 2019. ©2019 Association for Computational Linguistics

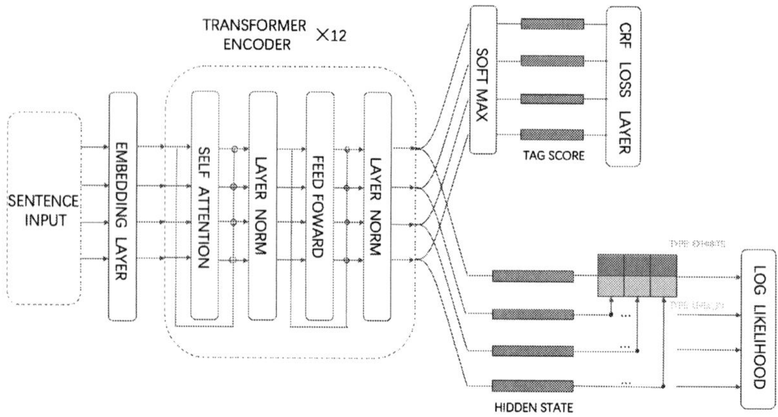

Figure 1: Diagram of Our Model

2.1 Input Feature and Representation

The input representation of each word w_i in sentence $S = \{w_1, w_2, ..., w_N\}$ consists of three parts: word vector, the embedding of features and positional encoding. A pre-trained word embedding using Skip-gram (Mikolov et al., 2013) model is used to map each word to a dense vector. The features we used are described in Table 1. Each feature is represented by a one-hot vector and pass a feature embedding layer. Positional encoding is added to make the model capture the relative and absolute position of each token (Vaswani et al., 2017). The three parts are concatenated and fed into transformer encoder.

2.2 Transformer Encoder

Transformer is widely used in various natural language processing task recently. We use transformer here to extract context features of each token. The encoder is composed of 12 layers. Each layer consists of a multi-head self attention sub-layer and point-wise fully connected feedfoward sub-layer, a residual connection is employed around each of the two sub-layers (Vaswani et al., 2017). The transformer is pre-trained using two novel unsupervised tasks including masked language model and next sentence predicting (Devlin et al., 2018) on the combination of BooksCorpus, English Wikipedia, PubMed and PubMed Central (PMC) corpus. The hyperparameters we use to pre-train are exactly the same as the $BERT_{BASE}$ of (Devlin et al., 2018). In fine-tuning stage, the output of the transformer encoder H_i will be fed into both mention recognition layer and relation extraction layer.

2.3 Mention Recognition Layer

Commonly, in named entity recognition, annotated data is tagged using BIO tagging schema in which each token is assigned into one of following tag: B means beginning, I means inside and O means outside of an entity mention. However this tagging schema is insufficient since some entity mentions in BB task are disjoint concepts with overlapping words. Taking the phrase "serotypes A, B and C" as an example, this phrase contains three disjoint Microorganism mentions: "serotypes A", "serotypes B" and "serotypes C". To handle these special mentions, we apply an alternative tagging schema which introduce 'H' and 'D' flag, where 'H' indicates the overlapping tokens and 'D' indicates discontinuous tokens. Figure 2 shows an annotation example. The tagging label set of this new tagging schema can be written as $\{\{GE, HA, PH, MI\} \times \{H, D\}\} \times \{B, I\} \bigcup \{O\}$.

We feed the final state H_i of each token to the softmax classification layer over the tagging set. The conditional random field (CRF) layer takes the sequence of output score vector V_i from the softmax classification layer. The tag prediction of w_i in sentence s is denoted as y_i^S, and further the CRF score of the tag predictions $y^S = \{y_1^S, y_2^S, ..., y_N^S\}$ is defined as follows:

$$score_{y^S} = E_{y^S} + T_{y^S} \qquad (1)$$

E represents emission score which can be defined

Feature Name	Description
Dot Flag Feature	Whether the word contains dot notations like "C. psittaci".
Capitalization Feature	Whether the first letter of the word is capitalized.
POS Tagging Feature	The output for the tokenized sentence of the POS tagging tool.
Dependency Parsing Feature	The output for the tokenized sentence of the dependency parsing tool.

Table 1: Input Features of Our Model and Their Description

as:

$$E_{y^S} = \sum_{i=1}^{N} V_i \qquad (2)$$

T represents transition score which can be defined as:

$$T_{y^S} = \sum_{i=1}^{N} TM_{y_{i-1}^S, y_i^S} \qquad (3)$$

where TM_{y_{i-1}, y_i} means the transition probability from tag y_{i-1} to y_i. The conditional probability $P(y|S)$ can be written as follows:

$$P(y^S|S) = \frac{e^{score_{y^S}}}{\sum_{y \in y^*} e^{score_y}} \qquad (4)$$

where y^* is the collection of all possible tag predictions for sentence s.

Serotypes	A,	B	and	C
HB-MI	DB-MI	DB-MI	O	DB-MI

Figure 2: Examples of BIOHD Tagging

2.4 Relation Extraction Layer

As depicted in Figure 1, the sequence of final state H_i is also fed into the relation extraction layer. We observe that each Microorganism entity may have multiple relations with entities of other three types. Moreover, all types of relation must contain a Microorganism entity. Thus we take the Microorganism entity as the center of relation prediction task.

The Microorganism entity which ends with the word w_i will be calculated the following score with another entity end with the word w_j:

$$R_{i,j,r} = \sigma(W_r f(H_r * Vi + T_r * Vj + b_r)) \qquad (5)$$

where H_r, T_r and b_r are parameter matrices associated with relation type r. The score $R_{i,j,r}$ represents probability that the Microorganism entity ends with words w_i has the relation r with another entity ends with words w_j. f is the activation function: relu. σ is used to normalize the probability.

2.5 Multi Task Training Objective

In training stage, we fine-tune the relation extraction layer and mention recognition layer simultaneously using a joint loss. The training loss defined by mention recognition layer can be written as:

$$L_{ner} = -log P(y^S|S) \qquad (6)$$

Moreover, the loss function of the relation extraction layer can defined as

$$L_{rel} = \sum_i \sum_j -log R_{i,j,r} \qquad (7)$$

The loss function of the whole system can be defined as

$$L = L_{ner} + L_{rel} \qquad (8)$$

3 Experiment and Result

In this section, we briefly introduce the dataset, evaluation metrics and the external resources that we use. We present our performance on different relation type with different metrics provided by organizers and comparison with other jointly information extraction methodology mentioned in Section 1 on development data.

3.1 Dataset Description

Bacteria Biotope task includes two types of documents: PubMed references (titles and abstracts) related to microorganism, extracts from full-text articles related to microorganisms living in food products.

The statistics of the dataset is shown in Table 2. The training and development data released for this task contains 133 and 66 files respectively, with gold standard annotations. Test data contains 32 files which are used to evaluate participation. The number of entity mentions in different file is unbalanced, ranging from 0 to 85.

Table 2: The Statistics of The Dataset: Number of Files, Relations and Entities

	File	Entity	Relation
train	133	2266	1127
development	66	1271	608
test	32	Unknown	Unknown

Table 3: Performance for Each Relation Type

	SER	precision	recall
All-types	0.954	0.509	0.351
Exhibits	0.982	0.492	0.449
Lived_in-geo	1.318	0.316	0.273
Lived_in-habitat	0.927	0.530	0.311

Table 4: Performance indicates statistically significant difference from our model, NTS, MHS, Copy RE and Pipeline.

	SER	precision	recall
Pipeline	1.472	0.231	0.294
NTS	1.456	0.261	0.288
MHS	1.183	0.381	0.302
Copy RE	1.128	0.376	0.291
Our model	0.947	0.493	0.339

3.2 External Resources

Here we introduce some external resources that we use in experiment. We use Google Word2vec tool to train word embeddings on corpora composed of PubMed, PubMed Central (PMC) corpus and English Wikipedia corpus. The LTP tool is used for sentence level dependency parsing and the NLTK tool is used for sentence tokenization and part of speech tagging.

3.3 Metric and Performance Comparison

Since the entity mentions which are potential arguments of each relation, are not given. In evaluation metrics (precision, recall), substitution errors are penalized. Morever, Slot Error Rate (SER) is taken as the main evaluation metric. Table 3 shows our results of different relation type.

We also evaluate some previous with famous jointly information extraction methodologies which are described in Section 1 on the BB 2019 development data for comparison:

NTS: Our implementation of (Zheng et al., 2017). Instead we use the tagging schema described in Section 2.3.

MHS: We use the code released by (Bekoulis et al., 2018) and train the model on the training data of BB rel+ner task.

Copy RE: Our implementation the sequence to sequence model using copy mechanism (Zeng et al., 2018). We train the model using the training data of BB rel+ner task.

Pipeline: The baseline method that we use includes two step separately: perform NER (Devlin et al., 2018) firstly, then perform relation extraction (Devlin et al., 2018) on the results of the NER task.

As shown in Table 4, our model achieves improvements on BB dataset comparing with the other four models. Particularly, our model significantly outperforms the **Pipeline** baseline by -0.525 SER.

3.4 Factor Analysis

We propose several strategies to improve the performance including feature engineering and utilizing the transformer encoder. To investigate the influence of these two factors, we conduct ablation study and list results on Table 5 .

"No" prefix in Table 5 means that we train and evaluate our model without the corresponding feature. "No Transformer Encoder" indicates that we replace the transformer with bi-directional lstm.

Results show that each feature listed in Table 1 plays a key role. Our model suffers serious performance degradation without any one of the four input features.

Table 5: Ablation Study

Model	SER	P	R
Our Model	0.947	0.493	0.339
No Dot Flag Feature	0.961	0.485	0.313
No Capitalization Feature	0.956	0.489	0.324
No POS Tagging Feature	0.949	0.499	0.335
No Dependency Parsing Feature	0.951	0.487	0.333
No Transformer Encoder	0.998	0.470	0.321

4 Conclusions

In this paper, we describe our participation in Bacteria Biotope rel+ner subtask. We propose a transfer multi-task learning framework to overcome data deficiency and fine-tune a joint entity and relation extraction model using multi-task training objective. Though we achieve the best performance in this subtask, we have some future directions to improve this work furthermore: adapting adversarial training or posterior regularization to improve the performance of our system.

References

Giannis Bekoulis, Johannes Deleu, Thomas Demeester, and Chris Develder. 2018. Joint entity recognition and relation extraction as a multi-head selection problem. *Expert Systems with Applications*, 114:34–45.

Jacob Devlin, Ming Wei Chang, Kenton Lee, and Kristina Toutanova. 2018. Bert: Pre-training of deep bidirectional transformers for language understanding.

Tomas Mikolov, Ilya Sutskever, Kai Chen, Greg S Corrado, and Jeff Dean. 2013. Distributed representations of words and phrases and their compositionality. In *Advances in neural information processing systems*, pages 3111–3119.

Sebastian Riedel, Limin Yao, and Andrew McCallum. 2010. Modeling relations and their mentions without labeled text. In *Joint European Conference on Machine Learning and Knowledge Discovery in Databases*, pages 148–163. Springer.

Ashish Vaswani, Noam Shazeer, Niki Parmar, Jakob Uszkoreit, Llion Jones, Aidan N. Gomez, Lukasz Kaiser, and Illia Polosukhin. 2017. Attention is all you need.

Xiangrong Zeng, Daojian Zeng, Shizhu He, Kang Liu, Jun Zhao, et al. 2018. Extracting relational facts by an end-to-end neural model with copy mechanism.

Suncong Zheng, Wang Feng, Hongyun Bao, Yuexing Hao, Zhou Peng, and Xu Bo. 2017. Joint extraction of entities and relations based on a novel tagging scheme.

YNU-junyi in BioNLP-OST 2019: Using CNN-LSTM Model with Embeddings for SeeDev Binary Event Extraction

Junyi Li, Xiaobing Zhou*, Yuhang Wu, Bin Wang
School of Information Science and Engineering
Yunnan University, Yunnan, P.R. China
*Corresponding author, zhouxb@ynu.edu.cn

Abstract

We participated in the BioNLP 2019 Open Shared Tasks: binary relation extraction of SeeDev task. The model was constructed using convolutional neural networks (CNN) and long short term memory networks (LSTM). The full text information and context information were collected using the advantages of C-NN and LSTM. The model consisted of two main modules: distributed semantic representation construction, such as word embedding, distance embedding and entity type embedding; and CNN-LSTM model. The F1 value of our participated task on the test data set of all types was 0.342. We achieved the second highest in the task. The results showed that our proposed method performed effectively in the binary relation extraction.

1 Introduction

The goal of Information Extraction (IE) (Finkel et al., 2005) is to transform textual information into structured information, and to focus on quickly locating and finding useful information in large amounts of data. Information Extraction (IE) (Fader et al., 2011) is also capable of mining useful data and hiding knowledge from a large number of corpus texts, which has led to some new research methods in many disciplines. For example, with the growing demand for key issues related to life and biology, many biological problems have fallen into the bottleneck due to inadequate methods. Biological information extraction (Bio-IE) emerges in time and attracts more and more researchers to solve problems. For instance, in the identification of named entities, the classification of relationships between proteins and the extraction of links between drugs. In addition, information extraction in the field of biology, especially event extraction, has entered people's views. This will be a far-reaching task and a major biological challenge for information extraction tasks.

The BioNLP Shared Task Series is a representative of biomolecular event extraction and has been held four times. This year is the fifth time that BioNLP has shared tasks. The topics in this series include fine-grained extraction, generalization to knowledge base construction. In addition, the scope of this task has become more extensive in each time. For instance, the BioNLP 2016 Shared Task(Nédellec et al., 2016) contained three separate parts, the Bacteria Biotope subtask (B-B3), the Seed Development subtask (SeeDev) and the Genia Event subtask (GE4). However, the BioNLP 2019 Open Shared Task contains seven separate parts, the Integrated structure, semantics and coreference subtask (CRAFT), the Pharma-CoNER task, the Active Gene Annotation Corpus subtask (AGAC), the BB3, the SeeDev and the Research Domain Criteria subtask(RDoc).

We mainly participated in the binary relation extraction task, which is part of the SeeDev task. The SeeDev task (Nédellec et al., 2013)(Chaix et al., 2016) aims to promote complex event extraction on regulations in plants from scientific articles. It focuses on events describing genetic and molecular mechanisms involved in seed development of the model plant, Arabidopsis thaliana. It involves n-ary and binary relation extraction. Meanwhile, the SeeDev task was proposed for the first time at BioNLP Shared Task 2016(Nédellec et al., 2016) (Mehryary et al., 2016). This 2019 edition is a rerun of the task, with an evaluation methodology more focused on the biological contribution.

Many teams participated in the BioNLP 2016 Shared Task(He et al., 2016). For example, VERSE uses a support vector machine (SVM) and k-fold cross-validation to identify the best parameters.(Lever and Jones, 2016) DUTIR uses a deep learning method that utilizes a convolutional neu-

Proceedings of the 5th Workshop on BioNLP Open Shared Tasks, pages 110–114
Hong Kong, China, November 4, 2019. ©2019 Association for Computational Linguistics

ral network(Li et al., 2016). Motivated by the previous study, based on CNN, we have integrated L-STM(Hochreiter and Schmidhuber, 1997) to solve the defect that convolutional neural networks can not obtain context information. After improving the method, we got good results.

The rest of our paper is structured as follows. Section 2 introduces models. Section 3 describes results and discussion. Conclusions are described in Section 4.

2 Model

The SeeDev-binary task can be thought of as a binary relationship extraction, which specifies whether there is interaction between the two entities. In relation extraction, the semantic and syntactic information of a sentence plays an important role. Traditional methods often require the design and extraction of complex features based on domain-specific knowledge (such as tree kernels and graphics kernels) to construct the model. As a result, this results in a much lower corpus-dependent generation capability. Therefore, we use CNN to replace complex manual design feature engineering, and learn the advanced function automation by modeling the word embedding and fully connected neural networks from the original input through convolution and pooling operations. Besides, we capture relative distance information and entity types as complementary features of the sentence. After that, we input the data processed by the CNN into the LSTM. Because CNN do not get good context information, and sometimes the connection between text contexts can help us do relation extraction more accurately. So, LSTM can get text context information, which allows us to get a better result in the end.

As shown in Figure 1, the model consists of two modules: distributed semantic representation construction, such as embedded characters, distance embedding and entity type embedding, and CNN-LSTM module. In the next section, we will introduce more details.

2.1 Data preprocessing

When doing data preprocessing, first we use the Stanford CoreNLP(Manning et al., 2014) tool to process the task's data. The text is divided into sentences and tokenized. Parts-of-speech and lemmas are identified and a dependency parse is generated for each sentence. Then, we further process the preprocessed data.

2.2 Embedding

We use the context of two entities to predict the type of relationship. In our task, the context is represented by words between two entities in a sentence. Then, by analyzing the data, we observe that different entities with different types have different mutual interaction probabilities if the entity types satisfy the relationship constraints. Therefore, the entity type of the two entities is the important factor of the predicted relationship type. In our model, entity types are seen as a complement to word embedding. In addition, we find that distance information usually plays an important role. The distance can capture the relative position between two entities. So, we concatenate the word embedding(Levy and Goldberg, 2014), type embedding(Su and Wang, 2011), and distance embedding(Cormode, 2003). We use the pre-trained word embedding.[1]

Then, we would introduce some formulas about word embedding, entity type embedding and distance embedding.

$$LT_W(S) =$$

$$[<W>_{E_1}, <W>_{W_1}, ..., <W>_{W_n}, <W>_{E_2}]$$

$$LT_{W,W^T}(S) =$$

$$[<W>_{E_1}, ..., <W>_{E_2}, <W^T>_{type(E_1)}]$$

$$LT_{W^d}(S) =$$

$$[<W^d>_{d(E_1,E_1)}, ..., <W^d>_{d(E_2,E_1)}, 0, 0]$$

where S stands for the sentences. E_1 and E_2 are the type 1 and type 2 respectively. W_1 stands for the first word. W is the word embedding table. W^T is type embedding table and W^d stands for the distance embedding table. $LT_W(S)$ is the representation of word embedding. $LT_{W,W^T}(S)$ is the representation type embedding. $LT_{W^d}(S)$ is the distance embedding. In the distance embedding, zero vector(0) is used to pad the sentence.

---

[1]https://github.com/cambridgeltl/BioNLP-2016

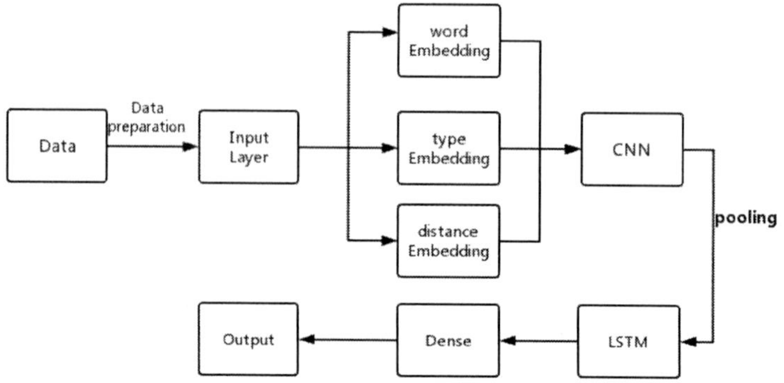

Figure 1: Our proposed CNN-LSTM based model

model	dropout	batch	epoch	F1
CNN	0.5	64	120	0.52
CNN-LSTM	0.5	64	120	0.60

Table 1: The F1 score of CNN and CNN-LSTM on the dev data set for SeeDev-binary task

Cluster	F1	Recall	Precision
Comparison	0.5	0.6	0.43
Function	0.25	0.19	0.35
Regulation	0.34	0.47	0.27
Genic Regulation	0.23	0.24	0.22
Composition	0.35	0.57	0.25
Interaction	0.22	0.16	0.33

Table 2: The F1, recall and precision of cluster on the test data set for SeeDev-binary task

Team	F1	Recall	Precision
MIC-CIS-1	0.373	0.511	0.295
YNU-junyi	**0.342**	**0.458**	**0.273**
Yunnan...	0.067	0.133	0.045
YNUBY	0.019	0.070	0.011

Table 3: The result of all types on the test data set for SeeDev-binary task

Team	F1	Recall	Precision
MIC-CIS-1	0.443	0.606	0.349
YNU-junyi	**0.394**	**0.528**	**0.314**
Yunnan...	0.135	0.267	0.090
YNUBY	0.074	0.274	0.043

Table 4: The result of ignoring types on the test data set for SeeDev-binary task

2.3 Model training

We run our model 5 times and use the maximum as the final result of the model. In all model runs, the dropout(Srivastava et al., 2014) is set to 0.5. We found that our loss function tends to stabilize when the epoch reaches around 120. So, we think that our model can converge at this time, so set epoch = 120. The batch size is set to 64. And, we use a pooling approach that combines average pooling and max pooling.

In this task, we choose the CNN-LSTM model to compare with a single CNN model. We find that the CNN-LSTM model works better than a single CNN model on development data set. So, we choose the CNN-LSTM model in the final submission.

3   Results and discussion

The SeeDev-binary task data sets consist of three parts which are the training set, the development set, and the test set. There are a total of 87 sections from 20 complete articles on Arabidopsis seed de-

Binary relation type	F1	Recall	Precision
Binds_To	0.31	0.28	0.35
Composes_Primary_Structure	0.34	0.44	0.28
Composes_Protein_Complex	0	0	0
Exists_At_Stage	0.14	0.1	0.25
Exists_In_Genotype	0.42	0.64	0.31
Interacts_With	0.09	0.06	0.19
Is_Involved_In_Process	0	0	0
Is_Localized_In	0.27	0.52	0.18
Is_Member_Of_Family	0.35	0.62	0.25
Is_Protein_Domain_Of	0.25	0.39	0.18
Occurs_In_Genotype	0.17	0.14	0.2
Occurs_During	0	0	0
Regulates_Accumulation	0.17	0.19	0.15
Regulates_Development_Phase	0.23	0.34	0.17
Regulates_Expression	0.22	0.25	0.19
Regulates_Molecule_Activity	0	0	0
Regulates_Process	0.43	0.66	0.32
Regulates_Tissue_Development	0	0	0
Transcribes_Or_Translates_To	0.34	0.38	0.32
Is_Linked_To	0.15	0.1	0.33
Is_Functionally_Equivalent_To	0.64	0.57	0.74
Has_Sequence_Identical_To	0.56	0.77	0.44

Table 5: Detailed results of our method on the test data set for SeeDev-binary task

velopment. This task defines 16 different types of entities and 22 different types of binary relationships.

Our method obtained F1 scores of 0.342 for all types and 0.394 for ignoring relation types and direction on the test set. In this task, the organizer gives the results of the evaluation obtained from three different evaluation conditions. Compared with 2016 BioNLP Shared Task, the organizer has added two more evaluations in order to have better biological contributions. These evaluation conditions are global results, relations by type cluster, and ignoring relation types and direction, respectively. We obtained a good score compared to the official results from different systems, and we ranked the second among all teams. It proves that our proposed method has good performance in binary relation extraction.

Table 2 shows the F1, recall and precision of cluster on the test data sets, and Table 3 shows the result of all types on the test data sets. Table 4 shows the result of ignoring types on the test data sets and Table 5 shows detailed results of our method on the test data set.

4 Conclusions

We use distributed semantic representation and CNN-LSTM model to extract the binary relationship between entities, then build a word embedding with rich semantic knowledge, distance embedding and entity type embedding to feed it into the CNN and learn the intrinsic relationship between the candidate entities. In the task, our F1-score of all types is 0.342, which indicates that our proposed method works efficiently in extraction of binary relations.

However, using only the original words embedded in CNN-LSTM may not be sufficient to understand the hidden information between words. Using our model to get this score does not mean that the model works well in other tasks.

In the future, we will continue to focus more on building rich distributed semantic embedding and we will improve our model by changing our model structure and adjusting paraments. In addition, we will explore various neural networks with multi-layer architectures, such as the attention mechanism and capsule networks, to solve binary relationships or event extraction problems.

References

Estelle Chaix, Bertrand Dubreucq, Abdelhak Fatihi, Dialekti Valsamou, Robert Bossy, Mouhamadou Ba, Louise Deléger, Pierre Zweigenbaum, Philippe Bessières, Loïc Lepiniec, and Claire Nedellec. 2016. Overview of the regulatory network of plant seed development (seedev) task at the bionlp shared task 2016. In *Proceedings of the 4th BioNLP Shared Task Workshop, BioNLP 2016, Berlin, Germany, August 13, 2016*, pages 1–11.

Graham Cormode. 2003. *Sequence distance embeddings*. Ph.D. thesis, University of Warwick.

Anthony Fader, Stephen Soderland, and Oren Etzioni. 2011. Identifying relations for open information extraction. In *Proceedings of the conference on empirical methods in natural language processing*, pages 1535–1545. Association for Computational Linguistics.

Jenny Rose Finkel, Trond Grenager, and Christopher Manning. 2005. Incorporating non-local information into information extraction systems by gibbs sampling. In *Proceedings of the 43rd annual meeting on association for computational linguistics*, pages 363–370. Association for Computational Linguistics.

Xinyu He, Lishuang Li, Jieqiong Zheng, and Meiyue Qin. 2016. Extracting biomedical event using feature selection and word representation. In *Proceedings of the 4th BioNLP Shared Task Workshop*, page 101, Berlin, Germany. Association for Computational Linguistics.

Sepp Hochreiter and Jrgen Schmidhuber. 1997. Long short-term memory. *Neural Computation*, 9(8):1735–1780.

Jake Lever and Steven JM Jones. 2016. VERSE: Event and relation extraction in the BioNLP 2016 shared task. In *Proceedings of the 4th BioNLP Shared Task Workshop*, pages 42–49, Berlin, Germany. Association for Computational Linguistics.

Omer Levy and Yoav Goldberg. 2014. Neural word embedding as implicit matrix factorization. In *Advances in neural information processing systems*, pages 2177–2185.

Honglei Li, Jianhai Zhang, Jian Wang, Hongfei Lin, and Zhihao Yang. 2016. DUTIR in BioNLP-ST 2016: Utilizing convolutional network and distributed representation to extract complicate relations. In *Proceedings of the 4th BioNLP Shared Task Workshop*, pages 93–100, Berlin, Germany. Association for Computational Linguistics.

Christopher Manning, Mihai Surdeanu, John Bauer, Jenny Finkel, Steven Bethard, and David McClosky. 2014. The stanford corenlp natural language processing toolkit. In *Proceedings of 52nd annual meeting of the association for computational linguistics: system demonstrations*, pages 55–60.

Farrokh Mehryary, Jari Björne, Sampo Pyysalo, Tapio Salakoski, and Filip Ginter. 2016. Deep learning with minimal training data: TurkuNLP entry in the BioNLP shared task 2016. In *Proceedings of the 4th BioNLP Shared Task Workshop*, pages 73–81, Berlin, Germany. Association for Computational Linguistics.

Claire Nédellec, Robert Bossy, and Jin-Dong Kim. 2016. Proceedings of the 4th bionlp shared task workshop. In *Proceedings of the 4th BioNLP Shared Task Workshop*.

Claire Nédellec, Robert Bossy, Jin-Dong Kim, Jung-Jae Kim, Tomoko Ohta, Sampo Pyysalo, and Pierre Zweigenbaum. 2013. Overview of bionlp shared task 2013. In *Proceedings of the BioNLP Shared Task 2013 Workshop*, pages 1–7.

Nitish Srivastava, Geoffrey Hinton, Alex Krizhevsky, Ilya Sutskever, and Ruslan Salakhutdinov. 2014. Dropout: a simple way to prevent neural networks from overfitting. *The Journal of Machine Learning Research*, 15(1):1929–1958.

Jiabao Su and Zhi-Qiang Wang. 2011. Sobolev type embedding and quasilinear elliptic equations with radial potentials. *Journal of Differential Equations*, 250(1):223–242.

Using Snomed to recognize and index chemical and drug mentions.

Pilar López-Úbeda, Manuel Carlos Díaz-Galiano,
Maria-Teresa Martín-Valdivia, L. Alfonso Ureña-López

Department of Computer Science, Advanced Studies Center in ICT (CEATIC)
Universidad de Jaén, Campus Las Lagunillas, 23071, Jaén, Spain
{plubeda, mcdiaz, maite, laurena}@ujaen.es

Abstract

In this paper we describe a new named entity extraction system. Our work proposes a system for the identification and annotation of drug names in Spanish biomedical texts based on machine learning and deep learning models. Subsequently, a standardized code using Snomed is assigned to these drugs, for this purpose, Natural Language Processing tools and techniques have been used, and a dictionary of different sources of information has been built. The results are promising, we obtain 78% in F1 score on the first sub-track and in the second task we map with Snomed correctly 72% of the found entities.

1 Introduction

Research in biology in the past decade has generated a large volume of available biological data. These texts usually contain information related to drugs, medications, chemicals, reactions, interactions, etc.

Named Entity Recognition (NER) of chemical compounds is receiving increased attention from researchers, as it may facilitate the application of information extraction to the pharmaceutical treatment of diseases. The recognition of pharmaceutical drugs and chemical entities is a critical step required for the subsequent detection of relations of chemicals with other biomedically relevant entities. Biomedical named entity recognition aims to find entities in biomedical texts, an invaluable function that becomes very important for further processing such as information retrieval, information extraction and knowledge discovery. At present, it has referred to kinds of domains, such as protein (Liu et al., 2005; Mitsumori et al., 2005; Tsuruoka and Tsujii, 2003), gene (Liu et al., 2005; Leser and Hakenberg, 2005) or drug (Campillos et al., 2008).

This challenge arises to address the task of recognizing chemicals and drugs. This task has already been studied by several workshops in English, but it is important to continue researching in other languages that have a lot of clinical information. Thanks to this challenge, we can continue studying one of the most widely spoken languages in the world: Spanish. The main aim is to promote the development of named entity recognition tools of practical relevance, that is chemical and drug mentions in non-English content, determining the current-state-of-the art, identifying challenges and comparing the strategies and results to those published for English data.

In terms of English, there were several challenges presented recently such as *CHEMDNER Task: Chemical compound and drug name recognition task* (Krallinger et al., 2015) and *JNLPBA* (Kim et al., 2004) that served to determine the state of the art methodology and systems performance in addition of providing valuable datasets for developing new systems (Tanabe et al., 2005). Some of the most important corpus in this domain are GENIA (Kim et al., 2003), CRAFT (Bada et al., 2012), CALBC (Rebholz-Schuhmann et al., 2010) corpora or SCAI corpus (Kolárik et al., 2008).

In this paper we introduce the participation of the SINAI group in the challenge named PharmaCoNER (Pharmacological Substances, Compounds and proteins and Named Entity Recognition). PharmaCoNER (Gonzalez-Agirre et al., 2019) is one of the workshops presented at BioNLP 2019 and consists of two tracks:

1.1 Track 1: NER offset and entity classification

In this first sub-track the main objective is to find the chemicals and drugs within the text. For its later evaluation it is necessary to write down the

Proceedings of the 5th Workshop on BioNLP Open Shared Tasks, pages 115–120
Hong Kong, China, November 4, 2019. ©2019 Association for Computational Linguistics

beginning and end position of the concept, as well as the appropriate label. The types of entities may be the following:

- *NORMALIZABLES*: those mentions of chemical compounds and drugs that can be standardized with a unique identifier from a database.

- *NO NORMALIZABLES*: those mentions of chemical compounds and drugs that cannot be normalized

- *PROTEÍNAS*: includes peptides, proteins, genes, peptide hormones and antibodies.

- *UNCLEAR*: for plants, oils, essences, plant principles and general formulations/compositions of various compounds.

1.2   Track 2: Concept indexing

The objective of the second task was to assign a unique identifier to each concept detected in the previous task. The Snomed (Systematized Nomenclature of Medicine) (National Health Service, 2019) terminology was used for this. Snomed is an international standard distributed by the International Health Terminology Standards Development Organisation (IHTSDO)[1], an organisation to which Spain belongs as a member.

2   Data collection

The used corpus was Spanish Clinical Case Corpus (SPACCC). This corpus contains a manually classified collection of sections of clinical cases derived from open-access Spanish medical journals. The corpus contains a total of 1000 clinical cases and 396,988 words.

The organizers provided us 500 documents for training, 250 validation documents and finally, 3751 test documents. The final collection had a total of 16504 sentences, with an average of 16.5 sentences per clinical case. The SPACCC corpus contains a total of 396,988 words, with an average of 396.2 words per clinical case.

3   Methodology

Our group has participated in both sub-tasks proposed by PharmaCoNER. In each sub-task we have sent 4 runs. For the first sub-track we have

created machine learning and deep learning approaches providing extra information with features. In the second sub-track we have used the outputs of the first task using a dictionary-based approach.

3.1   Track 1: NER offset and entity classification

3.1.1   Machine learning with CRF.

Conditional Random Fields (CRF) (Lafferty et al., 2001) are a probabilistic framework for the labeling or segmentation of sequential data. We used CRFsuite, the implementation provided by Okazaki (Okazaki, 2007), as it is fast and provides a simple interface for training and modifying the input features.

Similar to most machine learning-based systems, the token-level CRF requires a tokenization module at first. The tokenizer used is WordPunct-Tokenizer of the NLTK[2] library in Python.

Run 1. CRF + basic features.   For the first experiment, we incorporate to CRF some basic features of each word such as isLower, isUpper, isTitle, isDigit, isAlpha, isBeginOfSentence and isEndIfSentece.

Run 2. CRF + basic features + features based on medical terminology.   For this experiment, we decided to add a new feature to CRF using medical terminology to provide extra information for each word. This feature indicated if the word was contained in The Spanish Medical Abbreviation DataBase (AbreMES-DB), dictionary of chemicals, compounds, and drugs in Spanish (Nomenclator for Prescription) or Snomed in Spanish. Nomenclator for Prescription and AbreMES-DB are resources provided by the organizers and available on the workshop website[3]. On the other hand, Snomed was reduced using only the concepts of products and substances.

3.1.2   Deep learning with BiLSTM and CNN

For this sub-track, we present a hybrid model of bi-directional LSTMs and CNNs that learns both character and word-level features, based on the model of Chiu (Chiu and Nichols, 2016).

The first neural network, use the Convolutional Neural Network (CNN) to extract character features. For each word we employ a convolution and

---

[1]https://www.ihtsdo.org/

[2]https://www.nltk.org/
[3]http://temu.bsc.es/pharmaconer/index.php/resources/

a max layer to extract a new feature vector from the per character feature vectors such as character embedding and character type. This model also uses Bi-directional recurrent neural network with Long Short-Term Memory (BiLSTM) to transform word features into named entity.

The word embedding used for this task is Spanish Billion Word Corpus. The corpus for creating this embedding contain 1,000,653 words and the vector dimension is 300 (Cardellino, 2016).

Finally, the hyper-parameters used in these Neural Networks are those proposed by Chiu.

Run 3. BiLSTM + CNN + basic features. For this experiment, we used the model of Chiu (Chiu and Nichols, 2016). The Bi-LSTM take the concatenation of the output of CNN with the word embedding of each word.

Run 4. BiLSTM + CNN + basic features + features based on medical terminology. For the last experiment sent, we used the dictionary explained in the Run 2. In this case, the Bi-LSTM neural network used the CNN output, the word embedding of each word and if the word was contained in the new dictionary created.

3.2 Track 2: Concept indexing

Our approach is to create a large medical terminology dictionary to help map the named entity recognized in sub-track 1 with Snomed identifiers. The process for developing this task can be seen in Figure 1 and each step of the process followed is detailed below:

1. **Construction of the drug name dictionary.**

   At first, a drug name dictionary was build with different sources of knowledge related to chemicals, drugs and medicines. Our goal in creating this dictionary was to generate the maximum number of synonymous concepts. The sources of information used are explained below:

   (a) **Wikidata**: is a document-oriented database, focused on items, which represent topics, concepts, or objects. We downloaded from Wikidata all the elements that were *instances of*: *disease, gene, syndrome, protein, group or class of chemical substances, structural class of chemical compounds, medication, drug, chemical substance* and *chemical compound*. To make this query easy we used SPARQL and obtained the English and Spanish *alias* for each found object. We use both languages because most synonyms come with information in English.

   (b) **AbreMES-DB**: the Spanish Medical Abbreviation DataBase are extracted from the metadata of different biomedical publications written in Spanish, which contain the titles and abstracts.

   (c) **Nomenclator for prescription**[4]: is a medicine database designed to provide basic prescription information to healthcare information systems.

   (d) **Snomed**: we used the dictionary explained in Section 3.1.1, Snomed was reduced using only the concepts of products and substances in Spanish.

   (e) **Chemical symbols in Spanish**: abbreviated signs used to identify chemical elements of the periodic table and compounds.

   All these sources of information have something in common, they all contain synonyms, acronyms or other ways of referring to the same entity.

2. **Text pre-processing.**

   The second step of this architecture was to normalize the texts in order to make the matching of concepts. To do this we use the spaCy library because it is a free open source library for Natural Language Processing in Python. with *'es_core_news_sm'* module in Spanish. This pre-processing consists of:

   - Change the text to lower case.

   - Remove accents.

   - Use the lemma of each word with spaCy.

   - Remove punctuation marks.

   - Remove stop-word.

3. **Match with drug name dictionary.** At this point, we try to match the input text (recognized entity) with texts in the previously generated dictionary. If we can match them, then we will increase the list of possible synonyms

---

[4] https://cima.aemps.es/cima/publico/nomenclator.html

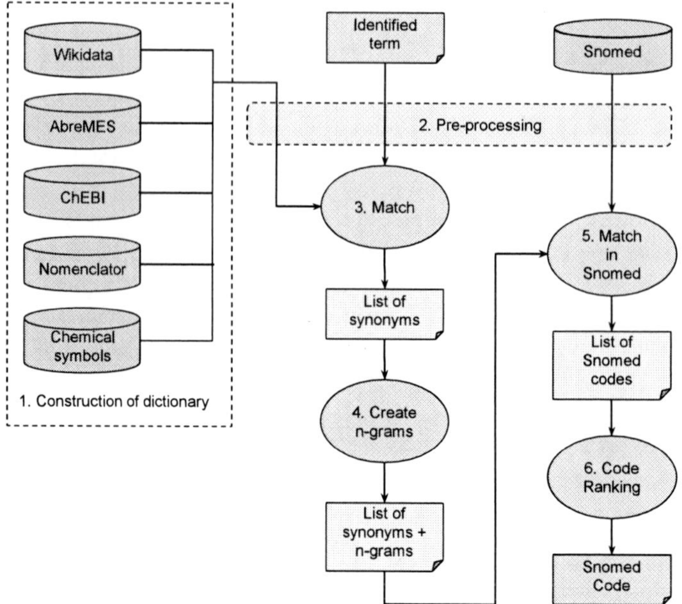

Figure 1: Snomed code indexing process.

to have more options to find the concept in Snomed.

4. **N-grams of terms.** After several tests with the development collection, we found that in some cases, the order of the multi-word concepts did not match well. For this reason we decided to create n-gram, where n is the size of the multi-word concept with all possible word combinations. This list of new n-grams was added to the list of possible synonyms of the concept.

5. **Match Snomed concepts.** Using the list of synonyms extracted from the previous steps, we try to match any of the synonyms with Snomed concepts. To get as many Snomed concepts as possible, we use a library called Hunspell[5].

Hunspell uses a special dictionary to which we have added Snomed concepts. With this library, we use the *hunspell_suggest* function where we can get similar words to the given word. This function will return many concepts of Snomed so later we must choose one of them.

6. **Ranking of concepts by Levenshtein distance.** Finally, we must choose a single Snomed concept. For this we use the Levenshtein distance (LD). LD is a measure of the similarity between two strings, which we will refer to as the source string and the target string. The distance is the number of deletions, insertions, or substitutions required to transform each synonym included with a Snomed concept. Lastly, we chose the Snomed concept that has the least distance with the input text.

In Table 1 we can see some examples of how the resources and tools applied in the architecture can contribute to the achievement of Snomed concept mapping.

4 Results

4.1 NER offset and entity type classification.

The first evaluation consist in the classical entity-based or instanced-based evaluation that requires that system outputs match exactly the beginning and end locations of each entity tag, as well as match the entity annotation type of the gold standard annotations.

The results obtained by our team for this sub-track are shown in Table 2.

---

[5]`http://hunspell.github.io/`

Resource	Input text	Snomed Term	Snomed Code
Wikidata	adriamicina	doxorrubicina	372817009
Chemical symbols	Na	sodio	39972003
AbreMES-DB	Hb	hemoglobina	38082009
Hunspell Library	6-Metil-Prednisolona	metilprednisolona	116593003

Table 1: Examples of Snomed concept indexing.

Run	Precision	Recall	F1
1	**0.92602**	0.61835	0.74154
2	0.88507	**0.69815**	**0.78058**
3	0.84404	0.64929	0.73397
4	0.85992	0.69653	0.76965

Table 2: Results of Track 1. NER offset and entity classification.

Run	Precision	Recall	F1
1	**0.87879**	0.55849	0.68295
2	0.85207	**0.63267**	**0.72616**
3	0.8335	0.57846	0.68295
4	0.82887	0.6184	0.70833

Table 3: Results of Track 2. Concept indexing.

In these results we can see that applying features in both methods (*Run 2* and *Run 3*) improves the base model (*Run 1* and *Run 3*). In the case of CRF the precision decreases but the recall increases and finally the F1 measure improves from 74% to 78%. For the RNN, in all the measures the use of new features improves, obtaining 76% of F1. For future occasions we will continue to exploit the use of new features in the different strategies.

4.2 Concept indexing.

For this sub-track the main objective was to index each document of the previous task and each concept detected with a unique Snomed code. Table 3 shows the evaluation of the systems for this sub-track.

The results are concordant to the previous task, we use the output of each Run of sub-track 1 to index the concepts detected with Snomed codes. For this reason we get 72% F1 score in *Run 2*, and in *Run 1* we obtain 87% precision.

We are still analyzing the results obtained, in this way, in the future we will know how we can improve the task of indexing with terminologies, which dictionaries to use or which Natural Language Processing (NLP) tools we can apply.

5   Conclusion and future work

The SINAI group presents its participation in the PharmaCoNER challenge. The first task was to find chemical and drug mentions in the text and assign a specific label. In the next task, the main objective was to index each found concept with a Snomed code.

For sub-track 1 we have developed four systems with different machine learning and deep learning approaches adding some relevant features obtained from the Snomed terminology in Spanish. The goal of sub-track 2 is to assign a unique identifier to each detected concept of sub-track 1 and for this we have developed a system based on a large dictionary of medical terminology according to the task, this dictionary provided us with a long list of synonyms for each entity to match with a Snomed code.

The results obtained have been as expected, adding extra information from Snomed terminology helps classifiers to detect relevant entities within medical texts. On the other hand, apply NLP techniques and tools and the creation of a medical dictionary has contributed to find synonyms for later assigning a single Snomed code. Using our methodology, we found the correct code for example to the input text *IgG* although in Snomed this concept is described as *immunoglobulin G*.

In future works we will continue working on machine learning approaches and different features of improvement. Specifically, we will create more sophisticated Neural Networks and explore different embeddings in Spanish. Normalization plays an important role in this track so we will use NLP to continue improving.

Acknowledgments

This work has been partially supported by Fondo Europeo de Desarrollo Regional (FEDER), LIVING-LANG project (RTI2018-094653-B-C21) and REDES project (TIN2015-65136-C2-1-R) from the Spanish Government.

References

Michael Bada, Miriam Eckert, Donald Evans, Kristin Garcia, Krista Shipley, Dmitry Sitnikov, William A Baumgartner, K Bretonnel Cohen, Karin Verspoor, Judith A Blake, et al. 2012. Concept annotation in the craft corpus. *BMC bioinformatics*, 13(1):161.

Monica Campillos, Michael Kuhn, Anne-Claude Gavin, Lars Juhl Jensen, and Peer Bork. 2008. Drug target identification using side-effect similarity. *Science*, 321(5886):263–266.

Cristian Cardellino. 2016. Spanish billion words corpus and embeddings.

Jason PC Chiu and Eric Nichols. 2016. Named entity recognition with bidirectional lstm-cnns. *Transactions of the Association for Computational Linguistics*, 4:357–370.

Aitor Gonzalez-Agirre, Montserrat Marimon, Ander Intxaurrondo, Obdulia Rabal, Marta Villegas, and Martin Krallinger. 2019. Pharmaconer: Pharmacological substances, compounds and proteins named entity recognition track. In *Proceedings of the BioNLP Open Shared Tasks (BioNLP-OST)*, pages 1–X, Hong Kong, China. Association for Computational Linguistics.

Jin-Dong Kim, Tomoko Ohta, Yuka Tateisi, and Jun'ichi Tsujii. 2003. Genia corpus—a semantically annotated corpus for bio-textmining. *Bioinformatics*, 19(suppl_1):i180–i182.

Jin-Dong Kim, Tomoko Ohta, Yoshimasa Tsuruoka, Yuka Tateisi, and Nigel Collier. 2004. Introduction to the bio-entity recognition task at jnlpba. In *Proceedings of the international joint workshop on natural language processing in biomedicine and its applications*, pages 70–75. Citeseer.

Corinna Kolárik, Roman Klinger, Christoph M Friedrich, Martin Hofmann-Apitius, and Juliane Fluck. 2008. Chemical names: terminological resources and corpora annotation. In *Workshop on Building and evaluating resources for biomedical text mining (6th edition of the Language Resources and Evaluation Conference)*.

Martin Krallinger, Florian Leitner, Obdulia Rabal, Miguel Vazquez, Julen Oyarzabal, and Alfonso Valencia. 2015. Chemdner: The drugs and chemical names extraction challenge. *Journal of cheminformatics*, 7(1):S1.

John Lafferty, Andrew McCallum, and Fernando CN Pereira. 2001. Conditional random fields: Probabilistic models for segmenting and labeling sequence data.

Ulf Leser and Jörg Hakenberg. 2005. What makes a gene name? named entity recognition in the biomedical literature. *Briefings in bioinformatics*, 6(4):357–369.

Hongfang Liu, Zhang-Zhi Hu, Jian Zhang, and Cathy Wu. 2005. Biothesaurus: a web-based thesaurus of protein and gene names. *Bioinformatics*, 22(1):103–105.

Tomohiro Mitsumori, Sevrani Fation, Masaki Murata, Kouichi Doi, and Hirohumi Doi. 2005. Gene/protein name recognition based on support vector machine using dictionary as features. *BMC bioinformatics*, 6(1):S8.

National Health Service. 2019. SNOMED International. http://www.snomed.org/.

Naoaki Okazaki. 2007. Crfsuite: a fast implementation of conditional random fields (crfs).

Dietrich Rebholz-Schuhmann, Antonio José Jimeno Yepes, Erik M Van Mulligen, Ning Kang, Jan Kors, David Milward, Peter Corbett, Ekaterina Buyko, Elena Beisswanger, and Udo Hahn. 2010. Calbc silver standard corpus. *Journal of bioinformatics and computational biology*, 8(01):163–179.

Lorraine Tanabe, Natalie Xie, Lynne H Thom, Wayne Matten, and W John Wilbur. 2005. Genetag: a tagged corpus for gene/protein named entity recognition. *BMC bioinformatics*, 6(1):S3.

Yoshimasa Tsuruoka and Jun'ichi Tsujii. 2003. Boosting precision and recall of dictionary-based protein name recognition. In *Proceedings of the ACL 2003 workshop on Natural language processing in biomedicine-Volume 13*, pages 41–48. Association for Computational Linguistics.

Bacteria Biotope at BioNLP Open Shared Tasks 2019

Robert Bossy, Louise Deléger, Estelle Chaix, Mouhamadou Ba and **Claire Nédellec**
MaIAGE, INRA, Université Paris-Saclay, 78350 Jouy-en-Josas, France
`robert.bossy@inra.fr louise.deleger@inra.fr`
`estelle.chaix@gmail.com mouhamadou.ba@inra.fr`
`claire.nedellec@inra.fr`

Abstract

This paper presents the fourth edition of the Bacteria Biotope task at BioNLP Open Shared Tasks 2019. The task focuses on the extraction of the locations and phenotypes of microorganisms from PubMed abstracts and full-text excerpts, and the characterization of these entities with respect to reference knowledge sources (NCBI taxonomy, OntoBiotope ontology). The task is motivated by the importance of the knowledge on biodiversity for fundamental research and applications in microbiology. The paper describes the different proposed subtasks, the corpus characteristics, and the challenge organization. We also provide an analysis of the results obtained by participants, and inspect the evolution of the results since the last edition in 2016.

1 Introduction

In this paper, we present the fourth edition[1] of the Bacteria Biotope (BB) task. The task was introduced in 2011. It has the ambition of promoting large-scale information extraction (IE) from scientific documents in order to automatically fill knowledge bases in the microbial diversity field (Bossy et al., 2012). BB 2019 is part of BioNLP Open Shared Tasks 2019[2]. BioNLP-OST is a community-wide effort for the comparison and evaluation of biomedical text mining technologies on manually curated benchmarks.

A large amount of information about microbes and their properties that is critical for microbiology research and development is scattered among millions of publications and databases (Chaix et al., 2019). Information extraction as framed by the Bacteria Biotope task identifies relevant entities and interrelationships in the text and map them to reference categories from existing knowledge resources. This information can thus be combined with information from other sources referring to the same knowledge resources. The knowledge resources used in the BB task are the NCBI taxonomy[3] (Federhen, 2011) for microbial taxa and the OntoBiotope ontology[4] (Nédellec et al., 2018) for microbial habitats and phenotypes. The large size of these resources relative to the small number of training examples reflects the real conditions of IE application development, whilst it challenges current IE methods. The lexical richness of the two resources partially offsets the difficulty.

Compared to the 2016 corpus that contained only scientific paper abstracts from the PubMed database (Deléger et al., 2016), the 2019 corpus is enriched with extracts from full-text articles. We introduced a new entity type (phenotype) and a new relation type (linking microorganisms and phenotypes). Phenotypes are observable characteristics such as morphology, or environment requirement (e.g. acidity, oxygen). It is very valuable information for studying the ability of a given microbe to adapt to an environment (Brbić et al., 2016). The definition of microorganism phenotype in the OntoBiotope ontology includes host interaction characteristics (e.g. symbiont) and community behavior and growth habit (e.g. epilithic). The task organization and the evaluation metrics remain unchanged.

2 Task Description

The representation scheme of the Bacteria Biotope task contains four entity types:

- *Microorganism*: names denoting microorganism taxa. These taxa correspond to microorganism branches of the NCBI taxon-

---

[1] `https://sites.google.com/view/bb-2019`
[2] `https://2019.bionlp-ost.org/`
[3] `https://www.ncbi.nlm.nih.gov/taxonomy`
[4] `https://tinyurl.com/OntoBiotope2019`

Proceedings of the 5th Workshop on BioNLP Open Shared Tasks, pages 121–131
Hong Kong, China, November 4, 2019. ©2019 Association for Computational Linguistics

omy. The set of relevant taxa is given on the BB task website.

- *Habitat*: phrases denoting physical places where microorganisms may be observed;

- *Geographical*: names of geographical places;

- *Phenotype*: expressions describing microbial characteristics.

The scheme defines two relation types:

- *Lives_in* relations which link a microorganism entity to its location (either a habitat or a geographical entity, or in few rare cases a microorganism entity);

- *Exhibits* relations which link a microorganism entity to a phenotype entity.

Arguments of relations may occur in different sentences. In addition, microorganisms are normalized to taxa from the NCBI taxonomy. Habitat and phenotype entities are normalized to concepts from the OntoBiotope ontology. We used the BioNLP-OST-2019 version of OntoBiotope available on AgroPortal [5]. We used the NCBI Taxonomy version as available on February 2, 2019 from NCBI website [6]. Copies of both resources can be downloaded from the task website. The microorganism part of the taxonomy contains 903,191 taxa plus synonyms, while the OntoBiotope ontology includes 3,601 concepts plus synonyms (3,172 for the Habitat branch and 429 for the Phenotype branch of the ontology).

Geographical entities are not normalized.

Figure 1 shows an example of a sentence annotated with normalized entities and relations.

As in the 2016 edition, we designed three tasks, each including two modalities, one where entity annotations are provided and one where they are not and have to be predicted.

2.1 Entity Normalization

The first task focused on entity normalization.

In the **BB-norm** modality of this task, participant systems had to normalize textual entity mentions according to the NCBI taxonomy for microorganisms and to the OntoBiotope ontology for habitats and phenotypes.

In the **BB-norm+ner** modality, systems had to recognize the mentions before normalizing them.

5 http://agroportal.lirmm.fr/ontologies/ONTOBIOTOPE
6 ftp://ftp.ncbi.nih.gov/pub/taxonomy

2.2 Relation Extraction

The second task focused on the extraction of the two types of relations— *Lives_in* relations among microorganism, habitat and geographical entities, and *Exhibits* relations between microorganism and phenotype entities.

In the **BB-rel** modality, participant systems only had to extract the relations, while in the **BB-rel+ner** modality they had to perform entity recognition in addition to relation extraction.

2.3 Knowledge Base Extraction

The goal of the third task is to build a knowledge base using the entities and relations extracted from the corpus. It can be viewed as the combination of the previous tasks, followed by a merging step. Participant systems must normalize entities and extract relations.

In the **BB-kb** modality, participant systems had to perform normalization and relation extraction with entity mentions being provided. In the **BB-kb+ner** modality, they had to perform entity recognition as well.

3 Corpus Description

3.1 Document Selection

The BB task corpus consists of two types of documents: PubMed references (titles and abstracts) related to microorganisms, and extracts from full-text articles related to beneficial microorganisms living in food products.

The PubMed references are the same as the 215 references of the Bacteria Biotope 2016 corpus. They were sampled from all PubMed entries indexed with a term from the Organisms/Bacteria subtree of the MeSH thesaurus. The full selection process is described in Deléger et al. (2016).

Full-text extracts were selected from scientific articles about microorganisms of food interest and annotated by microbiologist experts in the context of the Florilege project (Falentin et al., 2017). We reused and complemented this corpus for the BB task.

Because manual annotation is time-consuming and experts have limited time to dedicate to this task, they did not annotate the full articles. Instead, they chose the paragraphs and sentences they found the most informative in the articles. Thus, this part of the BB corpus is composed of 177 extracts of variable lengths (from one single

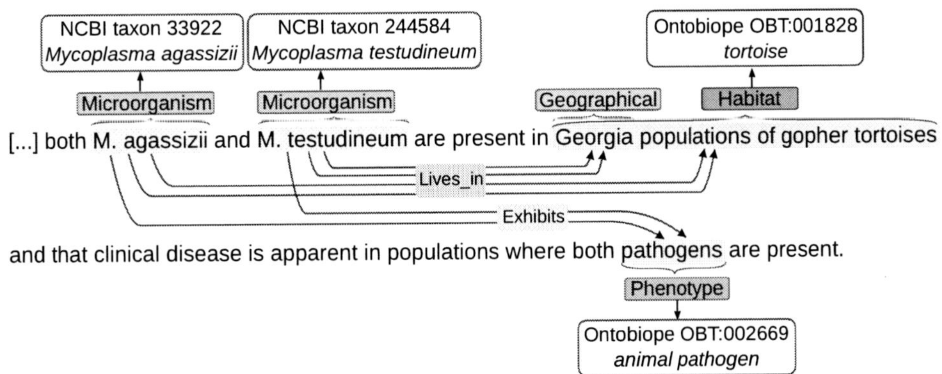

Figure 1: Annotation example

sentence to a few paragraphs) selected from 20 articles.

3.2 Annotation

The PubMed references were already annotated as part of the 2016 edition. We revised these annotations to add phenotype entities with their concept normalization and *Exhibits* relations. Habitat annotations were also revised to take into account the new and enriched version of the OntoBiotope ontology (compared to the 2016 version[7]).

We also extended the existing annotations of the full-text extracts of the Florilege project by assigning normalized concepts to the entities.

Annotation revision was performed by six annotators with backgrounds in biology, computer science and natural language processing. All documents were annotated independently by two annotators and disagreements were resolved through an adjudication phase. Detailed annotation guidelines (Bossy et al., 2019) were provided to the annotators and were regularly updated following issues raised during the annotation or adjudication phases.

The inter-annotator agreement was computed by evaluating one of the two annotations before adjudication against the other. Table 1 summarizes the inter-annotator agreement for named entities, normalization and relations. The metrics used for inter-agreement are the same as for the evaluation of predictions and thus are described below (5.1).

3.3 Descriptive Statistics

Table 2 gives the size of the corpus, in terms of documents, words, sentences and annotated ele-

[7] http://2016.bionlp-st.org/tasks/bb2/ OntoBiotope_BioNLP-ST-2016.obo

Named-entities (F1)	0.893
Normalization (semantic similarity)	0.974
Relations (F1)	0.786
BB-norm+ner evaluation (SER)	0.322
BB-norm+ner evaluation (F1)	0.823
BB-rel+ner evaluation (SER)	0.448
BB-rel+ner evaluation (F1)	0.765
BB-kb+ner evaluation	0.723

Table 1: Inter-annotator agreement metrics (SER stands for Slot Error Rate).

ments. The last row shows the number of unique relations in the whole corpus, i.e. the unique pairs of microorganism and habitat/phenotype concepts that are in a relation. The proportion is rather high (1,931 out of a total of 3,578 occurrences), which reflects the rich information content of the corpus.

Documents	392
Words	60,402
Unique words	12,566
Sentences	2,646
Entity mentions	7,232
Unique entity mentions	3,300
Concepts	1,072
Relations	3,578
Unique relations between concepts	1,931

Table 2: Global statistics of the corpus

In the following, we present more detailed statistics and highlight corpus characteristics that may be challenging for the participants.

3.3.1 Entities and Concepts

Table 3 shows the number of mentions, unique (lemmatized) mentions, concepts and average number of mentions per concept for each entity type. Habitat entities are the most frequent, followed by Microorganism entities. Geographical entities are very scarce.

There is much more variation in the expression of habitats and phenotypes than in that of microorganisms. There is an average of respectively 4 and 3.5 unique mentions per habitat and phenotype concept while microorganisms only have 1.9. Their proportion of unique entities out of all mentions is also higher (respectively 50.6% and 45.2% vs. 38.2% for microorganisms).

The proportion of direct mappings (i.e., exact string matches, taking into account lemmatization) between entity mentions and labels of concepts (from the NCBI taxonomy or the Onto-Biotope ontology) is displayed on Figure 2. It emphasizes once more the variability of Habitat and Phenotype entity expressions, with respectively 72.5% and 91.2% mentions that do not exactly match a concept label or synonym. Among exact matches, a small proportion of mentions are not actually normalized with the concept whose label they match. These are "contextual normalization" cases, i.e. entities are normalized with a more specific concept which can be inferred from the context. These often correspond to lexical coreference cases.

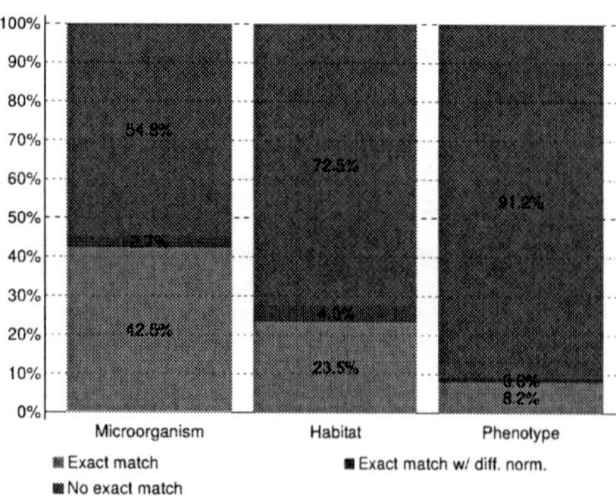

Figure 2: Exact matches between entity mentions and concepts. The *exact match* label refers to entities exactly matching the concept they are normalized with; the *exact match w/ diff. norm.* label refers to entities exactly matching a concept but normalized with a different one; the *no exact match* label refers to entities that do not match exactly a concept.

A distinctive feature of the BB task is that multiple concepts may be assigned to a given entity mention. Multiple normalization happens when two (or more) concepts can describe an entity and are all deemed necessary because each concept corresponds to a different aspect of the entity. An example of such a case is the Habitat entity *"diseased cow"* which is normalized by both the <cow> and <animal with disease> concepts. This is the case mainly for Habitat entities (8.7%), and rarely happens for Phenotype entities (0.6%) and Microorganism entities (only one occurrence).

Another characteristic of the corpus is the presence of nested entities (entities embedded in another larger entity) and discontinuous entities (entities split in several fragments). Both phenomena can be challenging for machine-learning methods and are often ignored. The proportion of discontinuous entities in the corpus is limited, with a total of 3.7%. Nested entities are more frequent (17.8% in total), especially for habitats. For instance, the Habitat entity *"cheese making factory"* also contains the smaller Habitat entity *"cheese"*.

3.3.2 Relations

Table 4 shows the number of relations for both *Lives_in* and *Exhibits* types, including intra-sentence and inter-sentence relations. Intra-sentence relations involve entities occurring in the same sentence while inter-sentence relations involve entities occurring in different sentences, not necessarily contiguous. Inter-sentence relations are known to be challenging for automatic methods. Their proportion in the corpus is not negligible (17.5% in total). An example can be seen in the following extract: <u>Vibrios</u> *[…] are ubiquitous to oceans, coastal waters, and estuaries. […] The bacterial pathogen is a growing concern in* <u>North America</u>. There is an inter-sentence relation between the two underlined entities.

3.3.3 Training, Development and Test Sets

The BB corpus is split into training, development and test sets. In practice, there are two test sets, one for the modalities involving entity recognition (the "+ner" sub-tasks) and one for the modalities where entity annotations are given. We kept the corpus division of the 2016 edition for the PubMed references. This was possible because the gold annotations of the test set were never released to the public. Then we split the Florilege full-text extracts using the same proportions as for

	Microorganism	Habitat	Phenotype	Geographical
Entity mentions	2,487	3,506	1,102	137
Unique entity mentions	950	1,774	498	78
Concepts	491	440	141	N/A
Unique mentions per concept (average)	1.9	4.0	3.5	N/A

Table 3: Statistics for each entity type

	Intra-sent.	Inter-sent.	Total
Lives_In	2,099 (79.8%)	532 (20.2%)	2,631
Exhibits	852 (90.0%)	95 (10.0%)	947
Total	2,951 (82.5%)	627 (17.5%)	3,578

Table 4: Statistics for each relation type

the PubMed references. Figure 3 shows the distribution of documents, entities, concepts and relations in the training, development and test sets of the BB-kb+ner task, as an example. The proportions are similar in all sub-tasks. Details for each sub-task can be found on the task website[8].

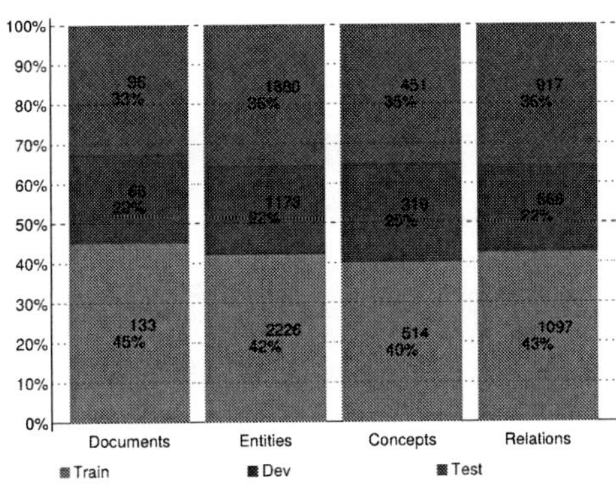

Figure 3: Distribution of documents, entities, concepts and relations in the training, development and test sets (BB-kb+ner task)

The proportion of concepts seen in the training set out of all concepts present in the knowledge resources is low for all entity types, which means that there is a large number of unseen examples (0.02% for microorganisms, 7.3% for habitats, and 15.6% for phenotypes). It emphasizes the need for methods that handle few-shot and zero-shot learning. Microorganisms have the lowest proportion, due to the large size of the microorganism taxonomies. However, the names of the microorganism entities show little variation in the corpus compared to habitat and phenotype types, and should be easier to recognize.

4 Supporting Resources

Supporting resources were made available to participants. They consist of outputs from state-of-the-art tools applied to the BB data sets (e.g., POS tagging, syntactic parsing, NER, word embeddings). We proposed in-house embeddings trained on selected relevant PubMed abstracts, and links to external embeddings (Pyysalo et al., 2013; Li et al., 2017) trained on PubMed and Wikipedia. The full list of tools and resources is available on the website.

5 Evaluation

5.1 Metrics

We used the same evaluation metrics as in the 2016 edition. The underlying rationale and formula of each score is detailed in Deléger et al. (2016); Bossy et al. (2013). Additionally we compute a variety of alternate scorings in order to distinguish the strengths of each submission. The evaluation tool was provided to participants[9].

Normalization accuracy is measured through a semantic similarity metric, and micro-averaging across entities. Relation extraction is measured with Recall, Precision, and F_1.

However for tasks where systems must recognize entities, we used the Slot Error Rate (SER) instead of F_1 in order to avoid sanctioning twice the inaccuracy of boundaries. The SER measures the amount of errors according to three types: insertions (false positives), deletions (false negatives), and substitutions (partial matches). The SER is normalized by the number of reference items. The higher the value the worse is the prediction, and there is no upper bound since insertions can exceed the number of items in the reference.

[8] https://sites.google.com/view/bb-2019/dataset/corpus-statistics

[9] https://github.com/Bibliome/bionlp-st

Confidence intervals were computed for each metric with the bootstrap resampling method (90%, n=100).

5.2 Baseline

We designed simple baselines for each sub-task in order to provide a comparison reference. We pre-processed the corpus with the AlvisNLP[10] engine, that performs tokenization, sentence splitting, and lemmatization using the GENIA tagger (Tsuruoka et al., 2005).

- BB-norm: we performed exact matching between lemmatized entities and the knowledge resources. When no match was found, we normalized habitats and phenotypes with the top-level concept of the Habitat and Phenotype ontology branches, and microorganisms with the high-level *<Bacteria>* taxon.

- BB-norm+ner: we used our exact matching approach on the lemmatized text of the documents instead of on given entity mentions.

- BB-rel: we used a simple co-occurrence approach, linking pairs of entities occurring in the same sentences.

- BB-rel+ner: we first detected entities using our exact matching strategy for microorganisms, habitats and phenotypes. For geographical entities, we used the Stanford Named Entity Recognition tool (Finkel et al., 2005). Then we linked entities occurring in the same sentences, as for the BB-rel task.

- BB-kb: we combined the BB-norm and BB-rel approaches.

- BB-kb+ner: we combined our BB-norm+ner method with our co-occurrence approach.

6 Outcome

6.1 Participation

The blind test data was released on the 22nd of July 2019 and participants were given until the 31st of July to submit their predictions. Each team was allowed two submissions to each sub-task.

Ten teams participated to all six sub-tasks and submitted a total of 31 runs. Table 5 details team affiliations. Teams are from five different countries in Europe, Asia, and North America. Six of the teams are affiliated to universities, three to industry companies, and one has a mixed university-industry affiliation.

Team	Affiliation
AliAI (Zhang et al., 2019)	Alibaba
Amrita_Cen	Amrita Vishwa Vidyapeetham
AmritaCen_healthcare	Amrita Vishwa Vidyapeetham
BLAIR_GMU (Mao and Liu, 2019)	George Mason University
BOUN-ISIK (Karadeniz et al., 2019)	Boğaziçi University & Işık University
MIC-CIS (Gupta et al., 2019)	Siemens AG & Ludwig Maximilian University of Munich
PADIA_BacReader (Deng et al., 2019)	Ping An Technology
UTU	University of Turku
whunlp (Xiong et al., 2019)	Wuhan University
Yuhang_Wu	Yunnan University

Table 5: Participating teams and their affiliations.

6.2 Participants' Methods and Resources

As in 2016, most methods are based on Machine Learning algorithms.

For named entity recognition, the CRF algorithm is still the most used (BLAIR_GMU), though sometimes combined with a neural network (MIC-CIS).

In 2016, the majority of participants used SVMs for relation extraction. In this edition nearly all participants used neural networks in a diversity of architectures: multi-layer perceptron (Yuhang_Wu), bi-LSTM (whunlp), AGCNN (whunlp). One participant predicted relations through filtered co-occurrences (BOUN-ISIK), and another by bagging SVM and Logistic Regression (BLAIR_GMU). Note that AliAI employed a multi-task architecture similar to BERT (Devlin

---

[10]https://bibliome.github.io/alvisnlp/

et al., 2019) to perform both named-entity recognition and relation extraction.

The normalization task was addressed in a more diverse manner. On one hand several distinct ML algorithms were used to discriminate entity categories: ensemble CNNs (PADIA_BacReader), kNN with reranking (BOUN-ISIK), or Linear Regression (BLAIR_GMU). On the other hand MIC-CIS employed an exact and an approximate matching algorithm.

Word embeddings trained with Word2Vec (Mikolov et al., 2013) on a domain-specific corpus (PubMed abstract, PMC articles) seem to be an universal resource since all but one submissions for any task used them. BLAIR_GMU used contextual embeddings based on BERT and XLNet (Yang et al., 2019).

Dependency parsing was used in every relation extraction submission, and also for normalization (BOUN-ISIK).

The most popular NLP tool libraries are Stanford CoreNLP (Manning et al., 2014) and NLTK (Bird et al., 2009). We also note that the Word-Piece segmentation is used even in systems that do not use BERT.

6.3 Results

In this section we report the results for all subtasks, and highlight notable results as well as a comparison with results obtained in 2016 in the third edition of the Bacteria Biotope task in BioNLP-ST 2016. The task site presents detailed results, including main and alternate metrics, as well as confidence intervals.

However comparison with 2016 is limited by the evolution of the task. On one hand the data set has increased approximately by 50%, and the annotations were revised and their quality improved. On the other hand the tasks were made harder because the schema was enriched with an entity type and a relation type, and the target taxa have been extended from *Bacteria* only to all microorganisms.

6.3.1 BB-norm and BB-norm+ner

The main results as well as the results for each entity type are shown in Tables 6 and 7. BOUN-ISIK and BLAIR_GMU obtained the best overall results for BB-norm, and MIC-CIS for BB-norm+ner.

The results for each entity type highlight different profiles. While BOUN-ISIK predicts accurate normalizations for habitat entities for BB-norm,

BLAIR_GMU predicts better normalizations for microorganism entities. PADIA_BacReader's predictions for habitats is on par with BOUN-ISIK, and their normalization of phenotype entities is outstanding.

As for BB-norm+ner, MIC-CIS consistently predicts the best entity boundaries and normalizations for all types.

In comparison to 2016, the state of the art for multi-word entity recognition and normalization, like habitats and phenotypes, has improved. We note that with the introduction of new taxa the recognition and normalization of taxa may have been rendered more difficult than anticipated since the results are lower than obtained in 2016.

6.3.2 BB-rel and BB-rel+ner

The results of BB-rel and BB-rel+ner are given in Tables 8 and 9 respectively. The table includes the scores obtained for each relation type, as well as the best results obtained in 2016.

The highest F-score for BB-rel was obtained by the whunlp submission, with AliAI as a very close contender. UTU, and very closely behind AliAI, obtained the highest Precision, whereas BOUN-ISIK the highest Recall. The Recall of the baseline prediction indicates the highest recall possible for relations contained in a single sentence. No participating system addresses cross-sentence relations, which appears to be the most productive lead to increase performance.

Most submissions outperform the best predictions of 2016 in at least one score, and five of the eleven submissions obtain a significantly higher F-score.

For BB-rel+ner, AliAI obtains the highest recall and precision, consistently for *Lives_In* and *Exhibits* relations. This submission also outperforms significantly the state of the art set in 2016.

6.3.3 BB-kb and BB-kb+ner

BLAIR_GMU is the only team to submit to the BB-kb and BB-kb+ner tasks, their results are shown in Table 10. The knowledge-base task and evaluation necessarily require end-to-end prediction systems that must perform named-entity recognition, entity normalization, relation extraction, as well as contributory tasks like POS-tagging, or coreference resolution. The limited scores obtained might be explained by the accumulation of errors by successive prediction steps.

Since the data of all sub-tasks comes from the

Team	All types	Habitats	Phenotypes	Microorganisms
Baseline	0.531	0.559	0.581	0.470
Best 2016	0.679	0.620		0.801
BOUN-ISIK-2	**0.679**	**0.687**	0.566	0.711
BLAIR_GMU-2	**0.678**	0.615	0.646	**0.783**
BOUN-ISIK-1	**0.675**	**0.687**	0.566	0.700
BLAIR_GMU-1	0.661	0.586	0.628	**0.783**
PADIA_BacReader-1	0.633	**0.684**	**0.758**	0.511
AmritaCen_healthcare-1	0.514	0.522	0.646	0.450

Table 6: Results for the BB-norm sub-task. The metric is the average of the semantic similarity between the reference and the predicted normalizations. Best scores are in bold font, several scores are in bold if their difference is not significant.

Team	All types	Habitat	Phenotype	Microorganism
Baseline	0.823	0.830	0.872	0.790
Best 2016	0.628	0.775		0.399
MIC-CIS-1	**0.716**	**0.728**	**0.747**	**0.686**
MIC-CIS-2	0.787	0.855	**0.759**	**0.715**
BLAIR_GMU-1	0.793	0.785	**0.775**	0.810
BLAIR_GMU-2	0.806	**0.722**	0.894	0.865
AmritaCen_healthcare-1	2.571	3.626	1.597	

Table 7: Results for the BB-norm+ner sub-task. The metric is the Slot Error Rate (lower is better) and takes into account false positives and negatives, entity boundary accuracy, and normalization accuracy. Best scores are in bold font, several scores are in bold if their difference is not significant.

| | Average | | | Lives_In | | | Exhibits | | |
Team	F1	Recall	Precision	F1	Recall	Prec.	F1	Recall	Prec.
Baseline	0.635	0.801	0.525	0.621	0.767	0.521	0.677	0.915	0.538
Best 2016				0.558	0.646	0.623			
whunlp-1	**0.664**	0.702	0.629	**0.643**	0.664	0.624	**0.725**	**0.829**	0.644
AliAI-1	**0.650**	0.620	**0.682**	**0.648**	0.606	**0.697**	0.654	0.667	0.642
Yuhang_Wu-1	0.605	0.670	0.551	0.593	0.645	0.549	0.640	0.752	0.556
BOUN-ISIK-1	0.603	**0.731**	0.514	0.592	**0.709**	0.508	0.640	**0.808**	0.530
BLAIR_GMU-2	0.594	0.650	0.548	0.578	0.618	0.543	0.642	0.752	0.560
BOUN-ISIK-2	0.575	0.601	0.552	0.562	0.562	0.561	0.613	0.729	0.529
UTU-2	0.550	0.474	0.655	0.495	0.417	0.610	**0.715**	0.662	**0.777**
BLAIR_GMU-1	0.549	0.496	0.617	0.526	0.463	0.609	0.619	0.603	0.636
UTU-1	0.529	0.428	**0.694**	0.505	0.403	**0.679**	0.603	0.510	**0.738**
Amrita_Cen-1	0.500	0.617	0.420	0.499	0.643	0.407	0.503	0.531	0.478
Amrita_Cen-2	0.493	0.610	0.414	0.491	0.642	0.397	0.505	0.502	0.507

Table 8: Results for the BB-rel sub-task. Best scores are in bold font, several scores are in bold if their difference is not significant.

Team	Average			Lives_In			Exhibits		
	SER	Recall	Prec.	SER	Recall	Prec.	SER	Recall	Prec.
Baseline	1.211	0.134	0.229	1.266	0.171	0.228	1.211	0.134	0.229
Best 2016				0.984	0.111	0.498			
AliAI-1	**0.954**	**0.351**	**0.509**	**0.941**	**0.309**	**0.520**	**0.982**	**0.449**	**0.492**
BLAIR_GMU-1	1.013	**0.330**	0.456	1.020	**0.325**	0.451	**0.996**	0.339	**0.468**
BLAIR_GMU-2	1.059	**0.331**	0.425	1.046	**0.320**	0.435	1.086	0.358	0.406
UTU-1	1.085	0.209	0.332	1.091	0.182	0.307	1.069	0.272	0.382
UTU-2	1.227	0.182	0.267	1.169	0.168	0.279	1.362	0.217	0.249

Table 9: Results for the BB-rel+ner sub-task (Prec. = Precision). Best scores are in bold font, several scores are in bold if their difference is not significant.

same pool of annotated documents, we were able to build a BB-kb prediction by combining the best predictions for the BB-norm and BB-rel tasks. The combination of the microorganism normalization by BLAIR_GMU, the habitat and phenotype normalization by PADIA_BacReader, and relations by whunlp yield a much higher precision. The best result for BB-kb+ner was obtained by combining the relation extraction of BLAIR_GMU and the normalization of MIC-CIS. The named entities concurrently predicted by the BB-norm+ner and BB-rel+ner systems were matched by maximizing the overlap segment.

Team	BB-kb	BB-kb+ner
Baseline	0.216	0.264
Combined	0.505	0.290
BLAIR_GMU-2	**0.308**	**0.269**
BLAIR_GMU-1	**0.291**	**0.259**

Table 10: Results for the BB-kb and BB-kb+ner sub-tasks. The metric is the average of the semantic similarity between the reference and the predicted normalizations for all relation arguments after removing duplicates at the corpus level. Best scores are in bold font, several scores are in bold if their difference is not significant.

7 Conclusion

The Bacteria Biotope Task arouses sustained interest with a total of 10 teams participating in the fourth edition. As usual, the relation extraction sub-tasks (BB-rel and BB-rel+ner) were the most popular, demonstrating that this task is still a scientific and technical challenge. The most notable evolution of participating systems since the last edition is the pervasiveness of methods based on neural networks and word embeddings. These systems yielded superior predictions compared to those in 2016. As mentioned previously, there is still much room for improvement in addressing cross-sentence relation extraction.

We also note a growing interest in the normalization sub-tasks (BB-norm and BB-norm+ner). The predictions improved for habitat entities, and are very promising for phenotype entities. However the generalization from bacteria-only taxa in 2016 to all microorganisms in this edition proved to pose an unexpected challenge.

Knowledge base population (BB-kb and BB-kb+ner) is the most challenging task, since it requires a wider set of capabilities. Nevertheless we demonstrated that the combination of other sub-task predictions allows to produce better quality knowledge bases.

To help participants, supporting resources were provided. The most used resources were pre-trained word embeddings, and general-domain named entities.

The evaluation on the test set will be maintained online[11] in order for future experiments to compare with the current state of the art.

Acknowledgments

The authors thank Léonard Zweigenbaum for his contribution to the manual annotation of the corpus.

The current affiliation of Estelle Chaix is the French Agency for Food, Environmental and Occupational Health & Safety (ANSES), 14 rue Pierre et Marie Curie, 94701 Maisons Alfort Cedex, France.

---

[11]`http://bibliome.jouy.inra.fr/demo/BioNLP-OST-2019-Evaluation/index.html`

References

Steven Bird, Ewan Klein, and Edward Loper. 2009. *Natural language processing with Python: analyzing text with the natural language toolkit.* O'Reilly Media.

Robert Bossy, Wiktoria Golik, Zorana Ratkovic, Philippe Bessières, and Claire Nédellec. 2013. BioNLP Shared Task 2013–an overview of the Bacteria Biotope task. In *Proceedings of the BioNLP Shared Task 2013 Workshop*, pages 161–169.

Robert Bossy, Julien Jourde, Alain-Pierre Manine, Philippe Veber, Erick Alphonse, Maarten Van De Guchte, Philippe Bessières, and Claire Nédellec. 2012. BioNLP shared task–the bacteria track. In *BMC bioinformatics*, volume 13, page S3. BioMed Central.

Robert Bossy, Claire Nédellec, Julien Jourde, Mouhamadou Ba, Estelle Chaix, and Louise Deléger. 2019. Bacteria biotope annotation guidelines. Technical report, INRA.

Maria Brbić, Matija Piškorec, Vedrana Vidulin, Anita Kriško, Tomislav Šmuc, and Fran Supek. 2016. The landscape of microbial phenotypic traits and associated genes. *Nucleic acids research*, page gkw964.

Estelle Chaix, Louise Deléger, Robert Bossy, and Claire Nédellec. 2019. Text mining tools for extracting information about microbial biodiversity in food. *Food microbiology*, 81:63–75.

Louise Deléger, Robert Bossy, Estelle Chaix, Mouhamadou Ba, Arnaud Ferre, Philippe Bessieres, and Claire Nedellec. 2016. Overview of the bacteria biotope task at BioNLP shared task 2016. In *Proceedings of the 4th BioNLP shared task workshop*, pages 12–22.

Pan Deng, Haipeng Chen, Mengyao Huang, Xiaowen Ruan, and Liang Xu. 2019. An ensemble CNN method for biomedical ontology alignment. In *Proceedings of the BioNLP Open Shared Tasks 2019 Workshop*.

Jacob Devlin, Ming-Wei Chang, Kenton Lee, and Kristina Toutanova. 2019. BERT: Pre-training of deep bidirectional transformers for language understanding. In *Proceedings of the 2019 Conference of the North American Chapter of the Association for Computational Linguistics: Human Language Technologies, Volume 1 (Long and Short Papers)*, pages 4171–4186, Minneapolis, Minnesota. Association for Computational Linguistics.

Hélène Falentin, Estelle Chaix, Sandra Derozier, Magalie Weber, Solange Buchin, Bedis Dridi, Stéphanie-Marie Deutsch, Florence Valence-Bertel, Serge Casaregola, Pierre Renault, Marie-Christine Champomier-Verges, Anne Thierry, Monique Zagorec, Francoise Irlinger, Céline Delbes, Sophie Aubin, Philippe Bessieres, Valentin Loux, Robert Bossy, Juliette Dibie, Delphine Sicard, and Claire Nédellec. 2017. Florilege : a database gathering microbial phenotypes of food interest. In *Proceedings of the 4th International Conference on Microbial Diversity 2017*, pages 221–227, Bari, Italy. Poster.

Scott Federhen. 2011. The NCBI taxonomy database. *Nucleic acids research*, 40(D1):D136–D143.

Jenny Rose Finkel, Trond Grenager, and Christopher Manning. 2005. Incorporating non-local information into information extraction systems by Gibbs sampling. In *Proceedings of the 43rd annual meeting on association for computational linguistics*, pages 363–370. Association for Computational Linguistics.

Pankaj Gupta, Usama Yaseen, and Hinrich Schütze. 2019. Linguistically informed relation extraction and neural architectures for nested named entity recognition in BioNLP-OST 2019. In *Proceedings of the BioNLP Open Shared Tasks 2019 Workshop*.

İlknur Karadeniz, Ömer Faruk Tuna, and Arzucan Özgür. 2019. BOUN-ISIK participation: An unsupervised approach for the named entity normalization and relation extraction of bacteria biotopes. In *Proceedings of the BioNLP Open Shared Tasks 2019 Workshop*.

Bofang Li, Tao Liu, Zhe Zhao, Buzhou Tang, Aleksandr Drozd, Anna Rogers, and Xiaoyong Du. 2017. Investigating different syntactic context types and context representations for learning word embeddings. In *Proceedings of the 2017 Conference on Empirical Methods in Natural Language Processing*, pages 2421–2431.

Christopher Manning, Mihai Surdeanu, John Bauer, Jenny Finkel, Steven Bethard, and David McClosky. 2014. The Stanford CoreNLP natural language processing toolkit. In *Proceedings of the 52nd Annual Meeting of the Association for Computational Linguistics: system demonstrations*, pages 55–60.

Jihang Mao and Wanli Liu. 2019. Integration of deep learning and traditional machine learning for knowledge extraction from biomedical literature. In *Proceedings of the BioNLP Open Shared Tasks 2019 Workshop*.

Tomas Mikolov, Kai Chen, Greg Corrado, and Jeffrey Dean. 2013. Efficient estimation of word representations in vector space. *arXiv preprint arXiv:1301.3781*.

Claire Nédellec, Robert Bossy, Estelle Chaix, and Louise Deléger. 2018. Text-mining and ontologies: new approaches to knowledge discovery of microbial diversity. *arXiv preprint arXiv:1805.04107*.

Sampo Pyysalo, Filip Ginter, Hans Moen, Salakoski Tapio, and Sophia Ananiadou. 2013. Distributional semantics resources for biomedical text processing. *Proceedings of LBM*, pages 39–44.

Yoshimasa Tsuruoka, Yuka Tateishi, Jin-Dong Kim, Tomoko Ohta, John McNaught, Sophia Ananiadou, and Jun'ichi Tsujii. 2005. Developing a robust part-of-speech tagger for biomedical text. In *Panhellenic Conference on Informatics*, pages 382–392. Springer.

Wuti Xiong, Fei Li, Ming Cheng, Hong Yu, and Donghong Ji. 2019. Bacteria biotope relation extraction via lexical chains and dependency graphs. In *Proceedings of the BioNLP Open Shared Tasks 2019 Workshop*.

Zhilin Yang, Zihang Dai, Yiming Yang, Jaime Carbonell, Ruslan Salakhutdinov, and Quoc V Le. 2019. XLNet: Generalized autoregressive pretraining for language understanding. *arXiv preprint arXiv:1906.08237*.

Qi Zhang, Chao Liu, Ying Chi, Xuansong Xie, and Xiansheng Hua. 2019. A multi-task learning framework for extracting bacteria biotope information. In *Proceedings of the BioNLP Open Shared Tasks 2019 Workshop*.

Linguistically Informed Relation Extraction and Neural Architectures for Nested Named Entity Recognition in BioNLP-OST 2019

***Usama Yaseen[1,2], *Pankaj Gupta[1,2], Hinrich Schütze[2]**

[1]Corporate Technology, Machine-Intelligence (MIC-DE), Siemens AG Munich, Germany
[2]CIS, University of Munich (LMU) Munich, Germany
{usama.yaseen,pankaj.gupta}@siemens.com

Abstract

Named Entity Recognition (NER) and Relation Extraction (RE) are essential tools in distilling knowledge from biomedical literature. This paper presents our findings from participating in BioNLP Shared Tasks 2019. We addressed Named Entity Recognition including nested entities extraction, Entity Normalization and Relation Extraction. Our proposed approach of Named Entities can be generalized to different languages and we have shown it's effectiveness for English and Spanish text. We investigated linguistic features, hybrid loss including ranking and Conditional Random Fields (CRF), multi-task objective and token-level ensembling strategy to improve NER. We employed dictionary based fuzzy and semantic search to perform Entity Normalization. Finally, our RE system employed Support Vector Machine (SVM) with linguistic features.

Our NER submission (team:MIC-CIS) ranked first in *BB-2019 norm+NER task* with standard error rate (SER) of **0.7159** and showed competitive performance on *PharmaCo NER* task with F1-score of **0.8662**. Our RE system ranked first in the *SeeDev-binary Relation Extraction Task* with F1-score of **0.3738**.

1 Introduction

Extracting knowledge from scientific articles is a challenging but very important problem. This becomes especially critical for biomedical literature which is growing at an increasing rate of at least 4% per year, as of June 2019 there are 30 Million documents in PubMed (Lu, 2011). Named Entity Recognition (NER) (Settles, 2004; Gupta et al., 2016; Lample et al., 2016) in the context of biomedical domain refers to the task of identifying the name of the biological entities e.g. name of a bacteria. Relation extraction[1] (RE) (Kambhatla,

---
* Equal Contribution
[1]Event extraction is treated as RE in this work

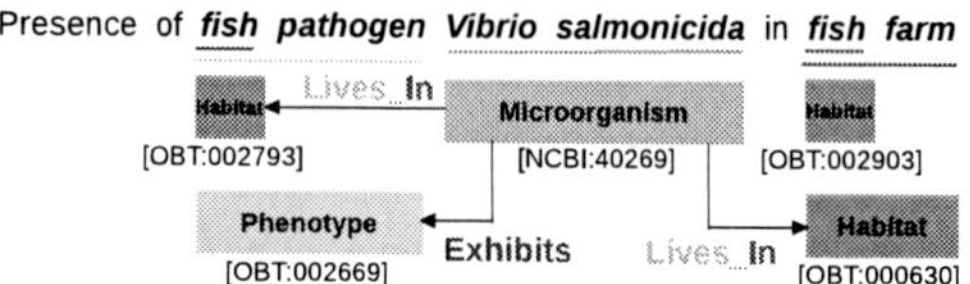

Figure 1: An illustration of (nested) NER + Normalization and Relation Extraction in Biomedical entities. Each rectangular box spans an entity, where the overlapping spans indicate nested entities. E.g., *fish* is a nested entity (a sub-concept) of type *Habitat* within the parent entity *fish pathogen* of type *Phenotype*. The identifiers (e.g. OBT:002669, NCBI:40269, etc.) refer to unique IDs in Biomedical databases (i.e., OBT → OntoBiotope Ontology and NCBI → NCBI Taxonomy), used to perform entity normalization (i.e., entity linking). The arrows indicate binary relationships.

2004; McDonald et al., 2005; Lever and Jones, 2016; Gupta et al., 2018) refers to identifying relations among biological entities (binary or n-ary).

Figure 1 illustrates an example of (nested) NER and RE consisting of five entities, where three entities participate in two distinct relationships. It is often required to link named entity(s) to a unique reference in database(s). For instance, one of the two occurrences of *fish* refers to *marine fish* while the second refers to a *farm fish*, where the two entities are linked (or normalized) to different identifiers (e.g., OBT:002793 and OBT:002903) in the biomedical database (e.g., OntoBiotope Ontology). The act of linking entities to standard entities with a unique identifier is known as *entity normalization* and is challenging as several entity mentions can correspond to the same standard entity (or unique identifier), e.g. *E. coli*, *Bacillus coli* and *Bacteriumcoli* refer to the standard entity *Escherichia coli* in the database. The linking process relies on knowledge base (KB) search (heuristic OR semantic) in order to resolve entities.

NER is a critical primitive step in the NLP

Proceedings of the 5th Workshop on BioNLP Open Shared Tasks, pages 132–142
Hong Kong, China, November 4, 2019. ©2019 Association for Computational Linguistics

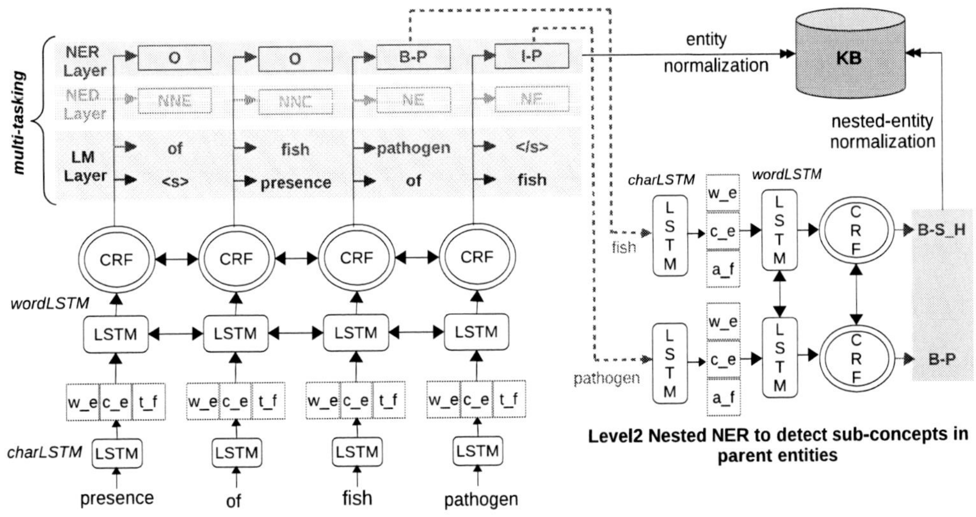

Figure 2: System Architecture for NER task, consisting of two bi-LSTM-CRF architectures: Level1 NER to detect parent entities and Level2 Nested NER to detect sub-concepts within the parent entities (output of Level1 NER). Here, w_e: a word embedding vector; c_e: an embedding vector for a word computed using character-level bi-directional LSTM; t_f: a vector of additional linguistic features; B_P: B_Pathogen; B-S_H: a sub-concept of type *Habitat* detected by the Level2 Nested NER run over the the parent entity.

pipeline as downstream tasks such as RE, text classification, Question Answering (QA) etc., depend on it. Even though several methods have been devised to engineer reliable NER systems; however, most of them don't explicitly address the extraction (or recognition) of nested entities, especially required in the biomedical domain. *Nested entity* is defined as an entity or sub-concept which is part of a longer entity (i.e., a parent). For instance in the Figure 1, *fish* is a nested entity as it is part of a parent entity *fish pathogen*. In this work, we have also investigated extracting nested entities via two bi-LSTM-CRF (Lample et al., 2016) networks: one for parent detection and another for nested entities with the parent entity.

2 Task Description and Contribution

We participate in the following three tasks organized by BioNLP workshop 2019: (1) **PharmaCoNER**: Recognition of pharmaceutical drugs and chemical entities in Spanish text. (2) **BB-norm+NER**: Recognition of *Microorganism*, *Habitat* and *Phenotype* entities and normalization with NCBI Taxonomy and OntoBiotope habitat concepts. (3) **SeeDev Binary RE**: Binary Relation extraction of genetic and molecular mechanisms involved in plant seed development.

Following are our multi-fold *contributions*:

1. To address NER tasks, we have employed neural network based sequence classifier, i.e., bi-LSTM-CRF and investigated multi-tasking of named entity detection (NED) and language modeling (LM). We further introduced hybrid loss including CRF and ranking. We also incorporated linguistic features such as POS, orthographic features, etc. We apply the proposed modeling approaches to both English and Spanish texts. Comparing with other systems, our submission (Team: MIC-CIS) is ranked 1st in *BB-norm+NER* task (Bossy et al., 2019) with standard error rate of 0.7159. In *PharmaCoNER* task (Gonzalez-Agirre et al., 2019), our submission scored F1-score of 0.8662.

2. To address RE task, we employed linguistic and entity features in SVM. Our submission (Team: MIC-CIS) is ranked 1st in SeeDev-binary RE task (Chaix et al., 2016) with F1-score of 0.3738.

The code to reproduce our results is available at: `https://github.com/uyaseen/bionlp-ost-2019`.

Features	Description
word-cap	capitalization features
POS	parts-of-speech tags
ortho	orthographic features
	e.g. *Egg Pulp, 97* encoded as *Ccc Ccccp nn*
tri-gram	tri-gram as features
five-gram	five-gram as features
length	length of the word
sdp-rel	dependency relation tag
alpha-features	detect if certain linguistic pattern occurred in the current word or the next word

Table 1: Word-level features for NER. The features are encoded as embeddings, except the *alpha* features that are represented as one-hot vector.

3 Methodology

In the following sections we discuss our proposed model for NER and RE.

3.1 Neural Architectures for NER

Figure 2 describes the architecture of our model, where we design two sequence taggers *Level1 NER* and *Level2 Nested NER* to extract parent and nested entities respectively. Furthermore, Level1 NER can be configured in two modes: (1) LSTM-CRF (Lample et al., 2016) with word embeddings (w_e), character embeddings (c_e) and token-level features (t_f) such as POS, capitalization features, word shape, etc. (refer to table 1 for the complete list of word level features) (2) *LSTM-CRF+Multi-task* that performs entity detection and language modelling as auxiliary tasks. Note that Level2Nested NER only operates on the parent entities detected by Level1 NER. The parent and nested entities are than normalized to unique identifiers in KB by our entity normalization algorithm.

3.1.1 BiLSTM-CRF

The input to LSTM is a sequence of word features $(\mathbf{w}_1, \mathbf{w}_2, \ldots, \mathbf{w}_n)$ and they compute a hidden state for each element in the sequence $(\mathbf{h}_1, \mathbf{h}_2, \ldots, \mathbf{h}_n)$. This hidden state can be used to jointly model tagging decisions using CRF (Lafferty et al., 2001). CRF imposes ordering constraints on the tagging decisions e.g. I_Habitat should always be preceded by B_Habitat. For an input sentence,

$$\mathbf{W} = (\mathbf{w}_1, \mathbf{w}_2, \ldots, \mathbf{w}_n),$$

we consider a matrix $\mathbf{P}$ of scores output by the bidirectional LSTM. The size of $\mathbf{P}$ is $n \times k$,

where k is the number of distinct tags, and $P_{i,j}$ corresponds to the score of the jth tag of the ith word in a sentence. For a sequence of predictions

$$\mathbf{y} = (y_1, y_2, \ldots, y_n),$$

we define its score to be

$$s(\mathbf{X}, \mathbf{y}) = \sum_{i=0}^{n} A_{y_i, y_{i+1}} + \sum_{i=1}^{n} P_{i, y_i}$$

where the matrix $\mathbf{A}$ express transition scores such that $A_{i,j}$ represents the score of a transition from the tag i to tag j. We add *start* and *end* tag to the set of possible tags, therefore, the size of A is $k + 2$. During training, we minimize the negative log-probability of the correct tag sequence:

$$\log(p(\mathbf{y}|\mathbf{X})) = s(\mathbf{X}, \mathbf{y}) - \log \left(\sum_{\widetilde{\mathbf{y}} \in \mathbf{Y_X}} e^{s(\mathbf{X}, \widetilde{\mathbf{y}})} \right)$$
$$= s(\mathbf{X}, \mathbf{y}) - \underset{\widetilde{\mathbf{y}} \in \mathbf{Y_X}}{logadd}\, s(\mathbf{X}, \widetilde{\mathbf{y}}), \quad (1)$$

$$loss_{CRF} = -\log(p(\mathbf{y}|\mathbf{X})) \quad (2)$$

3.1.2 Hybrid Loss: CRF + Ranking

We use a variant of ranking loss function proposed by dos Santos et al. (2015). Ranking maximizes the distance between the true label y^+ and the most competitive label c^-:

$$loss_{ranking} = max(0, 1 + (\gamma * (m^+ - y^+)) + (\gamma * (m^- + c^-))$$

where γ is the scaling factor that penalizes the predictions, m^+ and m^- are margins for correct and incorrect labels respectively. We follow Vu et al. (2016) to set the values of margins.

The hybrid loss function hence is the sum of CRF tagging loss and ranking loss:

$$loss_{hybrid} = loss_{CRF} + \alpha \cdot loss_{ranking}$$

where $\alpha \in [0, 1]$, weighs the contribution of ranking loss in the overall loss value. During training we minimize the hybrid loss and found it to improve the F1 score for both *BB-norm+NER* and *PharmaCoNER tasks*.

3.1.3 Multi-Tasking of Named Entity Recognition, Detection and Language Modelling

We employed auxiliary objectives of named-entity detection (NED) (Aguilar et al., 2017) and bidirectional language modelling (LM) (Rei, 2017) in our

---

Algorithm 1 Entity Normalization

    Input: NE, NE_Type
    Output: RF_ID
    Output: NE_PRED (Optional)
  1: RF_ID = None
  2: IF NE_Type == 'Microorganism':
  3:    found, RF_ID = exact_match(NE, NCBI)
  4:    if not found:
  5:      found, RF_ID = fuzzy_match(NE, NCBI)
  6:    return RF_ID
  7: ELSE
  8:    found, RF_ID = exact_match(NE, NCBI)
  9:    if not found:
10:      found, RF_ID = fuzzy_match(NE, NCBI)
11:    if found:
12:      # LABEL UPDATE !
13:      NE_PRED = 'Microorganism'
14:      return RF_ID, NE_PRED
15:    found, RF_ID = exact_match(NE, OBT)
16:    if not found:
17:      found, RF_ID = semantic_search(NE, OBT)
18: return RF_ID

---

model. Usually these auxiliary objectives acts as regularizes (Collobert and Weston, 2008) and improves the overall performance. With these multitasking objectives, for each word token our model predicts the NED tag, next word, previous word and the NER tag[2]. LM and NED layers in figure 2 realizes NED and LM objectives respectively. Note that Multi-tasking is only enabled at train time and requires no additional labelling.

3.1.4 Nested Entities

The dataset of *BB-norm+NER* task contains 17.4% nested entities [3] which cannot be extracted by standard Bi-LSTM CRF model. We employed two Bi-LSTM-CRF models: *Level1* NER model to detect parent entities and *Level2* Nested NER model to detect nested entities. Figure 2 (right) shows the architecture of Level2 Nested NER. The parent entities detected by Level1 NER are fed to Level2 Nested NER to detect nested entities in the parent entities. Level2 Nested NER has the same architecture as Level1 NER but without the multitasking objectives. It is easy to see that current architecture can only detect nested entities at level 2. The final output of model is the aggregation of parent entities and nested entities.

3.1.5 Entity Normalization

The goal of entity normalization (entity linking) is to map noisy predicted entities in text to canonical

---

[2] we used IOBES tagging scheme
[3] https://groups.google.com/d/msg/
bb-2019/A2MuFYiPQIY/9YtMmakeBQAJ

entities in knowledge base (KB). This is challenging because: (1) not all variations of textual forms for a canonical entity exists in the KB, (2) syntactic variations in the predicted entity mentions due to misspellings, abbreviations, acronyms and boundary errors.

For *BB-norm+NER* task, we used two Biomedical databases *OntoBiotope Ontology* and *NCBI Taxonomy*. OntoBiotope Ontology contains $3,602$ canonical forms of type *Habitat* and *Phenotype*. NCBI Taxonomy contains $1,082,401$ records for type *Microorganism*. We employed exact, fuzzy and semantic (embedding) search to perform entity normalization. Algorithm 1 illustrates the detailed steps of our algorithm, note that type and order of search depends on the predicted named entity type. We also employed *caching* to minimize pairwise comparisons and improve the overall run-time efficiency.

3.1.6 Post-processing for NER+norm

Our model (see Figure 2) employs CRF at decoding step to impose boundary ordering constraints on the predicted named entity types e.g. *I* should always be preceded by a *B* token. But our model does not always respect such ordering constraints and therefore, we resolve boundary inconsistencies at inference time to make the NER labels consistent. *Post-processing* column in the Table 3 illustrates the post-processing resolving inconsistent labels after the voting on majority labels, consider row *r3* where post-processing correctly imposes the semantics of boundary ordering by changing *I-Habitat* to *B-Habitat*.

3.2 Relation Extraction

Deep Learning based methods are state of the art in relation extraction (Wu and He, 2019; Wang et al., 2016) but they require large amount of labelled training data. In cases when enormous training data is not available than Kernel methods like Support Vector Machines (SVM) are an optimal choice. We employed SVM for performing relation extraction. One of the downsides of SVM is that they usually require lots of hand-crafted features to train properly. Table 2 lists computed general and entity features.

Our best model was trained with Radial Basis Function (RBF) Kernel with value of penalty parameter C determined by grid search for each dataset. We employed oversampling and classweight penalization to handle imbalanced data.

General Features	Description		Entity Features	Description
bow	bag-of-words (bow) representation of the complete sentence		entity-pos	position of entity in the bow representation
bow-partial	bow representation of the between context (i.e. word tokens between target entities) including three words to the target entities		entity-type	type of the entity mentions
bow-lemma	bow representation of the lemmatized tokens in the between context		dist-entities-cat	distance between target entities as categorical
pos-tags	part-of-speech tags		dist-entities	distance between target entities
sdp	shortest dependency path as bow		entity-count	count of entities in between context
sdp-len	length of shortest dependency path as scalar		entity-count-cat	count of entities in between context as categorical
sdp-rel	dependency relation tag		e1 type = e2 type	if type of e1 and e2 is same
emb-sdp	average embeddings of sdp		sdp-entity	sdp with entity as bow
keyword-vec	if current word is part of feature list of relations		entity-patterns	check if certain linguistic patterns occur in the vicinity of target entities

Table 2: General and Entity features used in Relation Extraction

	Tokens	Models			Voting	Post-
		M1	M2	M3		processing
r1	Presence	O	O	O	O	O
r2	of	O	O	B-H	O	O
r3	fish	I-H	B-H	I-H	I-H	B-H
r4	pathogen	I-H	I-P	I-P	I-P	B-P
r5	Vibrio	B-M	B-M	B-M	B-M	B-M
r6	salmonicida	I-M	O	I-M	I-M	I-M
r7	in	B-H	O	O	O	O
r8	fish	B-H	O	B-H	B-H	B-H
r9	farm	I-H	O	I-M	I-H	I-H
r10	.	O	O	O	O	O

Table 3: NER: Ensembling and Post-processing correcting individual models mistakes. Here, B, P and M refer to Habitat, Phenotype and Microorganism, respectively.

Surprisingly oversampling did not provide any performance improvement therefore, final models were trained only with higher class weights for minority classes. We did not normalize any input feature as it resulted in reduced performance.

In relation extraction participating entities are not known in advance, the usual practise is to test every valid pair of entities for a relation. We employed heuristic of *token counts* between entities to filter the probable invalid relations. The value of token counts was determined using cross-validation.

3.3 Ensemble Strategy

Bagging is a helpful technique to reduce variance without impacting bias of the learning algorithm. We employed a variant of *Bagging* (Breiman, 1996) which makes sure that every sample in the training set is part of the development set at least once and vice versa. We created three data folds and trained the model using optimal configuration on each fold, prediction on test involves majority voting among the *three* trained models.

The commonly used tagging schemes (BIO, BIOES etc.,) for NER contains information about the boundary of an entity along with the class of an entity, which is spitted by the model at each time-step. Due to this dual information in a single output, maximum voting is not trivial as models can not only disagree on the class but also on the boundary of an entity. Empirically we found that our model is better at predicting the class of an entity rather than the boundary of an entity, therefore, we followed the strategy *class determines the boundary*. In cases when voting results in a tie, we take the prediction of the *confident* model, we treat the model trained on original train/dev split as the confident model. We also experimented with an extreme version of ensembling where we aggregate the output of every model with distinct spans, as expected this improves the recall but with the cost of reduced precision. One possible optimization to this ensemble strategy is to only aggregate the non-overlapping spans to control reduction in precision without much decrease in recall, we will explore this as a future work. Table 3 shows the ensemble correcting individual model's erroneous predictions.

In case of ensemble for RE, we followed the straight forward approach of majority voting at sentence level for each test sample.

Task	Train	Dev	Test
Sentence Counts			
PharmaCo	8068	3748	3930
SeeDev	644	308	466
BB-norm+ner	822	413	735
PharmaCoNER Entities			
NORMALIZABLES	2304	1121	859
PROTEINAS	1405	745	973
UNCLEAR	89	44	34
NO_NORMALIZABLES	24	16	10
BB-norm+NER Entities			
Habitat	1118	610	-
Microorganism	739	402	-
Phenotype	369	161	-

Table 4: Dataset statistics for NER

Hyper-parameter	Value
NER	
learning rate	0.005
character (char) dimension	25
hidden unit::char LSTM	25
POS dimensions	$25^*, 50^+$
Ortho dimension	$25^*, 50^+$
hidden unit::word LSTM	$200^*, 100^+$
word embeddings dimension	$200^*, 100^+$
length dimension	10
sdp_rel	10
alpha_features	2
ranking loss::α	1.0
ranking loss::γ	1.0
RE	
kernel	RBF
class-weights	10.0

Table 5: Hyper parameter settings for NER and RE. * and + denote the optimal parameters for BB-norm+ner and PharmaCoNER respectively.

4 Experiments and Results

4.1 Dataset and Experimental Setup

Data: We employed bagging (discussed in section 3.3) to split the annotated corpus into 3-folds. We used pre-processed versions of datasets for BB-norm+NER[4] and SeeDev[5] provided by the organizers. This pre-processed version comes with sentence splitting, word tokenization and POS tagging.

PharmaCoNER: The dataset consists of four entity types with very few mentions of type *UNCLEAR* and *NO_NORMALIZABLES* as shown in table 4. Entities of type *UNCLEAR* are ignored in the evaluation of this shared task but we still treat them as regular entities.

BB-norm+NER: The dataset consists of three entity types with few mentions of type *Phenotype* (see table 4). The dataset also contains 3.6% *disconnected entities*[6], we did not employ any strategy to handle disconnected entities and instead treat them as separate (regular) entities.

SeeDev: The dataset consists of 22 binary relations among 16 entity types. The dataset is highly imbalanced with zero instances of type *Regulates_Molecule_Activity* and *Composes_Protein_Complex* in the default development set.

[4]https://sites.google.com/view/seedev2019/supporting-resources
[5]https://sites.google.com/view/bb-2019/supporting-resources
[6]https://groups.google.com/d/msg/bb-2019/A2MuFYiPQIY/9YtMmakeBQAJ

Experimental Setup: We found sub-word information to be very helpful in identifying entities and relations in biomedical domain and all our experiments used word embeddings trained using FastText (Bojanowski et al., 2017). For tasks in English language we used FastText embeddings trained on PubMed (Zhang et al., 2019). We don't employ any strategy for handling imbalanced classes for NER but have used class weighting by a factor of 10 for all positive classes for RE. Table 5 lists the best configuration of hyperparameters for all the tasks.

PharmaCoNER: We used *SPACCC_POS-TAGGER* (Soares and gonzalez agirre, 2019) for sentence splitting, word tokenization and POS tagging. We trained FastText embeddings on the following corpora: IBECS (Rodríguez, 2002), IULA-Spanish-English-Corpus (Marimon et al., 2017), MedlinePlus (Miller et al., 2000), PubMed (Lu, 2011), ScIELO (Goldenberg et al., 2007) and PharmaCoNer (Gonzalez-Agirre et al., 2019). We trained embeddings on two variants of corpora: (1) Include train and development set of PharmaCoNER (2) Include complete dataset of PharmaCoNER. We concatenated these two embeddings to provide complementary information and found them to empirically work better than the embeddings trained on individual corpora variant. We compute micro-F1 using the script

	Configration	PharmaCoNER			BB-norm+NER			
		P	**R**	**F1**	**P**	**R**	**F1**	**SER**
		Fold=1			**Fold=1**			
r1	*BiLSTM-CRF*	.884	.773	.824	.809	.474	.598	.576
r2	*+ word-emb*	.892	.857	.874	.831	.526	.644	.524
r3	*+ ortho*	.909	.846	.877	.823	.515	.633	.533
r4	*+ POS*	.906	.851	.877	.827	.523	.641	.526
r5	*+ multi-task*	.907	.851	.878	.806	.528	.638	.531
r6	*+ length*	-	-	-	**.842**	.487	.617	.545
r7	*+ ranking*	**.912**	**.860**	**.885**	.827	.535	.650	.520
r8	*+ search*	-	-	-	.810	**.600**	**.690**	**.489**
		Fold=2			**Fold=2**			
r9	*BiLSTM-CRF*	.915	.890	.902	.630	.400	.489	-
r10	*all features*	**.934**	**.889**	**.911**	**.719**	**.513**	**.599**	-
		Fold=3			**Fold=3**			
r11	*BiLSTM-CRF*	.899	.873	.886	.784	.699	.739	-
r12	*all features*	**.917**	**.877**	**.896**	**.813**	**.764**	**.788**	-

Table 6: Scores on dev set using different features on *PharmaCoNER* and *BB-norm+NER* tasks. Here, + signifies feature accumulation to the last row.

	Features	**P**	**R**	**F1**
r1	*bow-between*	.0	.0	.0
r2	*+ class-weights*	.214	.196	.205
r3	*+ entity-type*	.157	.589	.248
r4	*+ sdp-entity*	.204	.540	.296
r5	*+ emb-sdp*	.212	.479	.294
r6	*+ lemma*	**.220**	**.478**	**.301**

Table 7: Scores on dev set using different features on *SeeDev* task. Here, + signifies feature accumulation to the last row.

provided by the organizers on the dev set[7].

BB-norm+NER: For training NER model we compute macro-F1[8] (Tsai et al., 2006) on the dev set. NER and Entity normalization together are evaluated using *Standard Error Rate (SER)* (Bossy et al., 2015). During the entity normalization step, the fuzzy and semantic search can resolve an entity mention to multiple normalization identifiers. Our algorithm returns top 5 matched identifiers, however, we empirically found selecting the top most identifier gives superior performance.

SeeDev: We adopted two strategies to create negative relation instances for train and dev+test set: (1)*Train:* only consider sentences not participating in any positive relation (2) *Dev+Test:* consider all the sentences. Negative relation instances are always created only among the valid combination of entity types. We also employed an extended version of keywords match of Li et al. (2016) as a feature (referred as keyword vectors in table 2).

4.2 Results on Development Set

To investigate the impact of features we incrementally enabled them and observe the affect on performance on dev set.

NER: Table 6 shows the score on dev set for *PharmaCoNER* and *BB-norm+NER*. Observe that *FastText* embeddings (row r2) outperform randomly initialized embeddings (row r1) and con-

tribute to biggest performance boost for both datasets. Subsequently, *Orthographic* (row r3) and *POS* (row r4) features[9] improve the scores for PharmaCoNER but surprisingly lower the score for BB-norm+NER. In row r5, we perform multi-tasking with auxiliary task of NED leading to improvement only for PharmaCoNER. Next, we incorporate hybrid loss including ranking (row r7) which consistently improves the score on both datasets. In row r8, we employed Brute Force Search (discussed in section 4.3) that significantly reduce SER for BB-norm+NER. Finally, we create an ensemble of (r7, r10, r12) and (r8, r10, r12) on test set for PharmaCoNER and BB-norm+NER respectively.

RE: Table 7 shows the score on dev set for *SeeDev*[10]. In row r1, negative instances dominate the training set resulting in no learning. Observe that introduction of class weights (row r2) compensate the dominance of negative instances leading to F1 score of 0.205. Next, we added *entity-type* (row r3) and *sdp-entity* (row 4) features, both of these features significantly improves F1 score i.e. by an absolute value of more than 4.0. Subsequently, *emb-sdp* (row r5) and *lemma* (row r6) contribute to incremental improvements. Finally, we create an ensemble of *row r6* on all three data folds.

4.3 Analysis on Development Set

BB-norm+NER: We also explored approaching the problem of NER and entity normalization in a reverse manner by matching every entity mention from the biomedical databases (i.e. *NCBI Taxonomy* and *Ontobiotope*) in every sentence. This

---

9Additionally, we have employed document-topic proportion from neural topic models (Gupta et al., 2019a), however, no significant gains were observed.

10Results are only reported for standard data fold as it was not trivial to change evaluation script for non-standard folds.

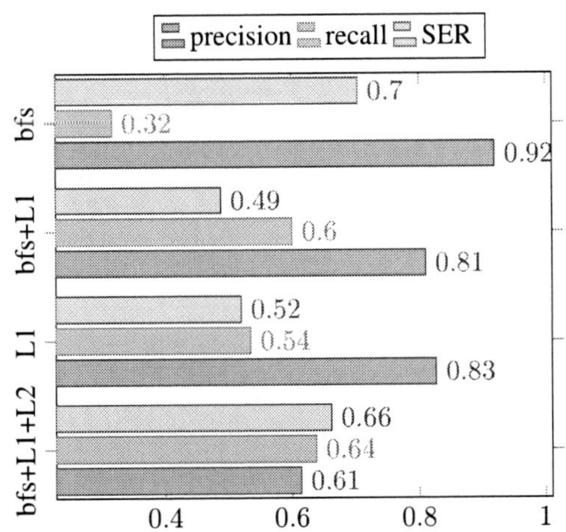

Figure 3: *BB-norm+NER*: Impact of brute-force search, Level1 NER and their aggregation on SER. Here bfs, L1 and L2 refer to *brute-force search, Level1 NER* and *Level2 Nested NER* respectively.

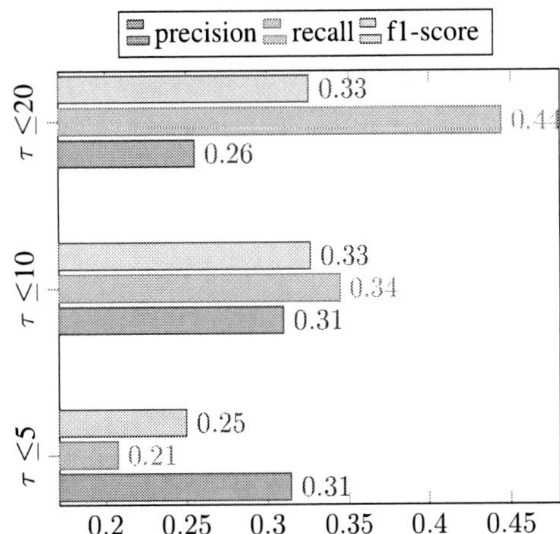

Figure 4: *SeeDev*: Impact of 'token counts between target entities' heuristic on system performance.

Task: SeeDev		Task: BB-norm+NER	
Team	P / R / $F1$	**Team**	P / R / SER
MIC-CIS	**.294** / **.511** / **.373**	**MIC-CIS-1**	**.624** / .433 / **.715**
YNU-junyi-1	.272 / .458 / .341	*MIC-CIS-2*	.560 / .449 / .786
Yunnan_University-1	.045 / .132 / .067	*BLAIR_GMU-1*	.496 / **.467** / .793
Yunnan_University-2	.020 / .132 / .035	*BLAIR_GMU-2*	.499 / .466 / .805
YNUBY-1	.011 / .070 / .019	*baseline-1*	.572 / .327 / .823

Table 8: Comparison of our system (MIC-CIS) with top-5 participants: Scores on Test set for SeeDev and BB-norm+NER

matching is indeed exhaustive search, we refer to it as *Brute-force search*. Figure 3 shows the comparison of: (1) *brute-force search* (2) *Level1 NER* (3) aggregation of *brute-force search* and *Level1 NER* (4) aggregation of *brute-force search, Level1 NER* and *Level2 NER*. Brute-force search yields high precision but a moderately low recall with SER value of 0.7. In comparison, Level1 NER has significantly higher recall with a little reduction in precision yielding SER value of 0.52. The aggregation of brute-force search and Level1 NER improves recall and lowers SER value to 0.49. Finally, aggregation of brute-force search, Level1 NER and Level2 NER results in a balanced precision and recall values but an overall higher value of SER. Our submission on test set employed aggregation of *brute-force search* and *Level1 NER*.

SeeDev: We employed the heuristic of token counts between target entities to filter potential negative relation instances. With this heuristic in place, we only consider sentences with entity distance less than or equal to threshold parameter τ. Figure 4 shows the impact of different values of τ on system performance. The value of $\tau \leq 20$ gives significant boost in precision with minor decrease in recall. Our submission employed the threshold value of $\tau \leq 20$ between entity tokens.

4.4 Comparison with Participating Systems

SeeDev: Table 8 (left) is the official result of SeeDev Shared Task. Our submission *MIC-CIS* achieves the best score among all participating systems with F1 score of **0.373** showing compelling advantage. The system attains the highest precision (0.294) and recall (0.511). Precision and recall are not balanced however, and our system need an improvement to bring down false positives.

BB-norm+NER: Table 8 (right) shows the comparison of performance among participating teams on BB-norm+NER test set. Our two submissions (MIC-CIS-1, MIC-CIS-2) ranked first and second with standard error rate (SER) of **0.7159** and *0.7867* respectively. The second submission employed Level2 NER to extract nested entities and hence has higher recall but with reduced precision. *MIC-CIS-1* has the highest precision 0.6242 and *MIC-CIS-2* has the recall close to the best recall of *BLAIR_GMU-1* with score *0.4676*. Precision and recall are not balanced, we hypothesize improvement in nested entities extraction and modelling discontinuous entities will improve the system recall.

5 Conclusion and Future Work

In this paper, we described our system with which we participate in *PharmaCoNER*, *BB-norm+NER* and *SeeDev* shared tasks. Our NER system employed linguistic features, multi-tasking via auxiliary objectives and hybrid loss including ranking loss to extract flat and nested entities in English and Spanish text. Our RE system employed SVM with linguistic features. Compared to other participating systems, our submissions are ranked 1st in BB-norm+NER and SeeDev task. Our system demonstrates competitive performance on PharmaCoNER with F1-score of 0.8662.

In future, we would like to explore improved modelling strategies for nested NER and discontinuous entities extraction. Further, in this work we only addressed intra-sentence RE, we would be interested to explore approaches for inter-sentence RE (Peng et al., 2017; Gupta et al., 2019b). Moreover, we would like to investigate interpretability of LSTMs for NER and RE (Gupta and Schütze, 2018).

Acknowledgment

This research was supported by Bundeswirtschaftsministerium (bmwi.de), grant 01MD19003E (PLASS, plass.io) at Siemens AG - CT Machine Intelligence, Munich Germany.

References

Gustavo Aguilar, Suraj Maharjan, Adrián Pastor López-Monroy, and Thamar Solorio. 2017. A multitask approach for named entity recognition in social media data. In *Proceedings of the 3rd Workshop on Noisy User-generated Text, NUT@EMNLP 2017, Copenhagen, Denmark, September 7, 2017*, pages 148–153.

Piotr Bojanowski, Edouard Grave, Armand Joulin, and Tomas Mikolov. 2017. Enriching word vectors with subword information. *TACL*, 5:135–146.

Robert Bossy, Louise Deléger, Estelle Chaix, Mouhamadou Ba, and Claire Nedellec. 2019. Bacteria Biotope at BioNLP Open Shared Tasks 2019. In *Proceedings of the BioNLP Open Shared Tasks 2019 Workshop*.

Robert Bossy, Wiktoria Golik, Zorana Ratkovic, Dialekti Valsamou, Philippe Bessières, and Claire Nedellec. 2015. Overview of the gene regulation network and the bacteria biotope tasks in bionlp'13 shared task. In *BMC Bioinformatics*.

Leo Breiman. 1996. Bagging predictors. *Machine Learning*, 24(2):123–140.

Estelle Chaix, Bertrand Dubreucq, Abdelhak Fatihi, Dialekti Valsamou, Robert Bossy, Mouhamadou Ba, Louise Deléger, Pierre Zweigenbaum, Philippe Bessières, Loïc Lepiniec, and Claire Nedellec. 2016. Overview of the regulatory network of plant seed development (seedev) task at the bionlp shared task 2016. In *Proceedings of the 4th BioNLP Shared Task Workshop, BioNLP 2016, Berlin, Germany, August 13, 2016*, pages 1–11.

Ronan Collobert and Jason Weston. 2008. A unified architecture for natural language processing: deep neural networks with multitask learning. In *Machine Learning, Proceedings of the Twenty-Fifth International Conference (ICML 2008), Helsinki, Finland, June 5-9, 2008*, pages 160–167.

Saul Goldenberg, Regina Célia Figueiredo Castro, and Fernando Redondo Moreira Azevedo. 2007. [scielo (scientific electronic library online) statistical data interpretation]. *Acta cirurgica brasileira*, 22 1:1–7.

Aitor Gonzalez-Agirre, Montserrat Marimon, Ander Intxaurrondo, Obdulia Rabal, Marta Villegas, and Martin Krallinger. 2019. Pharmaconer: Pharmacological substances, compounds and proteins named entity recognition track. In *Proceedings of the BioNLP Open Shared Tasks (BioNLP-OST)*, Hong Kong, China. Association for Computational Linguistics.

Pankaj Gupta, Yatin Chaudhary, Florian Buettner, and Hinrich Schütze. 2019a. Document informed neural autoregressive topic models with distributional prior. In *The Thirty-Third AAAI Conference on Artificial Intelligence, AAAI 2019, The Thirty-First Innovative Applications of Artificial Intelligence Conference, IAAI 2019, The Ninth AAAI Symposium on Educational Advances in Artificial Intelligence, EAAI 2019, Honolulu, Hawaii, USA, January 27 - February 1, 2019.*, pages 6505–6512.

Pankaj Gupta, Subburam Rajaram, Hinrich Schütze, and Thomas A. Runkler. 2019b. Neural relation extraction within and across sentence boundaries. In *The Thirty-Third AAAI Conference on Artificial Intelligence, AAAI 2019, The Thirty-First Innovative Applications of Artificial Intelligence Conference, IAAI 2019, The Ninth AAAI Symposium on Educational Advances in Artificial Intelligence, EAAI 2019, Honolulu, Hawaii, USA, January 27 - February 1, 2019.*, pages 6513–6520.

Pankaj Gupta, Benjamin Roth, and Hinrich Schütze. 2018. Joint bootstrapping machines for high confidence relation extraction. In *Proceedings of the 2018 Conference of the North American Chapter of the Association for Computational Linguistics: Human Language Technologies, NAACL-HLT 2018, New Orleans, Louisiana, USA, June 1-6, 2018, Volume 1 (Long Papers)*, pages 26–36.

Pankaj Gupta and Hinrich Schütze. 2018. LISA: explaining recurrent neural network judgments via layer-wise semantic accumulation and example to pattern transformation. In *Proceedings of the Workshop: Analyzing and Interpreting Neural Networks for NLP, BlackboxNLP@EMNLP 2018, Brussels, Belgium, November 1, 2018*, pages 154–164.

Pankaj Gupta, Hinrich Schütze, and Bernt Andrassy. 2016. Table filling multi-task recurrent neural network for joint entity and relation extraction. In *COLING 2016, 26th International Conference on Computational Linguistics, Proceedings of the Conference: Technical Papers, December 11-16, 2016, Osaka, Japan*, pages 2537–2547.

Nanda Kambhatla. 2004. Combining lexical, syntactic, and semantic features with maximum entropy models for extracting relations. In *Proceedings of the ACL 2004 on Interactive Poster and Demonstration Sessions*, ACLdemo '04, Stroudsburg, PA, USA. Association for Computational Linguistics.

John D. Lafferty, Andrew McCallum, and Fernando C. N. Pereira. 2001. Conditional random fields: Probabilistic models for segmenting and labeling sequence data. In *Proceedings of the Eighteenth International Conference on Machine Learning (ICML 2001), Williams College, Williamstown, MA, USA, June 28 - July 1, 2001*, pages 282–289.

Guillaume Lample, Miguel Ballesteros, Sandeep Subramanian, Kazuya Kawakami, and Chris Dyer. 2016. Neural architectures for named entity recognition. In *NAACL HLT 2016, The 2016 Conference of the North American Chapter of the Association for Computational Linguistics: Human Language Technologies, San Diego California, USA, June 12-17, 2016*, pages 260–270.

Jake Lever and Steven J. Jones. 2016. VERSE: event and relation extraction in the bionlp 2016 shared task. In *Proceedings of the 4th BioNLP Shared Task Workshop, BioNLP 2016, Berlin, Germany, August 13, 2016*, pages 42–49.

Chen Li, Zhiqiang Rao, and Xiangrong Zhang. 2016. Litway, discriminative extraction for different bioevents. In *Proceedings of the 4th BioNLP Shared Task Workshop, BioNLP 2016, Berlin, Germany, August 13, 2016*, pages 32–41.

Zhiyong Lu. 2011. Pubmed and beyond: a survey of web tools for searching biomedical literature. *Database*, 2011.

Montserrat Marimon, Jorge Vivaldi, and Núria Bel. 2017. Annotation of negation in the IULA Spanish clinical record corpus. In *Proceedings of the Workshop Computational Semantics Beyond Events and Roles*, pages 43–52, Valencia, Spain. Association for Computational Linguistics.

Ryan T. McDonald, Fernando C. N. Pereira, Seth Kulick, R. Scott Winters, Yang Jin, and Peter S. White. 2005. Simple algorithms for complex relation extraction with applications to biomedical IE. In *ACL 2005, 43rd Annual Meeting of the Association for Computational Linguistics, Proceedings of the Conference, 25-30 June 2005, University of Michigan, USA*, pages 491–498.

Naomi Miller, Eve-Marie Lacroix, and Joyce E. B. Backus. 2000. Medlineplus: building and maintaining the national library of medicine's consumer health web service. *Bulletin of the Medical Library Association*, 88 1:11–7.

Nanyun Peng, Hoifung Poon, Chris Quirk, Kristina Toutanova, and Wen-tau Yih. 2017. Cross-sentence n-ary relation extraction with graph lstms. *TACL*, 5:101–115.

Marek Rei. 2017. Semi-supervised multitask learning for sequence labeling. In *Proceedings of the 55th Annual Meeting of the Association for Computational Linguistics, ACL 2017, Vancouver, Canada, July 30 - August 4, Volume 1: Long Papers*, pages 2121–2130.

Joaquín Alberto Carballido Rodríguez. 2002. [the spanish bibliographic index of the health sciences (ibecs) and actas urológicas españolas]. *Actas urologicas espanolas*, 26 6:381–3.

Cícero Nogueira dos Santos, Bing Xiang, and Bowen Zhou. 2015. Classifying relations by ranking with convolutional neural networks. In *Proceedings of the 53rd Annual Meeting of the Association for Computational Linguistics and the 7th International Joint Conference on Natural Language Processing of the Asian Federation of Natural Language Processing, ACL 2015, July 26-31, 2015, Beijing, China, Volume 1: Long Papers*, pages 626–634.

Burr Settles. 2004. Biomedical named entity recognition using conditional random fields and rich feature sets. In *Proceedings of the International Joint Workshop on Natural Language Processing in Biomedicine and its Applications, NLPBA/BioNLP 2004, Geneva, Switzerland, August 28-29, 2004*.

Felipe Soares and Aitor gonzalez agirre. 2019. Spaccc-pos-tagger. Funded by the Plan de Impulso de las Tecnologas del Lenguaje (Plan TL).

Richard Tzong-Han Tsai, Shih-Hung Wu, Wen-Chi Chou, Yu-Chun Lin, Ding He, Jieh Hsiang, Ting-Yi Sung, and Wen-Lian Hsu. 2006. Various criteria in the evaluation of biomedical named entity recognition. *BMC Bioinformatics*, 7:92.

Ngoc Thang Vu, Pankaj Gupta, Heike Adel, and Hinrich Schütze. 2016. Bi-directional recurrent neural network with ranking loss for spoken language understanding. In *2016 IEEE International Conference on Acoustics, Speech and Signal Processing, ICASSP 2016, Shanghai, China, March 20-25, 2016*, pages 6060–6064.

Linlin Wang, Zhu Cao, Gerard de Melo, and Zhiyuan Liu. 2016. Relation classification via multi-level attention cnns. In *Proceedings of the 54th Annual Meeting of the Association for Computational Linguistics, ACL 2016, August 7-12, 2016, Berlin, Germany, Volume 1: Long Papers*.

Shanchan Wu and Yifan He. 2019. Enriching pre-trained language model with entity information for relation classification. *CoRR*, abs/1905.08284.

Yijia Zhang, Qingyu Chen, Zhihao Yang, Hongfei Lin, and Zhiyong Lu. 2019. Biowordvec, improving biomedical word embeddings with subword information and mesh. In *Scientific Data*.

An ensemble CNN method for biomedical entity normalization

Pan Deng, Haipeng Chen, Mengyao Huang, Xiaowen Ruan, Liang Xu

Ping An Technology (Shenzhen) Co., Ltd,

Shenzhen, People's Republic of China

{dengpan749, chenhaipeng867, huangmengyao666, ruanxiaowen571,

xuliang867}@pingan.com.cn

Abstract

Different representations of the same concept could often be seen in scientific reports and publications. Entity normalization (or entity linking) is the task to match the different representations to their standard concepts. In this paper, we present a two-step ensemble CNN method that normalizes microbiology-related entities in free text to concepts in standard dictionaries [1]. The method is capable of linking entities when only a small microbiology-related biomedical corpus is available for training, and achieved reasonable performance in the online test of the BioNLP-OST19 shared task Bacteria Biotope.

1 Introduction

With over 500K papers in the biomedical field published on average every year[2], it is important to promote efficient information retrieval and knowledge processing from the literatures automatically. Named entity recognition (NER), which extracts meaningful real-world objects from free text, and entity normalization (entity linking), which links ambiguous or varied extracted objects to standard concepts, are two fundamental natural language processing (NLP) tasks to approach the goal.

With many attempts made for general entity normalization (Hachey, Radford et al. 2013, Luo, Huang et al. 2015, Wu, He et al. 2018, Aguilar, Maharjan et al. 2019), biomedical entity linking faces more challenges handling entity variations, making it an enthralling field to be explored. Many studies endeavored to solve biomedical entity normalization issues have been published (Hanisch, Fundel et al. 2005, Leaman and Lu 2016, Cho,

Choi et al. 2017, Li, Chen et al. 2017, Luo, Song et al. 2018, Ji, Wei et al. 2019). Meanwhile, BioNLP Shared Tasks, one of the community-wide challenges that aim to find solutions for biomedical literature information retrieval, also addresses diverse tasks of entity linking (Bossy, Jourde et al. 2011, Bossy, Golik et al. 2013, Nédellec, Bossy et al. 2013, Chaix, Dubreucq et al. 2016, Deléger, Bossy et al. 2016). However, further investigations are required to improve the performance of the entity linking systems, especially when the available corpus is small.

Here, we present a two-step neural network-based ensemble method that links free text pre-annotated microbiology-related entities to standard concepts using semantic information from pre-trained word vectors. By integrating a perfect match method with a shallow CNN, our model's performance is comparable to the SOTA methods' performance when trained with a small biomedical corpus (2258 microbiology-related entities, or 1248 after de-duplication, from 198 microbiology related publications and reports) provided by the BioNLP-19 task Bacteria Biotope challenge.[3]

We have compared our ensemble model to both a baseline method, of which we linked free text entities to the standard concepts by vector distance (Manning, Raghavan et al. 2010), and ABCNN, one of the SOTA models that could be used for entity normalization (Yin, Schütze et al. 2016). In addition, the method was tested online, and the results indicated that our model achieved a reasonable performance for microbiology-related entities linking tasks with small corpora.

2 Related work

Entity normalization is a rich research field where diverse approaches have been proposed.

---

[1] Our code is available at:
https://github.com/OXPHOS/BioNLP

[2] http://dan.corlan.net/medline-trend/language/absolute.html
[3] https://sites.google.com/view/bb-2019/home

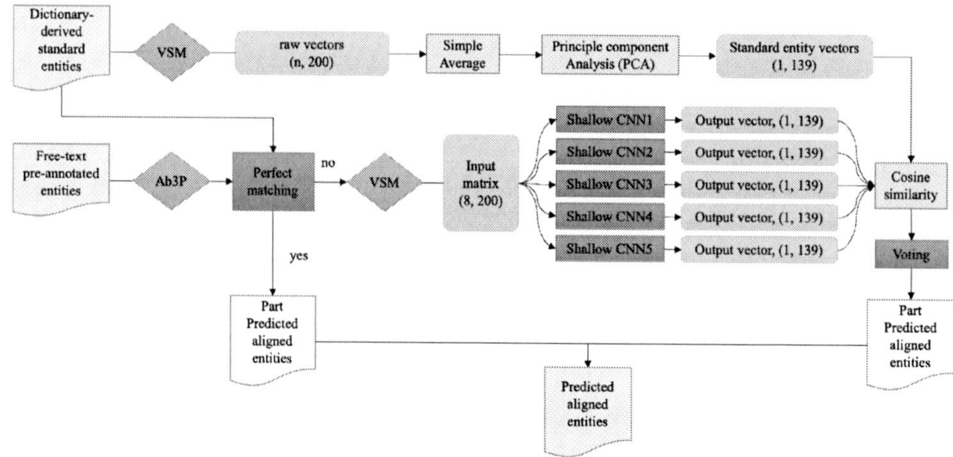

Figure 1 Model Architecture Overview. VSM: pre-trained word vector space model. Ab3P: an abbreviation detection tool developed specifically for biomedical concepts. CNN: convolutional neural network.

Most early studies utilized morphological similarity defined by editing distances between input terms and standard concepts to normalize the entities (Ristad and Yianilos 1998, Aronson 2001). Later, heuristic rules were incorporated to improve the performance (Hanisch, Fundel et al. 2005, Kang, Singh et al. 2012, Karadeniz and Özgür 2013, Tiftikci, Şahin et al. 2016, Cho, Choi et al. 2017). Among them, MaCallum et al. (2012), DNorm (Leaman, Islamaj Doğan et al. 2013) and LIMSI (Grouin and Moriceau 2016) introduced conditional random field (CRF) to the entity linking tasks, while TaggerOne (Leaman and Lu 2016) presented a NER and normalization joint system utilizing semi-markov models, and it has been adopted by an integrated bioconcept annotation and retrieval platform developed by NIH (Wei, Allot et al. 2019). However, many of the studies achieved good performance yet were limited for further improvements due to the common drawbacks of rule-based methods.

Approaches utilizing semantic information of the entities was made possible by the appearance of the word embedding technique. Word embedding projects words to vector spaces, where the cosine similarities between the vectors indicate their semantic similarities. The CONTES system (Ferré, Zweigenbaum et al. 2017) and the following HONOR system (Ferré, Deléger et al. 2018) performed entity linking tasks by minimizing the distances between embedded input terms and standard biomedical concepts. Karadeniz and Özgür (2019) proposed an unsupervised method

for entity linking tasks using word embeddings and a syntactic parser.

Meanwhile, neural networks have been combined with word embeddings to normalize biomedical entities. Limsopatham and Collier (2016) applied convolutional neural network (CNN) and recurrent neural network (RNN) to pre-trained word embeddings to normalize medical concepts in social media texts, and achieved the SOTA performance on several datasets. Li et al. (2017) utilized a CNN structure to rank the candidates generated by rule-based methods. Deep neural networks such as multi-view CNN and BERT have also been proposed to normalize biomedical entities (Luo, Song et al. 2018, Ji, Wei et al. 2019). However, their applications might be limited due to the requirement of large amount of data.

3 Models

Our model architecture is shown in **Figure 1**, where our major work is highlighted in blue and further discussed in Section 3.1-3.3.

To process the entities from the standard dictionary, let Y_i^d be the i-th entity from the dictionary, and $y_{ij} \in \mathbb{R}^k$ be the k-dimensional word vector of the j-th word in the entity Y_i^d. the embedded vector $\mathbf{y}_i$ of entity Y_i^d is defined as

$$\mathbf{y}_i = \text{PCA}(\frac{1}{n_i}\sum_{j=1}^{n_i} y_{ij}, n_{\text{component}} = 0.95),$$

where $n_i \in \mathbb{N}^+$ is the number of words present in a pre-trained vector space model (VSM) in the entity Y_i^d. The VSM was created from the biomedical scientific literature in the PubMed database and Wikipedia (Pyysalo 2013). $k = 200$. PCA was conducted to increase training efficiency, where $n_{\text{component}}$ is the reduction rate.

The processing of the entities from free text is described in the following section in detail. To be noted, abbreviations are commonly seen in free text from publications and reports. For example, CNS, standing for the central nervous system, might often be used in research literature over topics of neuroscience, and would be annotated as entities to be linked. However, abbreviations, mostly derived from phrases, are often absent from the pre-trained word vector spaces and would interfere with the model training. To solve this problem, we first converted potential abbreviations in free text pre-annotated entity list to their long forms with Ab3P. Ab3P is an abbreviation detection tool developed specifically for biomedical concepts. It reached 96.5% precision and 83.2% recall on 1250 randomly selected MEDLINE records as suggested by Sohn et al (2008).

The converted free text pre-annotated entities were then matched with dictionary-derived standard concepts by characters through a perfect match module (Section 3.1). The entities failed perfect matching were then fed to a set of shallow CNN models (Section 3.2) trained with bootstrap samples. Next, the outputs of the CNNs were mapped to standard entity vectors via cosine similarity. The standard entity vectors output from the voting classifier (Section 3.3) were predicted as the linked results of the input entities.

3.1 Perfect match

We noticed that some entities from the free text were able to match with the standard entities by characters after rule-based processing. These entities were then directly linked to the dictionary instead of being fed to the Word2Vector and CNN models. The rules we designed include:

- Hyphens were replaced with spaces.
- Characters except alphabetic letters and spaces were removed.
- Case-insensitive string matching was performed between the free text entities and standard entities.

3.2 Shallow CNN

The shallow CNN (**Figure 1**) was adapted from the previous ideas from Kim (2014) and Limsopatham and Collier (2016).

To start with, let X_i^t be the i-th input entity (which were provided by the task), and $x_{ij} \in \mathbb{R}^k$ be the k-dimensional word vector of the j-th word in the entity X_i^t, $k = 200$. The embedded matrix $\mathbf{x}_i$ of entity X_i^t is defined as

$$\mathbf{x}_i = x_{i1} \oplus x_{i2} \oplus \dots x_{in_i}.$$

Here $n_i \in \mathbb{N}^+$ is the number of words present in the pre-trained VSM (Pyysalo 2013) in the entity X_i^t. $\oplus$ is the concatenation operator. $\mathbf{x}_i$ is padded to length 8 as 98.8% of the input entities were composed of 8 or fewer words. For the entities with more than 8 words, average pooling was performed in prior with the pool size $= (w, 200)$ and step$= w$, where $w = \lceil n/8 \rceil$. In other words, simple average of the neighboring w words was calculated, so that the final embedded matrix would always have a length ≤ 8.

A temporal convolution kernel followed by a max-over-time pooling operation and a fully connected layer were applied to each $\mathbf{x}_i$. The output $\bar{\mathbf{y}}_i$ was then passed to a cosine similarity function to calculate the similarity scores between $\bar{\mathbf{y}}_i$ and each standard entity vector respectively. The standard concept with the highest score was predicted as the linked entity $\bar{\mathbf{y}}_i^e$.

3.3 Ensemble mechanism with voting

To reduce overfitting, we designed an ensemble method that combined 5 shallow CNNs with the concept of boosting (Valiant 1984). The 5 CNNs shared the identical architecture, but their weights were randomly initialized respectively. To increase the generalization capability of our model, the CNNs were fed with training data randomly subsampled with bootstrap method (Efron 1982), with the out-of-bag samples used for cross-validation.

The final normalized results were achieved with a majority-vote classifier over the outputs from the 5 shallow CNNs. If no majority output was present, the output from the network with the best cross-validation estimates would be chosen.

3.4 Baseline model

For each entity E_i^d in the standard dictionary and E_i^t in free text, the corresponding embedded

		Total Number			Number after de-duplication		
	Article Number	Habitat	Phenotype	Habitat+ Phenotype	Habitat	Phenotype	Habitat+ Phenotype
Training	133	1118	369	1487	627	176	803
Development	66	610	161	771	348	97	445
Test	97	924	252	1176	596	148	744

Table 1. Data Statistics

vector $\mathbf{v}_i^d$ and $\mathbf{v}_i^t$ are defined, respectively, as:

$$\mathbf{v}_i^d = \frac{1}{n_i}\sum_{j=1}^{n_i} v_{ij}^d, \; \mathbf{v}_i^t = \frac{1}{n_i}\sum_{j=1}^{n_i} v_{ij}^t,$$

where $n_i \in \mathbb{N}^+$ is the number of words present in the pre-trained VSM in the i-th entity.

Cosine similarity between each free text-dictionary entity pair was calculated. The free text entity was linked to the dictionary entity with the highest similarity score.

3.5 ABCNN

ABCNN (Yin, Schütze et al. 2016) is a state-of-the-art deep learning model for text similarity learning, which could also be applied for entity linking tasks. The model introduced attention mechanism into a pair of siamese architecture-based weight sharing CNNs (Bromley, Guyon et al. 1994).

For our purpose, we used a slight variant of a published ABCNN model[4]. In addition, attention mechanism could be applied to different layers of the CNN pair according to the original publication. Considering the data volume and the model complexity, we applied the attention mechanism to the input layer.

4 Experiments and Results

4.1 Data and resources

The biomedical corpus and pre-annotated entities were provided by BioNLP-OST19 task Bacteria Biotope. **Table 1** shows the detailed data statistics provided by the task. Two types of entities were involved in the task: *phenotype*, which describes microbial characteristics, and *habitat*, which describes physical places where microorganisms could be observed. Dictionary with 3602 standard concepts was also provided by the task. In the original dictionary, each concept is assigned to a unique ID, while its hierarchical information of its

direct parents is also listed. In our model, the hierarchical information is omitted.

Ab3P-detected abbreviations were provided as separate input files by the task organizers[5].

The 4 GB word vector space model was downloaded in binary format[6] and extracted with python package `gensim`.

4.2 Training

Our CNN model was trained using stochastic gradient descent optimizer with cosine proximity as the loss function. We randomly split 20% samples as validation dataset for each CNN and used early stopping criteria to determine the number of training epochs. The learning rate was fixed to 0.01. Batch size (2), kernel size (4) and filter number (5000) were determined by grid-search.

As expected with this small volume of data, extra convolution layers led to overfitting.

4.3 Held-out evaluation

We used precision metrics, the official metric of the challenge, to evaluate the performance of our model and the reference models on the held-out development dataset respectively.

	Habitat	Phenotype	Total
Ensemble CNN	0.624	0.615	0.622
Perfect match	0.863	0.937	0.869
Shallow CNN	0.526	0.572	0.538
ABCNN	0.244	0.134	0.221
baseline	0.207	0.101	0.184

Table 2. Performance of different models on development dataset (Precision)

As shown in **Table 2**, non-supervised baseline model yielded a precision score of 0.184 on the development dataset, while ABCNN yielded 0.221, which might be attributed to the small

---

training corpus. Our ensemble CNN model performed around 3 times better than both reference models with an average precision score of 0.622, indicating the efficiency of our model.

However, it should be noted that the perfect match module in our system had a remarkable higher precision score compared to the shallow CNN module, suggesting that the performance of the neural network could be further improved.

We analyzed the result of the shallow CNN module and concluded that 3 possible reasons might be associated with the performance: 1) Missing context. For example, our model normalized "children" to "child", while the provided label was "patient with infectious disease" in articles describing children with infectious disease. 2) Missing hierarchical information. For example, our model normalized "B cell" to "cell" instead of "lymphocyte", and the latter one was a more accurate description. Tackling the above two issues would require either the context or the hierarchical information of the standard concepts to be considered in the system. 3) Wrong match. For example, "cats" was normalized to "dog", suggesting that the networks were not trained well to normalize these words. However, we noticed that such errors mostly came with a majority vote of 2 or 1, which on the other hand demonstrated the power of the voting mechanism.

	Habitat	Phenotype
PADIA_BacReader	**0.684**	**0.758**
Challenge-provided baseline	0.559	0.581
AmritaCen healthcare	0.522	0.646
BLAIR_GMU[#]	0.615	0.646
BOUN-ISIK[#]	0.687	0.566

Table 3. Performance of different models on online test dataset (Precision). The performance of our model (PADIA_BacReader) is bolded. [#]: Best run of the submissions is considered.

4.4 Online test

The ensemble CNN model was then evaluated through online testing[7].

Our results showed a 12.5% and 17.7% precision increase in habitat and phenotype entity linking tasks respectively compared to the challenged-provided baseline model **(Table 3)**, where case-insensitive string matching was applied for

linking. In addition, it performed the best or among the best ones compared to models proposed by other participants, suggesting the advantages of our model. We did not test our own reference models online due to the limited number of submissions to the challenge.

5 Conclusions and Future direction

We introduced a two-step neural network-based ensemble method that linked microbiology-related biomedical entities extracted from free text to standard concepts. The shallow architecture and ensemble mechanism on top of a perfect-match morphological similarity method achieved reasonable predictions with limited training samples. The comparison with reference models suggested the efficiency of our model. In addition, our approach could be applied to other scenarios where semantic linking between entities is required as well.

Further improvement might be achieved once more semantic clues are incorporated, as we briefly discussed at the end of Section 4.3. The normalization deviation due to missing context clues did not only affect the performance of shallow CNN, but also affected the performance of perfect math as well. For example, though entity 'cell' has a perfect match in standard dictionary, it might be referred to 'lymphocytes' specifically in a research paper discussing about immunity. While some efforts have been made to preserve hierarchical information between concepts during entity linking (Ferré, Deléger et al. 2018), It would be interesting to investigate if knowledge graphs derived from the standard dictionaries and input corpus could contribute to the semantic-based entity normalization.

In addition, our model assigned the same weight to all the words present in the VSM, which might compromise the performance of the system. For example, only the word "children" is informative in the entity "children less than five years of age", as the entity is normalized to "child". The presence of other words might interfere with the normalization. Regarding this issue, syntactic parsers might be adopted for performance improvement.

---

[7] https://sites.google.com/view/bb-2019/prediction-submission

References

Aguilar, G., Maharjan, S., López-Monroy, A. P., & Solorio, T. (2019). A multi-task approach for named entity recognition in social media data. *arXiv preprint arXiv:1906.04135*.

Aronson, A. R. (2001). *Effective mapping of biomedical text to the UMLS Metathesaurus: the MetaMap program*. Paper presented at the Proceedings of the AMIA Symposium.

Bossy, R., Golik, W., Ratkovic, Z., Bessières, P., & Nédellec, C. (2013). *Bionlp shared task 2013–an overview of the bacteria biotope task*. Paper presented at the Proceedings of the BioNLP Shared Task 2013 Workshop.

Bossy, R., Jourde, J., Bessieres, P., Van De Guchte, M., & Nédellec, C. (2011). *BioNLP Shared Task 2011-Bacteria Biotope*. Paper presented at the Proceedings of BioNLP Shared Task 2011 Workshop.

Bromley, J., Guyon, I., LeCun, Y., Säckinger, E., & Shah, R. (1994). *Signature verification using a" siamese" time delay neural network*. Paper presented at the Advances in neural information processing systems.

Chaix, E., Dubreucq, B., Fatihi, A., Valsamou, D., Bossy, R., Ba, M., . . . Lepiniec, L. (2016). *Overview of the Regulatory Network of Plant Seed Development (SeeDev) Task at the BioNLP Shared Task 2016*. Paper presented at the Proceedings of the 4th bionlp shared task workshop.

Cho, H., Choi, W., & Lee, H. (2017). A method for named entity normalization in biomedical articles: application to diseases and plants. *BMC bioinformatics, 18*(1), 451.

Deléger, L., Bossy, R., Chaix, E., Ba, M., Ferre, A., Bessieres, P., & Nedellec, C. (2016). *Overview of the bacteria biotope task at bionlp shared task 2016*. Paper presented at the Proceedings of the 4th BioNLP shared task workshop.

Efron, B. (1982). The jackknife, the bootstrap and other resampling plans. *Journal of the American Statistical Association, 78*(384), 987. doi:10.1137/1.9781611970319

Ferré, A., Deléger, L., Zweigenbaum, P., & Nédellec, C. (2018). *Combining rule-based and embedding-based approaches to normalize textual entities with an ontology*. Paper presented at the Proceedings of the Eleventh International Conference on Language Resources and Evaluation (LREC-2018).

Ferré, A., Zweigenbaum, P., & Nédellec, C. (2017). *Representation of complex terms in a vector space structured by an ontology for a normalization task*. Paper presented at the BioNLP 2017.

Grouin, C., & Moriceau, V. (2016). *LIMSI at SemEval-2016 Task 12: machine-learning and temporal information to identify clinical events and time expressions*. Paper presented at the Proceedings of the 10th International Workshop on Semantic Evaluation (SemEval-2016).

Hachey, B., Radford, W., Nothman, J., Honnibal, M., & Curran, J. R. (2013). Evaluating entity linking with Wikipedia. *Artificial intelligence, 194*, 130-150.

Hanisch, D., Fundel, K., Mevissen, H.-T., Zimmer, R., & Fluck, J. (2005). ProMiner: rule-based protein and gene entity recognition. *BMC bioinformatics, 6*(1), S14.

Ji, Z., Wei, Q., & Xu, H. (2019). BERT-based Ranking for Biomedical Entity Normalization. *arXiv preprint arXiv:1908.03548*.

Kang, N., Singh, B., Afzal, Z., van Mulligen, E. M., & Kors, J. A. (2012). Using rule-based natural language processing to improve disease normalization in biomedical text. *Journal of the American Medical Informatics Association, 20*(5), 876-881.

Karadeniz, I., & Özgür, A. (2013). *Bacteria biotope detection, ontology-based normalization, and relation extraction using syntactic rules*. Paper presented at the Proceedings of the BioNLP Shared Task 2013 Workshop.

Karadeniz, I., & Özgür, A. (2019). Linking entities through an ontology using word embeddings and syntactic re-ranking. *BMC bioinformatics, 20*(1), 156.

Kim, Y. (2014). Convolutional neural networks for sentence classification. *arXiv preprint arXiv:1408.5882*.

Leaman, R., Islamaj Doğan, R., & Lu, Z. (2013). DNorm: disease name normalization with pairwise learning to rank. *Bioinformatics, 29*(22), 2909-2917.

Leaman, R., & Lu, Z. (2016). TaggerOne: joint named entity recognition and normalization with semi-Markov Models. *Bioinformatics, 32*(18), 2839-2846.

Li, H., Chen, Q., Tang, B., Wang, X., Xu, H., Wang, B., & Huang, D. (2017). CNN-based ranking for biomedical entity normalization. *BMC bioinformatics, 18*(11), 385.

Limsopatham, N., & Collier, N. (2016). *Normalising medical concepts in social media texts by learning semantic representation*. Paper presented at the Proceedings of the 54th Annual Meeting of the Association for Computational Linguistics (Volume 1: Long Papers).

Luo, G., Huang, X., Lin, C.-Y., & Nie, Z. (2015). *Joint entity recognition and disambiguation*. Paper presented at the Proceedings of the 2015 Conference on Empirical Methods in Natural Language Processing.

Luo, Y., Song, G., Li, P., & Qi, Z. (2018). *Multi-task medical concept normalization using multi-view convolutional neural network.* Paper presented at the Thirty-Second AAAI Conference on Artificial Intelligence.

Manning, C., Raghavan, P., & Schütze, H. (2010). Introduction to information retrieval. *Natural Language Engineering, 16*(1), 100-103.

McCallum, A., Bellare, K., & Pereira, F. (2012). A conditional random field for discriminatively-trained finite-state string edit distance. *arXiv preprint arXiv:1207.1406.*

Nédellec, C., Bossy, R., Kim, J.-D., Kim, J.-J., Ohta, T., Pyysalo, S., & Zweigenbaum, P. (2013). *Overview of BioNLP shared task 2013.* Paper presented at the Proceedings of the BioNLP Shared Task 2013 Workshop.

Pyysalo, S. G., Filip; Moen, Hans; Salakoski, Tapio; Ananiadou, Sophia. (2013). Distributional semantics resources for biomedical text processing. *Proceedings of LBM,* 39-44.

Ristad, E. S., & Yianilos, P. N. (1998). Learning string-edit distance. *IEEE Transactions on Pattern Analysis and Machine Intelligence, 20*(5), 522-532.

Sohn, S., Comeau, D. C., Kim, W., & Wilbur, W. J. (2008). Abbreviation definition identification based on automatic precision estimates. *BMC bioinformatics, 9*(1), 402.

Tiftikci, M., Şahin, H., Büyüköz, B., Yayıkçı, A., & Özgür, A. (2016). *Ontology-based categorization of bacteria and habitat entities using information retrieval techniques.* Paper presented at the Proceedings of the 4th BioNLP Shared Task Workshop.

Valiant, L. G. (1984). A theory of the learnable. *symposium on the theory of computing, 27*(11), 1134-1142. doi:10.1145/800057.808710

Wei, C. H., Allot, A., Leaman, R., & Lu, Z. (2019). PubTator central: automated concept annotation for biomedical full text articles. *Nucleic Acids Res, 47*(W1), W587-W593. doi:10.1093/nar/gkz389

Wu, G., He, Y., & Hu, X. (2018). Entity linking: an issue to extract corresponding entity with knowledge base. *IEEE Access, 6,* 6220-6231.

Yin, W., Schütze, H., Xiang, B., & Zhou, B. (2016). Abcnn: Attention-based convolutional neural network for modeling sentence pairs. *Transactions of the Association for Computational Linguistics, 4,* 259-272.

BOUN-ISIK Participation: An Unsupervised Approach for the Named Entity Normalization and Relation Extraction of Bacteria Biotopes

İlknur Karadeniz
Işık University
`ilknur.karadeniz`
`@isikun.edu.tr`

Ömer Faruk Tuna
Işık University
`218DCS8074`
`@isik.edu.tr`

Arzucan Özgür
Boğaziçi University
`arzucan.ozgur`
`@boun.edu.tr`

Abstract

This paper presents our participation at the Bacteria Biotope Task of the BioNLP Shared Task 2019. Our participation includes two systems for the two subtasks of the Bacteria Biotope Task: the normalization of entities (BB-norm) and the identification of the relations between the entities given a biomedical text (BB-rel). For the normalization of entities, we utilized word embeddings and syntactic re-ranking. For the relation extraction task, pre-defined rules are used. Although both approaches are unsupervised, in the sense that they do not need any labeled data, they achieved promising results. Especially, for the BB-norm task, the results have shown that the proposed method performs as good as deep learning based methods, which require labeled data.

1 Introduction

The amount of electronic resources in the biomedical domain and its rapid growth are major challenges for the scientists who make research in this domain. Text mining methods which aim to automatically extract useful information from the text of these electronic resources provide convenience to the researchers.

A number of shared tasks, including the BioNLP Shared Tasks, have been conducted with the goal of developing biomedical text mining methods. In 2011, the Bacteria Biotope Task has been conducted for the first time as a part of the BioNLP Shared Task targeting the extraction of useful information regarding bacteria and their habitats (Bossy et al., 2011). Since then, the participant teams of the following shared task series developed various solutions for the problem of bacteria biotopes (Bossy et al., 2015; Deleger et al., 2016).

The Bacteria Biotope Task of the BioNLP Shared Task 2019 (Bossy et al., 2019) is the final version of the tasks that have been conducted until now readdressing the problem of extraction of the information regarding the bacteria biotopes. This year's task has presented the opportunity to the participants to develop solutions for three sub-problems: normalization (BB-norm), relation extraction (BB-rel), and knowledge base extraction (BB-kb). For the BB-norm task of the Bacteria Biotope Task of the BioNLP Shared Task 2019, the participants are expected to develop systems to link the named entities (*Microorganism, Habitat*, and *Phenotype*) in a given text through a given ontology, when the entities are given with their boundaries. For instance, the sample sentence *"Atypical mycobacteria causing non-pulmonary disease in Queensland."* consists of the following mentions: *"mycobacteria"* microorganism mention, *"causing non-pulmonary disease"* phenotype mention, and *"pulmonary"* habitat mention, which should be normalized to the *"Mycobacteria"* term in the NCBI taxonomy, and *"human pathogen"* and *"lung"* terms in the Onto-Biotope ontology, respectively. For the BB-rel task of the Bacteria Biotopes Task of the BioNLP Shared Task 2019, the participants are required to extract the relations between the entities when the entities are given. There are two types of relations: *Lives_in* relation, which indicates a localization relation between a *Microorganism* entity and a *Habitat/Geographical* entity, and *Exhibits* relation, which indicates a property relation between a *Phenotype* entity and a *Microorganism* entity. For instance, the sample sentence above indicates two relations: a *Lives_in* relation between the *"Mycobacteria"* Microorganism entity and the *"Queensland"* Geographical entity, and an *Exhibits* relation between the *"Mycobacteria"* Microorganism entity and the *"causing non-pulmonary disease"* Phenotype entity.

We participated at the Bacteria Biotope Task in the BioNLP Shared Task 2019 with our system (named as the BOUN-ISIK system) and ob-

Proceedings of the 5th Workshop on BioNLP Open Shared Tasks, pages 150–157
Hong Kong, China, November 4, 2019. ©2019 Association for Computational Linguistics

tained promising results in the official evaluation. This paper presents our participating system for two sub-tasks: one for the BB-norm (Entity Normalization) sub-task and one for the BB-rel (Relation Extraction) sub-task. For the entity normalization sub-task, we utilized word embeddings and syntactic re-ranking to normalize the entities. On the other hand, for the relation extraction sub-task, we proposed a rule-based method. Although both systems are unsupervised, they achieved promising results. For the BB-norm sub-task, the official results of our system achieved state-of-the-art results on the BioNLP Shared Task 2019 Bacteria Biotope task test data set. The results have shown that our unsupervised approach, which does not require labeled data, performs as good as the deep learning based methods, which require labeled data.

1.1 Related Work

1.1.1 Named Entity Normalization

Among the previous series (2011, 2013, 2016) of the BioNLP Shared Task, the Bacteria Biotope Task in 2013 is the first shared task that addressed the problem of normalization of the entities in the bacteria biotopes domain. In 2013, the participant teams proposed rule-based methods and similarity-based methods. According to the official results of the Bacteria Biotope Task of 2013, for the habitat mention normalization, the best precision was obtained by the BOUN system, which utilized syntactic rules and shallow linguistic knowledge (Karadeniz and Ozgür, 2013; Karadeniz and Özgür, 2015).

In the following series of the Bacteria Biotopes task, the habitat mention normalization sub-task continued to attract the attention of the researchers. In the Bacteria Biotope task of the BioNLP Shared Task 2016, the best precision for the habitat normalization task was obtained by the BOUN system, which utilized both approximate string matching and cosine similarity of word-vectors weighted with Term Frequency-Inverse Document Frequency (TF-IDF) (Tiftikci et al., 2016).

After the Shared Tasks, the researchers continued to search for a solution for the problem of Bacteria Biotopes normalization (Ferré et al., 2017; Mehryary et al., 2017; Karadeniz and Özgür, 2019). Although promising results have been obtained by these approaches, the results showed that

there is still room for improvement for the normalization task of bacteria biotopes.

Besides the bacteria biotopes, there exist a significant amount of prior work on biomedical named entity normalization for different types of biomedical entities including genes/proteins (Morgan et al., 2008; Hakenberg et al., 2008; Wermter et al., 2009; Lu et al., 2011; Wei and Kao, 2011) and diseases (Leaman et al., 2013; Li et al., 2017). However, the need for manually annotated training data makes the adaptation of such methods to new entities difficult.

1.1.2 Relation Extraction

Several approaches, which consider the extraction of relations between various biomedical entities such as protein/protein (Giuliano et al., 2006; Airola et al., 2008; Choi, 2018), drug/drug (Segura-Bedmar et al., 2011; Kim et al., 2015), and gene/disease (Bravo et al., 2015) from biomedical text, have been presented in the literature. Relation extraction in the bacteria biotopes domain has also attracted considerable attention owing to the BioNLP Bacteria Biotope Shared Tasks.

Previous work in the bacteria biotopes domain consists of the extraction of relations between bacteria entities and habitat entities (Localization Relation Extraction) and of relations between two habitat entities (Part Of Relation Extraction). The participants of the BioNLP Shared Task 2011, which is the first shared task that addressed the relation extraction task of bacteria biotopes, utilized both machine learning and rule-based approaches for detecting the Localization and Part-of relations among bacteria and habitats (Bossy et al., 2011).

Sub-task 2 of the Bacteria Biotope (BB) Task in the BioNLP Shared Task 2013 also gave another opportunity to scientists to address the task of extracting the Localization and Part Of relations in the bacteria biotopes domain. For this sub-task, the best F-score (42%) was obtained by the TEES 2.1 system (Björne and Salakoski, 2013), which used support vector machine classification. After the shared task, a new sentence-level co-occurrence approach with an anaphora resolution component in order to handle relations that span multiple sentences has been developed in (Karadeniz and Özgür, 2015), which resulted in an improved F-score performance of 53% on Sub-task 2.

In the BioNLP Shared Task 2016, the VERSE

team (Lever and Jones, 2016) achieved the best F-score, which is 56%, on the relation extraction sub-task of Bacteria Biotopes by utilizing support vector machines.

2   Data Set

The data set, which was created by collecting titles and abstracts related to microorganisms from PubMed and extracts from full-text articles related to microorganisms living in food products, is provided by the BioNLP Shared Task 2019 BB Task organizers to the participants. The data set, consisting of 132 training, 67 development, and 97 test documents, was annotated by the bioinformaticians of the Bibliome team of MIG Laboratory at the Institut National de Recherche Agronomique (INRA).

For the training and development phases of BB-norm, document texts with manually annotated named entities and the concepts assigned to them through the OntoBiotope ontology (INRA, 2013) and NCBI taxonomy (NCBI, 2018) were provided, while in the test phase, only the entity boundaries and the entity types were given by the task organizers.

For the training and development phases of BB-rel, document texts with manually annotated *Microorganism*, *Habitat*, *Phenotype* and *Geographical* entities, as well as the *Lives_in* and *Exhibits* relations were provided, while in the test phase, document texts annotated only for *Microorganism*, *Habitat*, *Phenotype* and *Geographical* entities were given.

Since our system for the named entity normalization and relation extraction of bacteria biotopes is based on unsupervised approaches and does not require any labeled training data, the errors of the developed system are analyzed on the provided training and the development sets. The test set is used for the evaluation of the performance of the system.

3   Named Entity Normalization

In this section of the paper, the utilized methods for the BB-norm task are explained in detail. The BB-norm task includes the normalization of *Habitat* entities and *Phenotype* entities in a given set of documents through the Onto-Biotope ontology and the normalization of *Microorganism* entities through the NCBI Taxonomy.

The methods developed for the normalization of the named entities can be categorized into two according to the type of the entities: Habitat and Phenotype Normalization and Microorganism Normalization.

3.1   Habitat and Phenotype Entities

For the normalization of semantically meaningful entities such as Habitat and Phenotype entities, a two-step approach that we have previously proposed in (Karadeniz and Özgür, 2019) is adapted to this new data set. According to this approach, for the normalization of an entity mention, the top k semantically most similar ontology concepts are found at the first step using the word embedding representations of the entity mention and the ontology concepts. At the second step, these top k semantically most similar concepts are re-ranked according to a similarity metric that utilizes the constituency parses of the entity mention and ontology concept phrases. The resulting most similar ontology concept is assigned as the normalized concept for the corresponding mention. The details of this approach are explained in the following subsections.

3.1.1   Named Entity and Ontology Concept Representations

In the pre-processing step, the named entity mentions and the ontology concept names are tokenized, and the stop-words are removed from the mentions and the ontology concept names.

The intuition behind the adapted method is that semantically similar words have similar word vectors. Following this intuition, the semantic similarity between named entity mentions and ontology concept terms would be higher for the similar pairs, and lower for the dissimilar pairs, if the words can be converted into a machine processable format such as real-valued vectors.

After pre-processing, to convert each word into a real-valued vector, we utilized a pre-trained word embedding model (Chiu et al., 2016), which has been trained on PubMed by using the Word2Vec tool (Mikolov et al., 2013). The corresponding word vectors are obtained for each word by using this previously trained model. For the multi-word named entity mentions and ontology concept terms, the vector representations are obtained by averaging the real-valued vectors of their composing words.

3.1.2 Semantic Filtering

After the vector representations are obtained for each entity mention and for each ontology concept term, the semantic similarity between each pair is computed by using the cosine similarity. For each entity mention, the top k most similar ontology concepts are retained as candidates for further processing, i.e., for syntactic weighting based re-ranking. k is chosen as 5 based on the results obtained in our previous study (Karadeniz and Özgür, 2019).

3.1.3 Syntactic Re-ranking

For our re-ranking approach, the assumption is that the entity mentions are noun phrases and the most informative words in the mentions are the heads of the noun phrases. We used the Stanford Parser (version 3.8.0) (Klein and Manning, 2003) to obtain the corresponding head words of the entity mentions by providing the entity mentions as input and extracting the syntactic parses of the mentions as output. Next, the top level rightmost "noun" is searched in the tree structured syntactic parse and assigned as the head of the mention phrase.

The semantic similarities are recomputed using the mathematical formulation shown in Equation (1), which considers also the similarity between the head words of the entity mention and ontology concept pair. In Equation (1), $S_{RR}\ (m,\ c)$ is the final computed similarity between mention m and the candidate concept c, and S_S is the semantic similarity, in which m_{head} is the head word of the mention m and c_{head} is the head word of the concept c, $S_S\ (m,\ c)$ is the similarity between mention m and concept c computed as described in Section 3.1.1 , and w is a weighting parameter which can take values between 0 and 1. w is chosen as 0.25 based on the results reported in our previous study (Karadeniz and Özgür, 2019).

$$S_{RR}\ (m,\ c) = (w * S_S(m_{head},\ c_{head})) + ((1\text{-}w) * S_S(m,\ c)) \tag{1}$$

3.2 Microorganism Entities

The normalization of *Microorganism* entities component of our system is based on exact matching against the names and synonyms of the concepts in the NCBI taxonomy. Error analysis on the training and developments data sets revealed that applying some rules may improve the results. For instance, *"Escherichia coli"* has an exact match that can be successfully normalized to the referent concept with an ID *"562"* in the NCBI taxonomy. In the following parts of the document, although the *"E. coli"* mention indicates a clear reference to the same concept, it can not be normalized to the *"Escherichia coli"* concept with an exact matching approach. In this kind of cases, if an exact match does not exist, the previously mentioned similar entities in the text are searched. If a match is found, the same concept is assigned as the normalized concept for the corresponding mention *"E. coli"*. If there does not exist a match with the previously normalized concepts, the root concept with an ID *"2"* is assigned.

4 Relation Extraction

4.1 Localization Relation Extraction

Our system for the relation extraction sub-task is based on the naive assumption that the related entities for most of the relations appear within the same sentence. Therefore, firstly, the input texts are split into sentences using the NLTK library. For the extraction of *Lives_in* relations, all the sentences in the related document are searched to determine whether there exists a *Microorganism* entity and a *Habitat* entity or a *Microorganism* entity and a *Geographical* entity in the corresponding sentence. If there exists such a pair, this will be a sign of a *Lives_in* relation.

For any given sentence, there can be more than one *Habitat* entity and *Microorganism* entity. For this kind of sentences, two different approaches, which are called **smart matching** and **distributed matching**, are applied. In smart matching, each *Habitat* entity is paired with the closest *Microorganism* entity. In other words, the locations of each type of entities in the sentences are checked, and then the pairing process of the *Microorganism* and the *Habitat* entities are done based on the proximity criteria. In distributed matching, on the other hand, each *Habitat* entity is paired with every *Microorganism* entity in the sentence. Distributed matching can be seen as a type of N x N matching, while smart matching 1 x 1 matching. The performance of each approach is tested on the development data set. While there is slight increase in the precision, the recall is observed to decrease considerably for the smart matching method (see Table 1). As a result, the distributed matching approach is used in the final submission.

Table 1: Distributed vs Smart Matching for relation extraction. Precision, Recall, F-measure values for the development data set are reported.

	Distributed Matching	Smart Matching
Precision	0.491	0.576
Recall	0.785	0.515
F-measure	0.604	0.544

For the overlapping entities in which one entity contains another, some relations can be ignored. For instance, for the sample sentence *"An example of this fact is the presence of Psychrobacter DNA on the surface of Formaggio di Fossa cheeses"*, the *Habitat* entity *"surface of Formaggio di Fossa cheeses"*, *Habitat* entity *"Formaggio di Fossa cheeses"*, and *Habitat* entity *"cheeses"* are overlapping entities. In this case, it would not be appropriate to build three relations such as *"Psychrobacter"* - *"surface of Formaggio di Fossa cheeses"*, *"Psychrobacter"* - *"Formaggio di Fossa cheeses"*, and *"Psychrobacter"* - *"cheeses"*. Instead of extracting multiple relations, *"cheeses"* can be ignored and two relations between *"Psychrobacter"* - *"surface of Formaggio di Fossa cheeses"* and *"Psychrobacter"* - *"Formaggio di Fossa cheeses"* are extracted. This strategy, where the shortest overlapping entity is ignored, is called as the **soft filter** operation. On the other hand, the strategy when only the longest overlapping entity is retained and the remaining ones are ignored, is named as the **hard filter** operation. In hard filtering, *"Psychrobacter"* - *"Formaggio di Fossa cheeses"* and *"Psychrobacter"* - *"cheeses"* are ignored and only one relation between *"Psychrobacter"* - *"surface of Formaggio di Fossa cheeses"* is extracted. The performance of each approach is tested on the development data set (see Table 2).

Table 2: Soft Filter vs Hard Filter for relation extraction. Precision, Recall, F-measure values for the development data set are reported.

	Soft Filter	Hard Filter
Precision	0.584	0.575
Recall	0.768	0.639
F-measure	0.616	0.561

Since our rule-based system for relation extraction is based on the assumption that most of the relations appear within the same sentences, our system is not able to catch the relations that cross sentence boundaries. To overcome this problem, a new rule, which is called **remote matching**, is integrated into the system. According to this rule, if there exists only one entity type (Microorganism) in a sentence, and within a context window of three sentences there exists only one entity (Habitat or Geographical), then there is a relation between these two entities. The performance of the remote matching rule is tested on the development data set. The results show that the number of the predicted relations increased, which also led to an increase in recall. The obtained precision and recall values are 51.4% and 78.5%, respectively.

4.2 Exhibits Relation Extraction

Similar to the extraction of localization relations, for the extraction of *Exhibits* relations, all the sentences are searched for whether there exist a *Microorganism* entity and a *Phenotype* entity. The same rules that are explained in the previous subsection are applied for the extraction of the *Exhibits* relations.

5 Evaluation

In the BioNLP Shared Task 2019 Bacteria Biotopes normalization sub-task, entities are given with their boundaries in the text and the participants are required to predict the normalization of the entities. In the official evaluation, for each normalized Habitat/Phenotype entity, Wang similarity W (Wang et al., 2007) is calculated to measure the similarity between the reference concept and the predicted concept for the normalization. The performances of the submitted systems are evaluated with their Precision values, which are calculated as:

$$Precision = \sum S_p / N \qquad (2)$$

where S_p indicates the total Wang similarity W for all predictions (Deleger et al., 2016), and N is the number of predicted entities.

In the BioNLP Shared Task 2019 Bacteria Biotopes relation extraction sub-task, entities are given with their boundaries in the text and the participants are asked to predict the relations between the entities. The performances of the submitted systems are evaluated with their F1 (F-measure), recall and precision values.

5.1 Results of BB-norm

The official results obtained by our system and the other participants for the BB-norm sub-task are shown in Table 3. Our system (BOUN-ISIK-2) achieved the best performance with 67.9% Precision in the BB-norm sub-task (Entity Normalization).

Table 3: Comparison with the participant systems for the normalization task of bacteria biotopes. Precision values for the test data set are reported. k is set to 5 and w to 0.25 for the proposed system (BOUN-ISIK).

System	Precision
BOUN-ISIK-2 (Our system)	0.679
BLAIR_GMU-2	0.678
BOUN-ISIK-1 (Our system)	0.675
BLAIR_GMU-1	0.661
PADIA_BacReader-1	0.633
BASELINE-1	0.531
AmritaCen_healthcare-1	0.514

As the results in Table 4 demonstrate, our system performs significantly better than the other systems for the normalization of new Phenotype entities in the test set (Precision: 70.8%).

Table 4: Comparison with the participant systems for the normalization task considering only Phenotype entities. Precision values for the test data set are reported.

System	Phenotypes	Phenotypes (new in test)
BOUN-ISIK (Our system)	0.566	**0.708**
PADIA_BacReader-1	**0.758**	0.156
BASELINE-1	0.582	0.116
BLAIR_GMU-2	0.646	0.03
BLAIR_GMU-1	0.628	0.03
AmritaCen_healthcare-1	0.646	0.0

5.2 Results of BB-rel

The official results obtained by our system and the other participants for the BB-rel task are demonstrated in Table 5.

6 Conclusion

In this study, we presented two systems that are implemented in the scope of the BioNLP Shared Task 2019 - Bacteria Biotope Task. The aim of the first system is the normalization of the entity mentions in a biomedical text through the corresponding ontology, whereas the goal of the second

Table 5: Comparison with the participant systems for the relation extraction task of bacteria biotopes. F1, Recall and Precision values for the test data set are reported.

System	F1	Recall	Precision
whunlp-1	0.664	0.702	0.629
AliAI-1	0.650	0.620	0.682
BASELINE-1	0.635	0.801	0.525
Yuhang_Wu-1	0.605	0.670	0.551
BOUN-ISIK-1 (soft filter)	0.604	0.731	0.514
BLAIR_GMU-2	0.594	0.650	0.548
BOUN-ISIK-2 (hard filter)	0.575	0.601	0.552
BLAIR_GMU-1	0.549	0.496	0.617
UTU-2	0.550	0.474	0.655
UTU-1	0.529	0.428	0.694
Amrita_Cen-1	0.499	0.617	0.419
Amrita_Cen-2	0.493	0.610	0.414

system is the extraction of localization and property relations between the related entities when the entities are given. Both systems are unsupervised in the sense that they do not require domain-specific labeled data, while the normalization system makes use of word embeddings and syntactic re-ranking. According to the official evaluation, both of our systems achieved promising results, which have shown that the proposed methods are comparable to or better than the labeled data driven deep learning based approaches used in the shared task.

Acknowledgments

We would like to thank the BioNLP shared task organizers, especially, Robert Bossy for their help with the questions.

References

Antti Airola, Sampo Pyysalo, Jari Björne, Tapio Pahikkala, Filip Ginter, and Tapio Salakoski. 2008. A graph kernel for protein-protein interaction extraction. In *Proceedings of the workshop on current trends in biomedical natural language processing*, pages 1–9. Association for Computational Linguistics.

Jari Björne and Tapio Salakoski. 2013. Tees 2.1: Automated annotation scheme learning in the bionlp 2013 shared task. In *Proceedings of the BioNLP shared task 2013 workshop*, pages 16–25.

Robert Bossy, Louise Deléger, Estelle Chaix, Mouhamadou Ba, and Claire Nedellec. 2019. Bacteria Biotope at BioNLP Open Shared Tasks 2019. In *Proceedings of the BioNLP Open Shared Tasks 2019 Workshop*.

Robert Bossy, Wiktoria Golik, Zorana Ratkovic, Dialekti Valsamou, Philippe Bessieres, and Claire

Nédellec. 2015. Overview of the gene regulation network and the bacteria biotope tasks in bionlp'13 shared task. *BMC bioinformatics*, 16(10):S1.

Robert Bossy, Julien Jourde, Philippe Bessieres, Maarten Van De Guchte, and Claire Nédellec. 2011. Bionlp shared task 2011: bacteria biotope. In *Proceedings of the BioNLP Shared Task 2011 Workshop*, pages 56–64. Association for Computational Linguistics.

Àlex Bravo, Janet Piñero, Núria Queralt-Rosinach, Michael Rautschka, and Laura I Furlong. 2015. Extraction of relations between genes and diseases from text and large-scale data analysis: implications for translational research. *BMC bioinformatics*, 16(1):55.

Billy Chiu, Gamal Crichton, Anna Korhonen, and Sampo Pyysalo. 2016. How to train good word embeddings for biomedical nlp. *Proceedings of BioNLP16*, page 166.

Sung-Pil Choi. 2018. Extraction of protein–protein interactions (ppis) from the literature by deep convolutional neural networks with various feature embeddings. *Journal of Information Science*, 44(1):60–73.

Louise Deleger, Robert Bossy, Estelle Chaix, Mouhamadou Ba, Arnaud Ferre, Philippe Bessieres, and Claire Nedellec. 2016. Overview of the bacteria biotope task at bionlp shared task 2016. In *Proceedings of the 4th BioNLP Shared Task Workshop*, pages 12–22.

Arnaud Ferré, Pierre Zweigenbaum, and Claire Nédellec. 2017. Representation of complex terms in a vector space structured by an ontology for a normalization task. *BioNLP 2017*, pages 99–106.

Claudio Giuliano, Alberto Lavelli, and Lorenza Romano. 2006. Exploiting shallow linguistic information for relation extraction from biomedical literature. In *11th Conference of the European Chapter of the Association for Computational Linguistics*.

Jörg Hakenberg, Conrad Plake, Robert Leaman, Michael Schroeder, and Graciela Gonzalez. 2008. Inter-species normalization of gene mentions with gnat. *Bioinformatics*, 24(16):i126–i132.

INRA. 2013. *Onto-Biotope Ontology*. Accessed at December 2018.

Ilknur Karadeniz and Arzucan Ozgur. 2013. Bacteria biotope detection, ontology-based normalization, and relation extraction using syntactic rules. In *Proceedings of the BioNLP Shared Task 2013 Workshop*, pages 170–177.

İlknur Karadeniz and Arzucan Özgür. 2015. Detection and categorization of bacteria habitats using shallow linguistic analysis. *BMC bioinformatics*, 16(10):S5.

Ilknur Karadeniz and Arzucan Özgür. 2019. Linking entities through an ontology using word embeddings and syntactic re-ranking. *BMC bioinformatics*, 20(1):156.

Sun Kim, Haibin Liu, Lana Yeganova, and W John Wilbur. 2015. Extracting drug–drug interactions from literature using a rich feature-based linear kernel approach. *Journal of biomedical informatics*, 55:23–30.

Dan Klein and Christopher D Manning. 2003. Accurate unlexicalized parsing. In *Proceedings of the 41st Annual Meeting on Association for Computational Linguistics-Volume 1*, pages 423–430. Association for Computational Linguistics.

Robert Leaman, Rezarta Islamaj Doğan, and Zhiyong Lu. 2013. Dnorm: disease name normalization with pairwise learning to rank. *Bioinformatics*, 29(22):2909–2917.

Jake Lever and Steven JM Jones. 2016. Verse: Event and relation extraction in the bionlp 2016 shared task. In *Proceedings of the 4th BioNLP shared task workshop*, pages 42–49.

Haodi Li, Qingcai Chen, Buzhou Tang, Xiaolong Wang, Hua Xu, Baohua Wang, and Dong Huang. 2017. Cnn-based ranking for biomedical entity normalization. *BMC bioinformatics*, 18(11):385.

Zhiyong Lu, Hung-Yu Kao, Chih-Hsuan Wei, Minlie Huang, Jingchen Liu, Cheng-Ju Kuo, Chun-Nan Hsu, Richard Tzong-Han Tsai, Hong-Jie Dai, Naoaki Okazaki, et al. 2011. The gene normalization task in biocreative iii. *BMC bioinformatics*, 12(8):S2.

Farrokh Mehryary, Kai Hakala, Suwisa Kaewphan, Jari Björne, Tapio Salakoski, and Filip Ginter. 2017. End-to-end system for bacteria habitat extraction. *BioNLP 2017*, pages 80–90.

Tomas Mikolov, Ilya Sutskever, Kai Chen, Greg S Corrado, and Jeff Dean. 2013. Distributed representations of words and phrases and their compositionality. In *Advances in neural information processing systems*, pages 3111–3119.

Alexander A Morgan, Zhiyong Lu, Xinglong Wang, Aaron M Cohen, Juliane Fluck, Patrick Ruch, Anna Divoli, Katrin Fundel, Robert Leaman, Jörg Hakenberg, et al. 2008. Overview of biocreative ii gene normalization. *Genome biology*, 9(2):S3.

NCBI. 2018. *NCBI Taxonomy*. Accessed at December 2018.

Isabel Segura-Bedmar, Paloma Martinez, and Cesar de Pablo-Sánchez. 2011. Using a shallow linguistic kernel for drug–drug interaction extraction. *Journal of biomedical informatics*, 44(5):789–804.

Mert Tiftikci, Hakan Şahin, Berfu Büyüköz, Alper Yayıkçı, and Arzucan Özgür. 2016. Ontology-based categorization of bacteria and habitat entities using information retrieval techniques. In *Proceedings of the 4th BioNLP Shared Task Workshop*, pages 56–63.

James Z Wang, Zhidian Du, Rapeeporn Payattakool, Philip S Yu, and Chin-Fu Chen. 2007. A new method to measure the semantic similarity of go terms. *Bioinformatics*, 23(10):1274–1281.

Chih-Hsuan Wei and Hung-Yu Kao. 2011. Cross-species gene normalization by species inference. *BMC bioinformatics*, 12(8):S5.

Joachim Wermter, Katrin Tomanek, and Udo Hahn. 2009. High-performance gene name normalization with geno. *Bioinformatics*, 25(6):815–821.

Bacteria Biotope Relation Extraction via Lexical Chains and Dependency Graphs

Wuti Xiong[1], Fei Li[2], Ming Cheng[3], Hong Yu[2], Donghong Ji[1]
1. School of Cyber Science and Engineering, Wuhan University, China
2. Department of Computer Science, UMass Lowell, USA
3. Department of Medical Information,
The First Affiliated Hospital of Zhengzhou University, China
woody.xwt@gmail.com, dhji@whu.edu.cn

Abstract

In this article, we describe our approach for the Bacteria Biotopes relation extraction (BB-rel) subtask in the BioNLP Shared Task 2019. This task aims to promote the development of text mining systems that extract relationships between Microorganism, Habitat and Phenotype entities. In this paper, we propose a novel approach for dependency graph construction based on lexical chains, so one dependency graph can represent one or multiple sentences. After that, we propose a neural network model which consists of the bidirectional long short-term memories and an attention graph convolution neural network to learn relation extraction features from the graph. Our approach is able to extract both intra- and inter-sentence relations, and meanwhile utilize syntax information. The results show that our approach achieved the best F1 (66.3%) in the official evaluation participated by 7 teams.[1]

1. Chronic gastritis was recorded before treatment in all patients. Treatment reduced its activity and the presence of H. pylori.

   Relation: H. pylori *Live_in* Habitat patients

2. Atypical mycobacteria causing non-pulmonary disease in Queensland.

   Relation: mycobacteria *Live_in* Queensland
   Relation: mycobacteria *Exhibits* non-pulmonary disease

3. Erythromycin resistance was associated with Campylobacter coli.

   Relation: Campylobacter coli *Exhibits* Erythromycin resistance

Figure 1: Bacteria Biotopes relation examples. The Red, green and blue words denote Microorganism entities, Habitat entities and Phenotype entities respectively.

1 Introduction

The BioNLP Shared Task 2019 (Bossy et al., 2019) is a continuation of the previous efforts organized around the BioNLP Shared Task workshop series (Kim et al., 2009, 2011; Nédellec et al., 2013; Deléger et al., 2017). It aims to facilitate development and sharing of computational tasks of biomedical text mining and solutions to them. The Bacteria Biotope (BB) task is one of the six main tasks of the BioNLP Open Shared Tasks 2019. Three teams participated in the BB task when it was first organized in 2011. INRA Bibliome (Ratkovic et al., 2011) achieved the best Fscore of 45% with the Alvis system which used dictionary mapping, ontology inference and semantic analysis for NER, and co-occurrence-based rules for detecting relations between the entities. The 2013 BB task (Bossy et al., 2013) contained three

subtasks, the first one concerning recognition and normalization of bacteria and habitat entities, and the other two subtasks involving relation extraction. Four teams participated in these tasks, with the UTurku TEES system (Björne and Salakoski, 2013) achieving the first places with F-scores of 42% and 14%. Compared to the 2013 BB task, the 2016 BB task contains more subtasks and its subtask2 only concerned relation extraction. The team VERSE (Lever and Jones, 2016) achieved the best F-scores of 55.8% in the subtask2.

The Bacteria Biotopes relation extraction (BB-rel) in the BioNLP Shared Task 2019 aims to automatically extract Microorganism-Habitat or Microorganism-Phenotype relationships from biomedical literature. The BB-rel task follows the previous Bacteria Biotopes shared tasks, annotating directed binary relationships between Microorganism, Habitat and Phenotype entities. Fig-

---

[1]Code: https://github.com/woodyXwt/BB19-rel

Proceedings of the 5th Workshop on BioNLP Open Shared Tasks, pages 158–167
Hong Kong, China, November 4, 2019. ©2019 Association for Computational Linguistics

ure 1 shows some examples for each relationship. In the BB-rel task, not all the relations occur between two entities with the same sentence. In the preprocessing step, we found that there exist about one fourth of all relations whose argument entities are located in different sentences. Therefore, we need to build a model that does not only consider the entity relationship within one sentence, but also beyond the sentence boundary.

A lexical chain (Morris and Hirst, 1991) is a sequence of words which are semantically-similar or related. These words are related sequentially in the text, defining the topic of the text segment that they cover and establishing associations between sentences. Following this observation, some researchers have obtained success in many NLP tasks such as word sense induction(Tao et al., 2014), machine translation (Mascarell, 2017) and text (Stokes et al., 2004) segmentation. In the BB-rel dataset, the sentences where inter-sentence relations occur usually express the same topic or have semantic associations each other. These features usually appear as some related words which can form lexical chains. Following this observation, we propose a novel approach to build an inter-sentence dependency graph based on lexical chains.

In this paper, we propose a novel relation extraction method for the BB-rel task by incorporating dependency graphs and lexical chains into the neural network. As shown in Figure 1, inter-sentence relations are usually expressed in inter-related sentences, and these sentences may contain semantically-related words which can form lexical chains. We utilize these lexical chains and dependency graphs to build an inter-sentence dependency graph for inter-sentence relation extraction. Specifically, we utilize word embedding to find the semantic relationships of words that occur in different sentences for building reliable lexical chains. Then, we use the Stanford CoreNLP toolkit (Manning et al., 2014) to obtain sentence-level dependency and part-of-speech (POS) information, and build an inter-sentence dependency graph based on these information and lexical chains.

After that, we employ a neural network model which consists of the bidirectional long short-term memories and attention-guided graph convolutional neural networks to extract features from the inter-sentence dependency graph. The features are fed into a multi-layer perceptron (MLP) to classify the relation between an entity pair.

Our approach has two advantages. First, it is capable of extracting both intra-sentence and inter-sentence relations by connecting the dependency graphs of different sentences via lexical chains. Second, it is able to leverage syntax information. The results in the BB-rel task demonstrate the superiority of our method. It achieves the highest F1-score, the second highest precision and recall in the official evaluation.

2 Method

In this section, we first introduce our strategy of relation candidate generation. Then, the approach for constructing lexical chains is described. After that, we will introduce how to build inter-sentence dependency graphs. Lastly, the architecture of our neural network model is described.

2.1 Relation Candidate Generation

In the BB-rel dataset, if all candidate pairs (bacteria and habitat or phenotype) that occur in the document are enlisted as candidate training examples, the positive and negative examples will become very unbalanced because most entity pairs located beyond one sentence do not have any relation. Based on our observations, most entity pairs spanning more than two sentences have no relations between them. Therefore, we consider all entity pairs that span within two sentences as the candidates to generate training examples. The statistics of our dataset are summarized in Table 1.

2.2 Lexical Chain Construction

In previous work, there are mainly three approaches for constructing lexical chains. The first one utilized *WordNet* (Hirst and St-Onge, 1997) to capture the semantic relationship between words. The second approach (Remus and Biemann, 2013)

	Train	Dev
Lives_In	715	395
Exhibits	281	138
Total relatonships	996	533
Intra-sentence relationships	885	467
Inter-sentence relationships	111	66

Table 1: BB-rel data statistics on the training and development set.

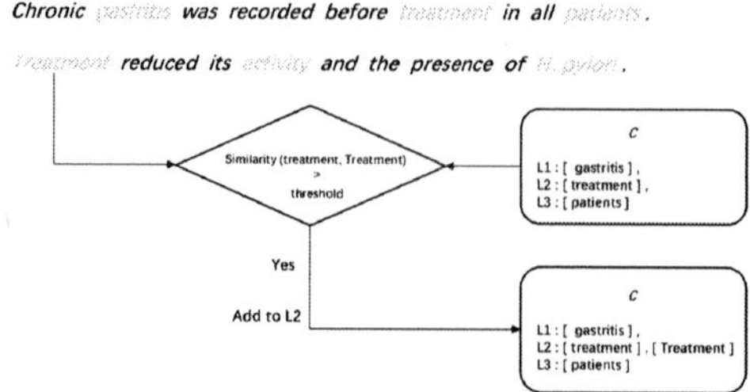

Figure 2: Process of lexical chain construction. Orange words denote nouns. C is the set of lexical chains. The similarity here refers to the cosine similarity between word vectors. We set the threshold to 0.5.

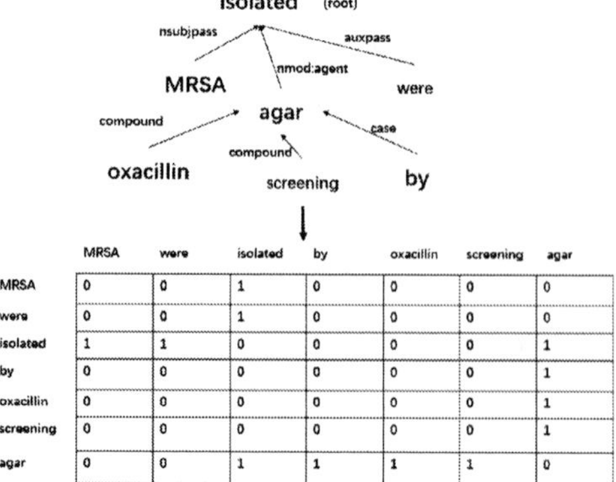

	MRSA	were	isolated	by	oxacillin	screening	agar
MRSA	0	0	1	0	0	0	0
were	0	0	1	0	0	0	0
isolated	1	1	0	0	0	0	1
by	0	0	0	0	0	0	1
oxacillin	0	0	0	0	0	0	1
screening	0	0	0	0	0	0	1
agar	0	0	1	1	1	1	0

Figure 3: An example of the dependency graph and its corresponding adjacent matrix. If there is a dependency relation between the node i and j in the dependency graph, the value of the element M_{ij} in the adjacent matrix is 1.

automatically extracted lexical chains using statistical methods . Another approach (Li et al., 2017) is based on semantic word vectors. In this paper, we assume that lexical relationships can be captured by calculating the similarity of their semantic vectors. To compute similarities, we use 200-dimensional pre-trained word vectors released by Pyysalo et al. (2013). Moreover, we only consider nouns for constructing the lexical chains since they usually contain relevant information.

Given a sentence, we first use the Stanford CoreNLP toolkit (Manning et al., 2014) to obtain POS tags for each word. Then we pick those words whose POS tags belonging to *N= (NN,NNP,NNS)* as candidates for chain construction. We take one candidate at a time and check where it should be placed. Assuming that C is the set of lexical chains, we add each candidate w to C according to the following steps (Figure 2):

- *Step 1*: each noun is treated as a candidate w. If C is empty, we will create a new lexical chain in C and add the current candidate w into it.

- *Step 2*: for the current candidate w, we traverse all the lexical chains in C and compute the similarity between the last word of each lexical chain and the current candidate w. If the similarity surpasses a predefined threshold, the current candidate w will be attached to the corresponding lexical chain.

- *Step 3*: if the current candidate w cannot be attached to any existing lexical chain, we will create a new lexical chain for it.

2.3 Dependency Graph Construction

In this section, we propose an approach to build an inter-sentence dependency graph by lexical chains. For an entity pair that occurs within the same sentence, we directly use their sentence dependency graph. If two entities occur in different sentences, we construct their dependency graph by lexical chains. We design two rules to build an inter-sentence graph. Here we define the following notations: C is the set of lexical chains, A and B are nouns belonging to sentence s_1 and sentence s_2, respectively.

- *Rule 1*: if A and B exist in the same chain of C, we will add an edge between A and B to build an inter-sentence dependency graph.

- *Rule 2*: if A and B do not appear in the same lexical chain, we will use the root nodes of two sentences to build the dependency inter-sentence graph.

Then we convert the dependency graph into an adjacency matrix. An example of such process is shown in Figure 3. Give a sequence $S = \{s_1, s_2, ..., s_n\}$, we considered its dependency graph as an undirected graph, which can be converted into an adjacent matrix. If there is a dependency relation between nodes i and j in the dependency graph, the element M_{ij} in the adjacent matrix is assigned with 1.

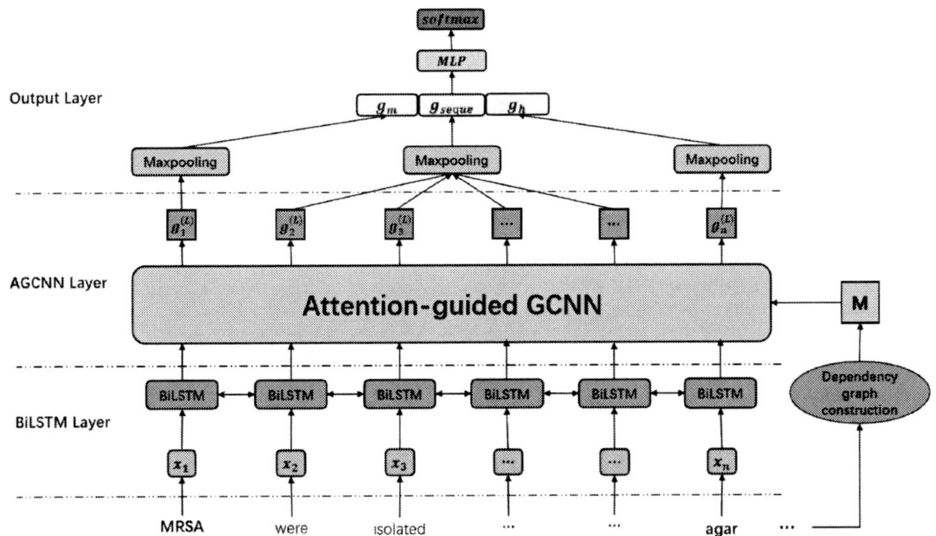

Figure 4: The architecture of our model. The input sentence is "MRSA were isolated by oxacillin screening agar" with a Microorganism entity "MRSA" and a Habitat entity "oxacillin screening agar". M denotes the adjacency matrix.

2.4 Neural Network Model

2.4.1 BiLSTM Layer

Figure 4 shows the neural network architecture of our model. It uses the words and POS tags as input. We adopt the 200-dimensional word embeddings and 20-dimensional POS tag embeddings. The final representation for the token is the concatenation x_i of the word embedding s_i and the POS tag embedding p_i. We initialize our word embeddings with the pre-trained biomedical embeddings (Pyysalo et al., 2013) and randomly initialize the POS tag embeddings.

After obtaining the word representation sequence $x = \{x_1, x_2, ..., x_n\}$, we leverage bidirectional LSTMs (Hochreiter, 1998) to encode the context information into each word. The forward and backward hidden states ($\overrightarrow{h_i}$ and $\overleftarrow{h_i}$) will be concatenated, formalized as $h_i = [\overrightarrow{h_i} \odot \overleftarrow{h_i}]$.

2.4.2 Attention-Guided GCNN Layer

We employ the attention-guided graph convolutional neural network (AGCNN) (Guo et al., 2019a) to incorporate the dependency information into word representations, which is composed of M identical blocks. Each block has three types of layers: attention-guided layer, densely connected layer, linear combination layer.

In the attention guided layer, we first update the representation of the node using a graph convolution network (GCNN) (Zhang et al., 2018). For an L-layer GCNN, we denotes the inputs in the first layer as $g_1^{(0)}, ..., g_n^{(0)}$ and the outputs in the last layer as $g_1^{(L)}, ..., g_n^{(L)}$. The $g_i^{(l)}$ denotes the output vectors of the node i in the l-th layer. The convolution operation in the l-th layer can be written as:

$$g^l = \sigma(\sum_{j=1}^{n} \tilde{M}_{ij}, W^l g^{l-1}/d_i + b^l), \quad (1)$$

where W^l is a linear transformation, b^l is a bias term, and σ is a nonlinear function (e.g., $ReLU$). The $\tilde{M}$ can be computed by $M + I$, where $I \in \mathbb{R}^{n \times n}$ is an identity matrix and $d_i = \sum_{j=1}^{n} \tilde{M}_{ij}$ is the degree of node i in the dependency graph. Intuitively, during the graph convolution of each layer, each node gathers all the information of its neighboring nodes in the graph.

After the L-layer graph convolution operation, we transform the original dependency graph into a fully connected edge-weighted graph by constructing N (N is a hyper-parameter) attention-guided adjacency matrix. Each attention-guided adjacency matrix $\tilde{A}$ corresponds to a completely connected graph. In this paper, we use the multi-head attention(Vaswani et al., 2017) to calculate $\tilde{A}$, which allows the model to focus on information from different representation sub-spaces. The output is computed as a weighted sum of values, where the weight is calculated by the function of the query and the corresponding key.

$$\tilde{A}^{(t)} = softmax(QW_i^Q \times (KW_i^K)^T/\sqrt{d})V, \tag{2}$$

where Q and K are both equal to the collective representation h^{l-1} at layer $l-1$ of the model. The projections are parameter matrices $W_i^Q \in \mathbb{R}^{d\times d}$ and $W_i^K \in \mathbb{R}^{d\times d}$. $\tilde{A}^{(t)}$ is the t-th attention guided adjacency matrix corresponding to the t-th head.

Following (Guo et al., 2019b), we employ the dense connection (Huang et al., 2017) into the our model to capture more structural information on the large graph. We concatenate the initial node representation $h_j^{(l)}$ and the node representations $g_j^{(1)}, ..., g_j^{(l-1)}$ produced in layer $1, ..., l-1$:

$$h_j^{(l)} = [x_j; g_j^{(1)}, ..., g_j^{(l-1)}], \tag{3}$$

Each densely connected layer has L sub-layers. The dimensions of these sub-layers d_{hidden} are decided by L and the input feature dimension d. In our model, we use $d_{hidden} = d/L$.

Then we use N separate dense connection layers to modify the computation of each layer as follows (for the t-th matrix $\tilde{A}^{(t)}$):

$$g_{t_i}^l = \rho(\sum_{j=1}^{n} \tilde{A}^{(t)} W_t^l h_i^l + b_t^l), \tag{4}$$

where $t = 1, ..., N$ and t selects the weight matrix and bias term associated with the attention guided adjacency matrix $\tilde{A}^{(t)}$. The column dimension of the weight matrix increases by d_{hidden} per sub-layer, i.e., $W_t^l \in \mathbb{R}^{d_{hidden}\times d^{(l)}}$ where $d^{(l)} = d + d_{hidden}(l-1)$.

Finally, we use linear combination layer to integrate representations from N different densely connected layers. Formally, the output of the linear combination layer is defined as:

$$g_{comb} = W_{comb}g_{out} + b_{comb}, \tag{5}$$

where g_{out} is the output by concatenating outputs from N separate densely connected layers, i.e., $g_{out} = [g^{(1)}; ...; g^{(N)}] \in \mathbb{R}^{d\times d}$. $W_{comb} \in \mathbb{R}^{d\times d}$ is a weight matrix and b_{comb} is a bias vector for the linear transformation.

2.4.3 Output Layer

We treat the BB-rel task as a classification task. $S = [s_1, ..., s_n]$ denotes a sequence, s_i is the i-th token, M_e and H_e denote Microorganism and Habitat or Phenotype entities. The entities may consist of several tokens, namely $[s_{e_1}, ..., s_{e_n}]$ and $[s_{h_1}, ..., s_{h_n}]$. The goal of the BB-rel task is to predict whether there is a "Live_in" or "Exhibits" relationship between the entities H_e and M_e.

After applying the attention-guided GCNN layer to the input word vectors, we obtain the representation for each word. The sequence representation can be obtained using the following equation:

$$g_{seque} = f(g_1, ..., g_n), \tag{6}$$

where $g_1, ..., g_n$ denotes the outputs of the the the attention-guided GCNN layer and $f : R^{d\times n} \rightarrow R^d$ is a max-pooling function. Since we also observed that the entity information is often critical for BB-rel extraction, the entity representations M_e and H_e are also used, given by:

$$g_m = f(g_{m_1}, ..., g_{m_n}),$$
$$g_h = f(g_{h_1}, ..., g_{h_n}). \tag{7}$$

Inspired by (Santoro et al., 2017; Lee et al., 2017), we obtained the final feature for BB-rel extraction by feeding the sequence and entity representations into a multi-layer perceptron (MLP):

$$g_{final} = MLP([g_{seque}; g_m; g_h]), \tag{8}$$

where "[]" denotes the concatenation operation. Finally, g_{final} is fed into a softmax layer to compute the probability distribution over all classes. During training, our model uses the cross-entropy loss:

$$loss(\theta) = -\sum_{j=1}^{J} logP(y_j|S_j), \tag{9}$$

where J denotes the size of the training set $S = \{(S_1, y_1), ..., (S_J, y_J)\}$ and y_j denotes the gold answer of the j-th training instance. $P(y_j|S_j)$ denotes the probability that S_j belongs to y_j, which is calculated as $P(y_j|S_j) = softmax(g_{final})$.

3 Experiments

3.1 Evaluation Metrics

We send the prediction results of our model on the test set to the task organizer for evaluation. The

Hyper-parameter	Value
Number of heads N	2
Block number M	2
Word emb size	200
POS emb size	20
LSTM hidden size	300
BiSTM layer	2
GCNN layer	2
GCNN output size	200
Dropout of GCNN	0.5
Multi-head attention head	3
Sublayers	5
d_{hidden}	300
Epoch	100
Decay rate	0.9
Learning rate	0.5
Optimizer	sgd
MLP layer	1

Table 2: Hyper-parameter setting.

Team Name	P	R	F1
Amrita_Cen	41.9	61.7	49.9
UTU	47.3	65.5	55.5
BLAIR_GMU	54.7	64.9	59.4
BOUN-ISIK	51.3	**73.1**	60.3
Yuhang_Wu	55.1	67.0	60.4
AliAI	**68.2**	62.0	64.9
Our method	62.9	70.2	**66.3**

Table 3: The official results of the BB-rel task.

performances of our model were evaluated by the standard evaluation measures: precision (P), recall (R) and F1-score (F1).

3.2 Hyper-parameter

The hyper-parameter setting is listed in Table 2. We tuned hyper-parameters based on the development set.

3.3 Official Results

The official results on the test set are shown in Table 3. There are totally 7 teams participating in the BB-rel task. Each team could submit up to 2 predictions. We report the top results for all teams. As we can see, our method achieved the highest F1 (66.3%), and the second highest precision (62.9%) and recall (70.2%).

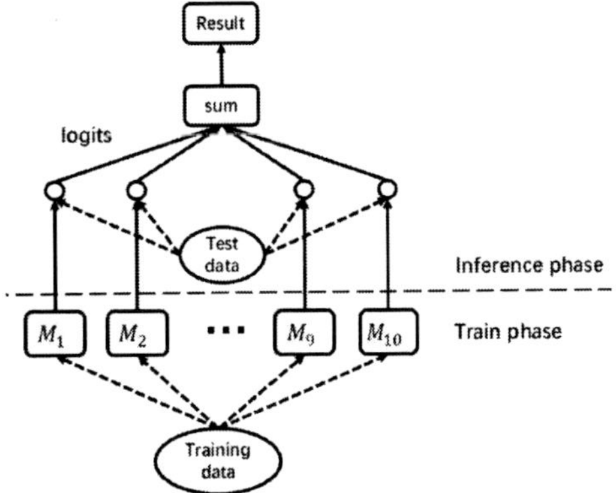

Figure 5: Ensemble training and inference.

3.4 Ensemble Training and Inference

In relation extraction tasks, the ensemble training and inference have proven to be an effective way to improve performance of the neural network model (Mehryary et al., 2016; Lim and Kang, 2018). Following previous work (Lim and Kang, 2018), we improve performance of our model using the ensemble training and inference. We sum the output probabilities (logits) of ensemble members, which are generated using the same neural network model but different weight initialization.

As shown in Figure 5, M1 to M10 are the models using the same structure and hyper-parameters. In the training phase, we independently trained each ensemble member with different initialized parameters. When inferring a relation for an easy sample, the trained ensemble members make relatively consistent predictions. When inferring for a difficult sample, the trained ensemble members may make different predictions. We incorporate the voting results of 10 ensemble members to produce final results.

To investigate the effectiveness of ensemble training and inference, we conducted the following experiment on the development set. First, we run five times of our model and average the results as the final result of the single model as shown in Table 4. Second, we run one time for the ensemble training and inference. The results show that the approach using ensemble training and inference achieved relatively balanced precision and recall, thus yielding a better F1.

Method	P	R	F1
Single	59.1	69.3	63.8
Ensemble	63.1	68.4	65.7

Table 4: Effects of ensemble training and inference.

	Relation	P	R	F1
Intra	Live_in	61.6	60.0	60.8
	Exhibits	73.4	80.6	76.8
	Total	64.8	65.2	65.0
Intra+Inter	Live_in	59.5	63.7	61.5
	Exhibits	72.8	82.4	77.3
	Total	63.1	68.4	65.7

Table 5: Results of recognizing inter- and intra-sentence relations.

3.5 Results of Recognizing Inter- and Intra-Sentence Relations

In this section, we discuss the performance of our model in Intra- and inter-sentence relation. As shown in Table 5, we obtained an F1-score of 65.0 when we only evaluated the intra-sentence relationships. When we evaluated both intra- and inter-sentence relationship, F1-score, Recall increase by 0.7% and 3.2% respectively. But Precision drops by 1.7%. We can also see from the table that the performance of "Exhibits" relation is better than the performance of the "Live_in" relation. Because most of the "Exhibits" relation happen within a sentence and have a certain pattern.

3.6 Effects of Lexical Chains

In order to verify the effectiveness of constructing inter-sentence dependency graphs by lexical chains, we also conducted related experiments on development set. The experimental results are shown in Table 6. "lexical chains" denotes the model employing the proposed method that constructs inter-sentence dependency graphs by lexical chains. "root nodes" denotes the model where the inter-sentence dependency graphs are built using root nodes. Table 6 shows the performance comparison of the "lexical chains" method and the "root nodes" method on the development set. The "lexical chains" method obtained better perfor-

Method	P	R	F1
Root nodes	62.7	67.3	64.9
Lexical chains	63.1	68.4	65.7

Table 6: Effects of lexical chains.

Figure 6: Examples of false positives. The Red and green words denote Microorganism and Habitat entities respectively.

mance than the "root nodes" model. This demonstrates our idea is effective. The relevant sentences are usually expressed using relevant words. These relevant words found by lexical chains can be used as the associations to connect the dependency graphs of different sentences. Therefore, we can build an effective representation for an inter-sentence entity pair.

3.7 Error Analysis

In this section, we manually analyzed what cases lead to false positives, since those are more critical than false negatives. Figure 6 shows some examples of false positives. The most of false positives are caused by overlapping target entities. For example, there is a "Live_in" relation between "Listeria sp." and "chicken nugget processing plant", but there is no "Live_in" relation between "Listeria sp." and "chicken" or "chicken nugget". The reason for these errors is that the model is confused by overlapping entities with similar context.

4 Related Work

In the natural language processing community, there are a number of related competitions and tasks (Wei et al., 2015; Nédellec et al., 2013; Deléger et al., 2016). Most prior work focused on extracting the relations within one sentence, and ignored the relations beyond one sentence.

In the NLP community, it has proven to be effective to combine linguistic features with neural networks for relation extraction (Zhou et al., 2015; Miwa and Bansal, 2016). Bunescu et al. (2005) demonstrated that the relationship of an entity pair can be captured along their shortest dependency path in the dependency graph because the words on the shortest dependency path concentrate the most relevant information and diminish redundant information. Following this observation, several studies (Xu et al., 2015; Liu et al., 2015) achieved outstanding performance by combining shortest dependency paths with various neural networks. As deep learning develops, some attention-based neural architectures (Zhou et al., 2016; Lin et al., 2016) have been proposed for relation classification and show the state-of-the-art performance. But with a few exceptions, almost all related work only focused on intra-sentence relation extraction, without considering the inter-sentence relations.

Recent work has explored some approaches to consider inter-sentence relations, such as Graph LSTMs (Peng et al., 2017), self-attention (Verga et al., 2018), Graph CNNs (Sahu et al., 2019). However, none of these work investigated lexical chains for inter-sentence relation extraction. In the future, we will evaluate our approach on some large-scale datasets for intra- and inter-sentence relation extraction (Yao et al., 2019).

5 Conclusion

In this paper, we describe our approach used for participating the Bacteria Biotope task at BioNLP-OST 2019. Our approach achieved very competitive performance in the official evaluation. We found that the idea using lexical chains to build inter-sentence dependency graphs is effective. Moreover, ensemble training and inference can improve the performance of our model. The attention-guided graph convolution neural network performs well in extracting Bacteria Biotope relations. However, our approach is not specific to Bacteria Biotope relation extraction, and it can be applied to other relation extraction tasks.

Acknowledgments

This work is supported by the Young Scientists Fund of the National Natural Science Foundation of China (No. 61802350), the National Natural Science Foundation of China (No. 61772378), the Major Projects of the National Social Science Foundation of China (No. 11&ZD189), the National Key Research, Development Program of China (No. 2017YFC1200500).

References

Jari Björne and Tapio Salakoski. 2013. TEES 2.1: Automated annotation scheme learning in the bionlp 2013 shared task. In *Proceedings of the BioNLP Shared Task 2013 Workshop, Sofia, Bulgaria, August 9, 2013*, pages 16–25.

Robert Bossy, Louise Deléger, Estelle Chaix, Mouhamadou Ba, and Claire Nedellec. 2019. Bacteria Biotope at BioNLP Open Shared Tasks 2019. In *Proceedings of the BioNLP Open Shared Tasks 2019 Workshop*.

Robert Bossy, Wiktoria Golik, Zorana Ratkovic, Philippe Bessières, and Claire Nédellec. 2013. Bionlp shared task 2013 - an overview of the bacteria biotope task. In *Proceedings of the BioNLP Shared Task 2013 Workshop, Sofia, Bulgaria, August 9, 2013*, pages 161–169.

Razvan C Bunescu and Raymond J Mooney. 2005. A shortest path dependency kernel for relation extraction. In *EMNLP*, pages 724–731.

Louise Deléger, Robert Bossy, Estelle Chaix, Mouhamadou Ba, Arnaud Ferre, Philippe Bessieres, and Claire Nedellec. 2016. Overview of the bacteria biotope task at bionlp shared task 2016. In *Proceedings of the 4th BioNLP shared task workshop*, pages 12–22.

Louise Deléger, Robert Bossy, Estelle Chaix, Mouhamadou Ba, Arnaud Ferré, Philippe Bessières, and Claire Nédellec. 2017. Overview of the bacteria biotope task at bionlp shared task 2016. In *Bionlp Shared Task Workshop-association for Computational Linguistics*.

Zhijiang Guo, Yan Zhang, and Wei. Lu. 2019a. Attention guided graph convolutional networks for relation extraction. In *ACL*.

Zhijiang Guo, Yan Zhang, Zhiyang Teng, and Wei Lu. 2019b. Densely connected graph convolutional networks for graph-to-sequence learning. *TACL*, 7:297–312.

Graeme Hirst and David St-Onge. 1997. Lexical chains as representations of context for the detection and correction of malapropisms. *Lecture Notes in Physics*, 728(9):123–149.

Sepp Hochreiter. 1998. The vanishing gradient problem during learning recurrent neural nets and problem solutions. *International Journal of Uncertainty, Fuzziness and Knowledge-Based Systems*, 06(02):–.

Gao Huang, Zhuang Liu, Laurens van der Maaten, and Kilian Q. Weinberger. 2017. Densely connected convolutional networks. In *CVPR*, pages 2261–2269.

J D Kim, S. Pyysalo, T. Ohta, R. Bossy, N. Nguyen, and J. Tsujii. 2011. Overview of bionlp shared task 2011. In *Bionlp Shared Task Workshop*.

Jin Dong Kim, Tomoko Ohta, Sampo Pyysalo, Yoshinobu Kano, and Jun'Ichi Tsujii. 2009. Overview of bionlp'09 shared task on event extraction. In *Workshop on Current Trends in Biomedical Natural Language Processing: Shared Task*.

Kenton Lee, Luheng He, Mike Lewis, and Luke Zettlemoyer. 2017. End-to-end neural coreference resolution. In *EMNLP*, pages 188–197.

Jake Lever and Steven J. Jones. 2016. VERSE: event and relation extraction in the bionlp 2016 shared task. In *Proceedings of the 4th BioNLP Shared Task Workshop, BioNLP 2016, Berlin, Germany, August 13, 2016*, pages 42–49.

Liunian Li, Xiaojun Wan, Jin-ge Yao, and Siming Yan. 2017. Leveraging diverse lexical chains to construct essays for chinese college entrance examination. In *IJCNLP*.

Sangrak Lim and Jaewoo Kang. 2018. Chemical-gene relation extraction using recursive neural network. *Database*, 2018:bay060.

Yankai Lin, Shiqi Shen, Zhiyuan Liu, Huanbo Luan, and Maosong Sun. 2016. Neural relation extraction with selective attention over instances. In *ACL*, pages 2124–2133.

Yang Liu, Furu Wei, Sujian Li, Heng Ji, Ming Zhou, and WANG Houfeng. 2015. A dependency-based neural network for relation classification. In *ACL*, pages 285–290.

Christopher D. Manning, Mihai Surdeanu, John Bauer, Jenny Finkel, Steven J. Bethard, and David Mcclosky. 2014. The stanford corenlp natural language processing toolkit. In *Meeting of the Association for Computational Linguistics: System Demonstrations*.

Laura Mascarell. 2017. Lexical chains meet word embeddings in document-level statistical machine translation. In *Discourse in Machine Translation (DiscoMT)*.

Farrokh Mehryary, Jari Björne, Sampo Pyysalo, Tapio Salakoski, and Filip Ginter. 2016. Deep learning with minimal training data: Turkunlp entry in the bionlp shared task 2016. In *Proceedings of the 4th BioNLP Shared Task Workshop, BioNLP 2016, Berlin, Germany, August 13, 2016*, pages 73–81.

Makoto Miwa and Mohit Bansal. 2016. End-to-end relation extraction using lstms on sequences and tree structures. In *ACL*, pages 1105–1116.

Jane Morris and Graeme Hirst. 1991. Lexical cohesion computed by thesaural relations as an indicator of the structure of text. *Computational Linguistics*, 17(1):21–48.

Claire Nédellec, Robert Bossy, Jin-Dong Kim, Jung-Jae Kim, Tomoko Ohta, Sampo Pyysalo, and Pierre Zweigenbaum. 2013. Overview of bionlp shared task 2013. In *Proceedings of the BioNLP Shared Task 2013 Workshop*, pages 1–7.

Claire Nédellec, Robert Bossy, Jin Dong Kim, Jung Jae Kim, Tomoko Ohta, Sampo Pyysalo, and Pierre Zweigenbaum. 2013. Overview of bionlp shared task 2013. In *Bionlp Shared Task Workshop*.

Nanyun Peng, Hoifung Poon, Chris Quirk, Kristina Toutanova, and Wen-tau Yih. 2017. Cross-sentence n-ary relation extraction with graph LSTMs. *Transactions of the Association for Computational Linguistics*, 5:101–115.

S. Pyysalo, F. Ginter, H. Moen, T. Salakoski, and S. Ananiadou. 2013. Distributional semantics resources for biomedical text processing. In *Proceedings of LBM 2013*, pages 39–44.

Zorana Ratkovic, Wiktoria Golik, Pierre Warnier, Philippe Veber, and Claire Nédellec. 2011. Bionlp 2011 task bacteria biotope - the alvis system. In *Proceedings of BioNLP Shared Task 2011 Workshop, Portland, Oregon, USA, June 24, 2011*, pages 102–111.

Steffen Remus and Chris Biemann. 2013. Three knowledge-free methods for automatic lexical chain extraction. In *NAACL*.

Sunil Kumar Sahu, Fenia Christopoulou, Makoto Miwa, and Sophia Ananiadou. 2019. Inter-sentence relation extraction with document-level graph convolutional neural network. In *Proceedings of the 57th Annual Meeting of the Association for Computational Linguistics*, pages 4309–4316, Florence, Italy. Association for Computational Linguistics.

Adam Santoro, David Raposo, David G. T. Barrett, Mateusz Malinowski, Razvan Pascanu, Peter W. Battaglia, and Tim Lillicrap. 2017. A simple neural network module for relational reasoning. In *NIPS*, pages 4967–4976.

Nicola Stokes, Joe Carthy, and Alan F. Smeaton. 2004. Select: a lexical cohesion based news story segmentation system. *Ai Communications*, 17(17):3–12.

Qian Tao, Donghong Ji, and Congling Xia. 2014. Word sense induction using lexical chain based hypergraph model. In *Coling*.

Ashish Vaswani, Noam Shazeer, Niki Parmar, Jakob Uszkoreit, Llion Jones, Aidan N. Gomez, Lukasz Kaiser, and Illia Polosukhin. 2017. Attention is all you need. In *NIPS*, pages 5998–6008.

Patrick Verga, Emma Strubell, and Andrew McCallum. 2018. Simultaneously self-attending to all mentions for full-abstract biological relation extraction. In *Proceedings of the 2018 Conference of the North American Chapter of the Association for Computational Linguistics: Human Language Technologies, Volume 1 (Long Papers)*, pages 872–884.

Chih-Hsuan Wei, Yifan Peng, Robert Leaman, Allan Peter Davis, Carolyn J Mattingly, Jiao Li, Thomas C Wiegers, and Zhiyong Lu. 2015. Overview of the biocreative v chemical disease relation (cdr) task. In *Proceedings of the fifth BioCreative challenge evaluation workshop*, volume 14.

Yan Xu, Lili Mou, Ge Li, Yunchuan Chen, Hao Peng, and Zhi Jin. 2015. Classifying relations via long short term memory networks along shortest dependency paths. In *EMNLP*, pages 1785–1794.

Yuan Yao, Deming Ye, Peng Li, Xu Han, Yankai Lin, Zhenghao Liu, Zhiyuan Liu, Lixin Huang, Jie Zhou, and Maosong Sun. 2019. DocRED: A large-scale document-level relation extraction dataset. In *Proceedings of the 57th Annual Meeting of the Association for Computational Linguistics*, pages 764–777, Florence, Italy. Association for Computational Linguistics.

Yuhao Zhang, Peng Qi, and Christopher D. Manning. 2018. Graph convolution over pruned dependency trees improves relation extraction. In *EMNLP*.

Hao Zhou, Yue Zhang, Shujian Huang, and Jiajun Chen. 2015. A neural probabilistic structured-prediction model for transition-based dependency parsing. In *ACL*, pages 1213–1222.

Peng Zhou, Wei Shi, Jun Tian, Zhenyu Qi, Bingchen Li, Hongwei Hao, and Bo Xu. 2016. Attention-based bidirectional long short-term memory networks for relation classification. In *ACL*, pages 207–212.

Gap in pagination due to unavailable paper.

Pages 168-173

CRAFT Shared Tasks 2019 Overview — Integrated Structure, Semantics, and Coreference

William A Baumgartner Jr.[1], Michael Bada[1], Sampo Pyysalo[2], Manuel R. Ciosici[3],
Negacy Hailu[1], Harrison Pielke-Lombardo[1], Michael Regan[4] and Lawrence Hunter[1]

[1]University of Colorado Anschutz Medical Campus
[2]Turku NLP Group, University of Turku, Finland
[3]UNSILO A/S
[4]University of New Mexico

Abstract

As part of the BioNLP Open Shared Tasks 2019, the CRAFT Shared Tasks 2019 provides a platform to gauge the state of the art for three fundamental language processing tasks — dependency parse construction, coreference resolution, and ontology concept identification — over full-text biomedical articles. The structural annotation task requires the automatic generation of dependency parses for each sentence of an article given only the article text. The coreference resolution task focuses on linking coreferring base noun phrase mentions into chains using the symmetrical and transitive identity relation. The ontology concept annotation task involves the identification of concept mentions within text using the classes of ten distinct ontologies in the biomedical domain, both unmodified and augmented with extension classes. This paper provides an overview of each task, including descriptions of the data provided to participants and the evaluation metrics used, and discusses participant results relative to baseline performances for each of the three tasks.

1 Introduction

With its multiple layers of annotation, the Colorado Richly Annotated Full Text (CRAFT) corpus provides a unique foundation for integrating natural language processing (NLP) tasks involving structure, semantics, and coreference. As part of the BioNLP Open Shared Tasks 2019, the CRAFT corpus was used for the evaluation of three fundamental NLP tasks: dependency parse construction, coreference resolution, and ontology concept annotation. Each of these tasks is a foundational element to many NLP systems and their performances can propagate downstream and directly affect overall system accuracy. Dependency parses have been successfully employed for information extraction, e.g. from clinical records (Gupta et al.,

2018), relation extraction, e.g. identifying protein post-translational modifications (Sun et al., 2017), and used as features for machine learning tasks, e.g. gene mention detection (Smith and Wilbur, 2009), among other uses. By linking noun phrases to a referent entity, coreference systems serve as annotation multipliers, amplifying results of entity recognition systems (Cohen et al., 2017), and have been shown to improve information extraction in biomedical text (Choi et al., 2016). The concept annotation task, also known as named entity recognition (NER), is a prerequisite for many biomedical NLP applications. Its importance is buttressed by the many previous shared tasks that have included aspects of NER (Hirschman et al., 2005; Smith et al., 2008; Krallinger et al., 2013) . Measuring the state of the art of these foundational tasks will inform the BioNLP community by resetting the performance benchmarks and demonstrating optimal methodologies.

The CRAFT Shared Tasks (CRAFT-ST) 2019 mark the inaugural use and subsequent release of thirty articles annotated in CRAFT that had previously been held in reserve. All 97 articles and accompanying annotations of the CRAFT corpus are now available in the public domain. To augment the results of the CRAFT-ST 2019, and to account for the relatively low participation rate, baseline systems for each task were evaluated in the same manner as the participant systems. The CRAFT-ST 2019 made use of the CRAFT v3.1.3 release[1]. Original task descriptions are available on the CRAFT-ST website [2]. An integrated scoring platform capable of supporting the evaluation of all three sub tasks of the CRAFT-ST 2019 is

---

[1]https://github.com/UCDenver-ccp/
CRAFT/releases/tag/v3.1.3;
doi:10.5281/zenodo.3460908
[2]https://sites.google.com/view/
craft-shared-task-2019

Proceedings of the 5th Workshop on BioNLP Open Shared Tasks, pages 174–184
Hong Kong, China, November 4, 2019. ©2019 Association for Computational Linguistics

also available as a standalone system[3], and as a pre-built Docker container[4].

2 The CRAFT Structural Annotation Task

For the structural annotation task (CRAFT-SA), participants were asked to automatically parse full-length biomedical journal articles of the CRAFT Corpus into dependency structures for each sentence. The CRAFT-SA task targets dependency parses as opposed to constituency parses in order to emphasize differences that directly affect the meaning of a parsed sentence; differences in constituent parse conventions can result in parse differences that do not affect the resultant meaning of a parsed sentence (Clegg and Shepherd, 2007).

There have been previous shared tasks in the general domain NLP community to evaluate dependency parse construction using both the CoNLL-X (Buchholz and Marsi, 2006) and CoNLL-U (Zeman et al., 2018) file formats. Although the dependency parses initially distributed with the CRAFT corpus more closely resemble the older CoNLL-X format, the CRAFT dependency data was transformed into a quasi-CoNLL-U format to allow the input provided to participants to be only the text of the documents making for a more realistic scenario compared to the CoNLL-X shared tasks which required participants to match gold standard tokenization for evaluation purposes.

2.1 Data

2.1.1 Data preparation – CoNLL-X

The dependency parses distributed as part of the CRAFT corpus are automatically derived (Choi and Palmer, 2012) from the manually annotated Penn Treebank style data, which identifies the syntactic structure of each sentence. During the course of data preparation and testing, several updates were made to the Treebank data. The constituency parses for two sentences that were missing from the Treebank data were added. Also, in cases where the automatically derived dependency parse contained multiple ROOT nodes, the corresponding syntactic parse was edited, usually by dividing into multiple sentences, to ensure each

dependency parse contained only a single ROOT node. Once the errors were fixed and the CoNLL-X formatted data was finalized, the data was transformed into a quasi-CoNLL-U form.

2.1.2 Data preparation – CoNLL-U

The CoNLL-U format[5] is a revised version of the CoNLL-X format that adds a number of features such as universal part-of-speech tags, language-specific part-of-speech tags, and a standardized multi-language dependency format. It includes representations of the original raw text in addition to its segmented and tokenized form. This is required for training systems that address sentence boundary detection and tokenization as part of extracting syntactic dependencies from raw text.

The CoNLL-X representation of the CRAFT dependency parses was converted into CoNLL-U format using scripts that 1) introduce document, paragraph, and sentence boundary markers and include the original untokenized text of each sentence, 2) supplement the Penn Treebank part-of-speech tags with their corresponding universal tags following the mapping proposed by the Universal Dependencies (UD) project[6], and 3) introduce morphological features based on the same part-of-speech migration guide. Spacing and paragraph information is added to the CRAFT CoNLL-U files by aligning the CoNLL-X files with the raw text for each article.

We note that while the resulting data is in the CoNLL-U format and includes UD part-of-speech tags and features, it retains the Stanford Dependency structure and labels from the CoNLL-X files and thus, does not fully conform to the UD representation in terms of its content.

2.2 Scoring

Scoring of the CRAFT-SA task made use of the scoring software provided for the CoNLL 2018 Shared Task (Zeman et al., 2018). Dependency parse performance is measured using three metrics, LAS, MLAS, and BLEX. We provide brief definitions of these metrics in the following and refer to Zeman et al. (2018) for details.

2.2.1 LAS

The Labeled Attachment Score (LAS) metric is the de facto standard metric for evaluating de-

---

pendency parsing performance, and is commonly defined simply as the fraction of tokens for which the predicted head and dependency relation type (label) match the gold standard, i.e. `#correct/#tokens`. In the CoNLL 2018 setting applied in the CRAFT-SA task, this definition is generalized to account for cases where the predicted tokenization does not fully match the gold standard tokenization, and LAS is defined over aligned predicted (`pred-tokens`) and gold standard tokens (`gold-tokens`) as the harmonic mean (F1-score) over the precision `#correct/#pred-tokens` and recall `#correct/#gold-tokens`.

2.2.2 MLAS

The Morphology-aware Labeled Attachment Score (MLAS) is a modification of LAS that focuses on content words – ignoring e.g. punctuation and determiners – while also taking into account the part-of-speech, aspects of morphology, and associated function words. For a predicted token to be considered correct according to the MLAS criteria, it must match the gold standard values for the head and dependency label (as in LAS), and also the universal POS tag, selected morphological features (e.g. `Case`, `Number`, and `Tense`) and function words attached with particular dependency relations (e.g. `aux` and `case`). Similarly to LAS, MLAS is defined for system-predicted tokenization in terms of precision, recall and F1-score.

2.2.3 BLEX

Like MLAS, the Bilexical Dependency Score (BLEX) is a modification of LAS that focuses on content words, emphasizing lemmas instead of morphology. A predicted token is correct according to BLEX criteria if it matches the head, dependency relation, and lemma of the corresponding gold token. BLEX accounts for differences between the predicted and gold tokenization similarly to LAS and MLAS.

2.3 Baseline system

SyntaxNet (Andor et al., 2016), a transition-based neural network framework built using TensorFlow was used as the baseline system for the structural annotation task. The system was composed of two models of similar architecture: a part of speech (POS) tagger and a dependency parser. The Python NLTK punkt (Bird et al., 2009) sentence

Team	LAS	MLAS	BLEX
T013 - Run 1	65.994	0	45.618
T013 - Run 2	69.318	0	54.798
T014 - Run 1	89.695	85.549	86.631
T014 - Run 2	89.65	85.441	86.596
T014 - Run 3	89.536	85.318	86.545
Baseline	56.68	44.22	0.0

Table 1: Results showing the average score over all test documents for each metric from the structural annotation (dependency parse construction) task for all participating teams.

tokenizer was used to segment the articles into sentences which where used as input to the POS tagger model to generate POS annotations. The dependency parser model uses the POS annotations as input and generates dependency parses for each sentence. Each of the models was trained using the CRAFT training data as a gold standard.

2.4 Results

Two teams submitted five runs in total for the CRAFT-SA task (Table 1). Team T013 used the SpaCy dependency parser with (Run1) and without (Run2) the OGER NER system to test whether adding semantic information in the form of named entities can improve resultant dependency parses. In the case of this evaluation, the incorporation of an NER system caused a drop in performance, however this decrease in performance is confounded by tokenization differences resulting from their system grouping entities as single tokens. Using a neural approach and custom biomedical word embeddings, Team T014 demonstrated state of the art performance in dependency parsing over biomedical text, achieving high marks for all submitted runs. Both submitted systems out-performed the baseline by a large margin.

3 The CRAFT Coreference Resolution Task

Coreference resolution, linking strings of text that have the same referent, is a challenging NLP task that offers potential benefit to downstream tasks if done successfully. The challenge arises in linking strings of text over long distances across a document, or possibly between documents. The benefit of doing so can be substantial as coreference resolution has the ability to amplify results of upstream tasks such as concept recognition, thereby potentially improving the performance of downstream tasks, e.g. information extraction, that require explicitly represented entities. It has been estimated that successful coreference resolution would inherently add over 106,000 additional concept annotations to the CRAFT corpus through referent linkages (Cohen et al., 2017).

Coreference resolution is an active area in the NLP research community, and the most relevant previous shared task on coreference resolution is the CoNLL-2012 Shared Task (Pradhan et al., 2012), which evaluated identity chains curated in the OntoNotes project (Hovy et al., 2006). The OntoNotes corpus consists of text from conversational speech, broadcast conversations, broadcast news, magazine articles, newswire, and web data in three languages (English, Arabic, and Chinese), covering 1M words per language. The CRAFT corpus presents some unique challenges to the coreference resolution task. While slightly smaller than the OntoNotes corpus in regards to word count (1M), 620k works is still substantial, and scientific text is a domain not covered in OntoNotes explicitly. Further, CRAFT equals the highest median token count (24.0) per sentence (news wire) and the second highest median sentence count per document (318 vs. 565 for broadcast conversations) in the OntoNotes corpus. The combination of longer sentences and more sentences per document allows for an increase in the potential distances between coreference mentions within the sentences themselves and within each document. Adding further complexity to the task is CRAFT's use of discontinuous mentions, i.e. coreference mentions that have intervening text (see example of a discontinuous mention in Figure 1). Discontinuous mentions comprise 5.7% of all identity chain mentions in the CRAFT corpus. This is the first task on coreference resolution that allows for discontinuous mentions as far as the task organizers are aware.

3.1 Data

Annotation of the identity chains in the CRAFT corpus is described in (Cohen et al., 2017). For the purposes of the CRAFT-CR task, the strings of text (referred to as `mentions` below) that are linked to form coreference chains must exist in the same document, but can be localized any distance from one another. Some mentions may be

Statistic	Training	Test
Min IC length	2	2
Max IC length	187	157
Median IC length	3	2
Average IC length	4.77	4.70
Total IC	16,302	7,185
IC per document	243.3	239.5
Total mentions	77,755	33,749
Discont. mentions	4,485	1,845

Table 2: Descriptive statistics of the coreference resolution annotations in the CRAFT training and test sets. IC = identity chain

found to be adjacent while others may exist only in the document title and conclusion, for example. Two types of coreference have been resolved for all base noun phrases in the CRAFT corpus. Identity chains link mentions of the same referent, and can span the entire document. Apposition relations link adjacent noun phrases that have the same referent and are not linked by a copula. The CRAFT-CR task focuses on reproducing the manually curated identity chains.

3.1.1 Data preparation

During the course of data preparation for the CRAFT-CR task, some errors in the coreference annotations were discovered, and subsequently fixed. The most common error involved two identity chains sharing a single base noun phrase mention. Each shared mention was manually reviewed, and the two identity chains were merged in cases where the chains were deemed to be about the same referent. In cases where the presence of a shared mention in one chain was clearly an error, it was removed and the identity chains remained distinct. The CRAFT-CR training and test data are summarized in Table 2.

3.1.2 Data format

The CRAFT-CR task makes use of the CoNLL-2011/2012 data format for representing identity chains[7], with a modification to enable representation of discontinuous mentions. Discontinuous mentions are denoted by the addition of a character or characters (non-digit) after the chain identifier (integer) as depicted in Figure 1.

---

[7]See the `*_conll File Format` heading: `http:// conll.cemantix.org/2012/data.html`

```
48141 0 7   high JJ  - ... - (64a)
48141 0 8   and  CC  - ... - -
48141 0 9   low  JJ  - ... - (65
48141 0 10  IOP  NN  - ... - (64a)|65)
```

Figure 1: Sample representation of two coreference mentions, `high..IOP` and `low IOP`. Note the use of the character a in the chain identifier (`64a`) to indicate a discontinuous mention for the `high..IOP` mention. Empty columns 7-11 have been elided for figure layout consideration.

3.2 Scoring

There are a wide range of coreference resolution scoring metrics available. For historical purposes, the five reference metrics (MUC, B^3, CEAFE, CEAFM, BLANC) of Pradhan et al. (2014) are used to score the CRAFT-CR task. Due to their apparent unreliability and their low agreement rate, the Link-based Entity-Aware (LEA) metric proposed by Moosavi and Strube (2016) is also used to measure coreference system performance. The LEA metric was designed specifically to address the shortcomings of the previously used metrics. By taking into account all coreference links and evaluating resolved coreference relations instead of resolved mentions, the LEA metric accurately assesses recall and precision.

The coreference scoring implementations were modified in two ways for the CRAFT-CR task. First, because the CRAFT-CR data allows for mentions with discontinuous spans, the implementations were augmented to take as input the modified CoNLL-Coref 2011/2012 file format. Second, the implementations were updated to allow overlapping mentions to match instead of enforcing strict mention boundary matching. This option was added to allow for a slightly more flexible, permissive evaluation. The augmented implementations of all metrics used in the CRAFT-CR task have been made publicly available[8].

3.3 Baseline system

For comparison purposes, we evaluated the Berkeley coreference resolution system using the CRAFT-CR task test data (Durrett and Klein, 2013). The Berkeley system is an english coreference system predicated on learning using simple, but large numbers of lexicalized features.

This baseline evaluation made use of the built-in preprocessing machinery for sentence splitting, tokenization, and parsing, and their pre-trained CoNLL 2012 model. Prior to evaluation, results from the Berkeley system were post-processed to adjust for some system idiosyncrasies, e.g. replacing "-LRB-" in the 'word' column with the "(" or "[" that is found in the actual text, and then the coreference information was mapped onto the gold standard tokenization provided with the test data.

3.4 Results

One team submitted three runs for evaluation in the CRAFT-CR task (Table 3). They augmented the state-of-the-art end-to-end neural coreference resolution system of Lee et al. (2017) by incorporating extra syntactic features including grammatical number agreements between mentions, as well as semantic features using MetaMap to identify entity mentions. They also investigated the use of PubMed word vectors (Chiu et al., 2016) (Run1) and SciBERT word vectors (Beltagy et al., 2019) (Run2, Run3) as inputs to their model. As implemented, the system of Team T010 performed admirably compared to the baseline. F-scores are in line with some previous coreference systems used on CRAFT (Cohen et al., 2017), thus emphasizing the challenge of coreference resolution in general, and of coreference resolution over biomedical text in particular. While the baseline system and Run1 of the participant system produced on average shorter chains than those in the evaluation set (p<0.01, Mann-Whitney U test), Run2 and Run3 of the participant system were both able to generate distributions of coreference chain lengths that were not significantly different from the evaluation set (Run2: p=0.94, Run3: p=0.79, Mann-Whitney U test) suggesting that inclusion of the SciBERT embeddings helps to achieve the proper chain length distribution.

4 The CRAFT concept annotation task

Concept annotation has been a mainstay in BioNLP shared tasks dating back to the very first BioCreative, which involved the detection of gene/protein mentions in abstracts and their subsequent normalization to gene identifiers from model organism databases (Hirschman et al., 2005). Detecting biomedical concepts is a foundational NLP task, and performance of this task

---

[8]`https://github.com/bill-baumgartner/reference-coreference-scorers`; doi:10.5281/zenodo.3462790

Metric	Run	P_M	R_M	F_M	P_{CR}	R_{CR}	F_{CR}
B^3	T010 - Run 3	0.731	0.578	0.646	0.517	0.384	0.440
	Baseline	0.552	0.294	0.384	0.379	0.195	0.257
B^3_{APM}	T010 - Run 3	0.779	0.615	0.687	0.538	0.406	0.462
	Baseline	0.685	0.364	0.476	0.435	0.224	0.296
BLANC	T010 - Run 3	0.731	0.578	0.646	0.506	0.473	0.489
	Baseline	0.552	0.294	0.384	0.413	0.193	0.263
$BLANC_{APM}$	T010 - Run 3	0.779	0.616	0.688	0.513	0.480	0.496
	Baseline	0.686	0.365	0.476	0.447	0.209	0.284
CEAFE	T010 - Run 3	0.731	0.578	0.646	0.454	0.354	0.398
	Baseline	0.552	0.294	0.384	0.334	0.195	0.247
$CEAFE_{APM}$	T010 - Run 3	0.779	0.615	0.688	0.484	0.377	0.424
	Baseline	0.685	0.364	0.476	0.393	0.230	0.290
CEAFM	T010 - Run 3	0.731	0.578	0.646	0.555	0.439	0.490
	Baseline	0.552	0.294	0.384	0.429	0.228	0.298
$CEAFM_{APM}$	T010 - Run 3	0.779	0.615	0.688	0.574	0.453	0.507
	Baseline	0.685	0.365	0.476	0.487	0.259	0.338
LEA	T010 - Run 3	0.731	0.578	0.646	0.475	0.345	0.400
	Baseline	0.552	0.294	0.384	0.335	0.171	0.226
LEA_{APM}	T010 - Run 3	0.779	0.615	0.687	0.491	0.360	0.415
	Baseline	0.685	0.364	0.476	0.376	0.193	0.255
MUC	T010 - Run 3	0.731	0.578	0.646	0.644	0.511	0.570
	Baseline	0.552	0.294	0.383	0.450	0.233	0.307
MUC_{APM}	T010 - Run 3	0.779	0.616	0.688	0.665	0.527	0.588
	Baseline	0.685	0.365	0.476	0.530	0.275	0.362

Table 3: Results for the coreference resolution task. Runs achieving highest coreference F-score are shown. The APM subscript indicates that partial mention matches were allowed. P_M: mention precision; R_M: mention recall; F_M: mention F-score; P_{CR}: coreference precision; R_{CR}: coreference recall; F_{CR}: coreference F-score

impacts many potential downstream applications. Mapping textual mentions of ontology concepts presents its own set of challenges. Well-known among these are conceptual synonymy, by which a given represented concept may be indicated by multiple unique textual mentions, and textual polysemy, by which a given text string may refer to multiple represented concepts. Particularly prevalent in the biomedical literature are acronyms and other abbreviations of represented concepts. Additionally, some ontologies employ standard patterns for concept labels, but some of these may result in long, complex labels that are infrequently seen in the literature (Ogren et al., 2005; Funk et al., 2014).

The CRAFT corpus is uniquely positioned to gauge the state of the art in ontological concept recognition as it comprises over 159,000 concept annotations spanning ten ontologies from the Open Biomedical Ontologies (OBO) (Smith et al., 2007) collection. Participants in the CRAFT con-

cept annotation (CRAFT-CA) task were provided the plain-text version of each article and a file containing each ontology in the OBO format[9]. The CRAFT-CA task was further subdivided into two subtasks. The first subtask involved recognition of concepts in the original OBO files. The second subtask involved the recognition of concepts in the original OBO files augmented with *extension classes*, which are classes created by CRAFT developers but defined in terms of proper OBO classes. These extension classes were created for various reasons[10]: Some were created to capture mentions of concepts different from, but corresponding to, concepts represented in the ontologies, e.g., functionally defined entities corresponding to represented molecular functionalities. Oth-

---

[9]`https://github.com/owlcollab/oboformat`

[10]`https://github.com/UCDenver-ccp/CRAFT/blob/master/concept-annotation/README.md`

ers are semantically broadened forms of the represented concepts, while others were created to unify classes from different ontologies that were semantically equivalent so that there would not be multiple concept annotations for the same text spans if disparate annotation sets are aggregated.

4.1 Data

Concept annotations in the CRAFT corpus span ten Open Biomedical Ontologies (Smith et al., 2007), including the Chemical Entities of Biomedical Interest (ChEBI) ontology (Degtyarenko et al., 2007), the Cell Ontology (CL) (Bard et al., 2005), the Biological Process (GO_BP), Cellular Component (GO_CC) and Molecular Function (GO_MF) subontologies of the Gene Ontology (Ashburner et al., 2000), the Molecular Process Ontology (MOP)[11], the NCBI Taxonomy (NCBITaxon) (Federhen, 2011), the Protein Ontology (PR) (Natale et al., 2010), the Sequence Ontology (SO) (Eilbeck et al., 2005), and the Uberon cross-species anatomy ontology (UBERON) (Mungall et al., 2012). Note that concept annotations in the CRAFT corpus are permitted to have discontinuous spans with intervening text; e.g., for the phrase `somatic and germ cells`, the combination of the two substrings `somatic` and `cells` is annotated with the concept for `somatic cells` (CL:0002371) even though `somatic` and `cells` are not adjacent to one another in the text. There are over 2,300 concept annotations with discontinuous spans in the CRAFT corpus. The ontologies provided for the CRAFT-CA task were the same versions used during the annotation of CRAFT. As with the other tasks, the data is divided into a training set consisting of 67 full-text articles from the PMC Open Access subset, and a test set of 30 full-text articles chosen using identical selection criteria. Concept annotation of the CRAFT articles is described in detail in Bada et al. (2012) and Bada et al. (2017). Summary statistics showing total annotation counts for the ten ontologies used in the CRAFT corpus are shown in Table 4.

4.1.1 Data preparation

Some minor concept annotation errors were discovered and addressed during preparation for the CRAFT-CA task. These errors included an NCBITaxon concept that was found to not exist in the version of the NCBI Taxonomy used to annotate CRAFT, as well as some erroneous extension class prefixes used in the GO_MF extended ontology file. Errors were addressed prior to the commencement of the shared tasks.

4.1.2 Data format

The CRAFT corpus is distributed with a script that can convert its native annotation format to a variant of the BioNLP format[12] which is used for both input and output for the CRAFT-CA task. This format captures span information, the concept identifier, and the covered text for each annotation (See Figure 2).

4.2 Scoring

The method of Bossy et al. (2013) was used to measure performance of the concept annotation systems with respect to the CRAFT corpus. This method employs a hybrid measure taking into account both the degree to which the predicted annotation boundaries match the reference, as well as a similarity metric for scoring the concept match. The boundary match uses the modified Jaccard index scheme described in Bossy et al. (2012), which allows for flexible matching but prefers exact matches. The concept similarity metric of Wang et al. (2007) is used to score the predicted concepts. As suggested by Bossy et al. (2013), the weight factor, w, was set to 0.65, which ensures that ancestor/descendant predictions always have a greater value than sibling predictions, while root predictions never yield a similarity greater than 0.5. An implementation of the scoring algorithm has been made publicly available[13].

4.3 Baseline system

We evaluated a baseline system on the CRAFT-CA data to use as a comparison for the participant-submitted runs. The baseline system is a two-stage machine learning system proposed in Hailu (2019) and trained only on the CRAFT corpus. The first stage makes use of NERSuite (Cho et al., 2010) to detect concept mention spans using a conditional random field (CRF) model. The CRF model was trained as described in Okazaki (2007), and uses as features words, parts of speech, and constituency parse information within a window of three tokens

---

[11]`http://obofoundry.org/ontology/mop.html`

[12]`http://2013.bionlp-st.org/file-formats`

[13]`https://github.com/UCDenver-ccp/craft-shared-tasks`; doi:10.5281/zenodo.3460928

Ontology	Training	Test		Ontology	Training	Test
CHEBI	4,548 (18)	2,200 (14)		CHEBI_EXT	11,915 (38)	5,248 (19)
CL	4,043 (244)	1,749 (175)		CL_EXT	6,275 (249)	2,872 (175)
GO_BP	9,280 (493)	3,681 (272)		GO_BP_EXT	13,954 (526)	5,847 (287)
GO_CC	4,075 (80)	1,184 (14)		GO_CC_EXT	8,495 (150)	3,217 (30)
GO_MF	375 (0)	94 (0)		GO_MF_EXT	4,070 (28)	1,822 (20)
MOP	240 (0)	101 (0)		MOP_EXT	386 (0)	111 (0)
NCBITaxon	7,362 (2)	3,101(0)		NCBITaxon_EXT	7,592 (2)	3,219 (0)
PR	17,038 (84)	6,409 (44)		PR_EXT	19,862 (110)	7,932 (44)
SO	8,797 (108)	3,446 (45)		SO_EXT	24,955 (182)	9,136 (72)
UBERON	12,269 (235)	6,551 (118)		UBERON_EXT	14,910 (255)	7,416 (133)

Table 4: Total and discontinuous (in parentheses) concept annotation counts by ontology for both the 67 article training and 30 article test sets.

```
T1    CL:0000540 83 89      neuron
T2    CL:0002613 239 247;259 265    striatal ... neuron
T3    CL:0002613 434 442;451 457    striatal ... neuron
T4    CL:0000540 703 709    Neuron
```

Figure 2: Sample annotations demonstrating the BioNLP format used as input and output for the CRAFT-CA task. Note the presence of two annotations with discontinuous spans. The document identifier is indicated in the filename for each annotation file.

Ontology	Submission	Proper OBO				OBO + extension			
		SER	P	R	F1	SER	P	R	F1
CHEBI	T013 - Run 3/1	0.34	0.79	0.75	0.77	0.27	0.84	0.79	0.81
	Baseline	0.44	0.91	0.59	0.72	0.29	0.89	0.73	0.80
CL	T013 - Run 3/2a	0.56	0.68	0.62	0.65	0.35	0.77	0.67	0.72
	Baseline	0.53	0.83	0.48	0.61	0.33	0.79	0.67	0.73
GO_BP	T013 - Run 3/1	0.30	0.83	0.78	0.80	0.29	0.81	0.81	0.81
	Baseline	0.39	0.83	0.64	0.72	0.29	0.84	0.74	0.79
GO_CC	T013 - Run 1/2a	0.39	0.77	0.75	0.76	0.20	0.92	0.83	0.87
	Baseline	0.44	0.88	0.60	0.71	0.20	0.93	0.83	0.88
GO_MF	T013 - Run 2/2a	0.04	0.99	0.96	0.98	0.39	0.82	0.68	0.74
	Baseline	0.07	0.99	0.92	0.95	0.45	0.82	0.56	0.66
MOP	T013 - Run 3/2a	0.27	0.81	0.94	0.87	0.34	0.89	0.73	0.79
	Baseline	0.43	0.87	0.65	0.75	0.36	0.88	0.72	0.79
NCBITaxon	T013 - Run 3/2a	0.05	0.97	0.97	0.97	0.077	0.98	0.93	0.96
	Baseline	0.07	0.99	0.93	0.96	0.07	0.99	0.94	0.96
PR	T013 - Run 3/1	0.68	0.50	0.59	0.54	0.73	0.49	0.46	0.47
	Baseline	0.69	0.60	0.40	0.48	0.62	0.61	0.45	0.52
SO	T013 - Run 3/2a	0.16	0.90	0.88	0.89	0.13	0.92	0.91	0.92
	Baseline	0.21	0.91	0.82	0.86	0.18	0.92	0.85	0.89
UBERON	T013 - Run 1/2a	0.37	0.77	0.71	0.74	0.39	0.77	0.69	0.73
	Baseline	0.41	0.84	0.61	0.70	0.36	0.86	0.66	0.75

Table 5: Aggregate concept annotation results evaluated per ontology against the 30 CRAFT test documents. For Team T013, their highest scoring run is displayed based on SER. Run identifiers indicate (proper OBO/OBO_EXT). Note that Run 2a is an unofficial run as it was submitted after the deadline, however since there were no other teams participating, Run 2a is included in the official results. SER = Slot Error Rate; P = Precision; R = Recall; F1 = F1-score.

upstream and downstream of each concept mention. The second stage links each textual mention identified by the CRF to an ontology identifier using a stacked Bi-LSTM approach implemented by the OpenNMT system (Klein et al., 2018). By modeling concept normalization as sequence-to-sequence translation at the character level, the baseline system maps characters in the text spans identified in the first stage to characters in ontology identifiers to normalize concepts.

4.4 Results

One team submitted three runs to the CRAFT-CA task (Table 5). They used variants of two systems, one a modified ontology-specific BioBert[14] model with (Run3) and without (Run1) input from the OGER NER system (Furrer et al., 2019) and with weights pretrained on PubMed using identifiers from the ontologies as the tag set, and the other a BiLSTM with ontology pretraining (Run2). With regard to overall system performances, marked improvement in recognition of concepts from CHEBI, GO_BP, GO_MF, and SO was observed compared to past evaluations using the CRAFT public dataset (Funk et al., 2014). However, it is important to note that past evaluations were performed on CRAFT v1/2 concept annotations, whereas the testing of this shared task was performed on v3 concept annotations, which constitute a major update of the concept annotations relative to those of v1/2 (including first usage of extension classes), so we do not believe it is safe to directly compare evaluations performed on these substantially different versions of the concept annotations. The BioBert approach augmented with the OGER NER system (Run3) generally outperforms the other approaches when normalizing to proper OBO concepts, whereas the BiLSTM approach is generally better when the extension classes are used.

Neither the baseline system, nor any of the submitted runs identified annotations with discontinuous spans. Though annotations with discontinuous spans make up only a small percentage (1.46%) of the overall annotations, their exclusion from system output could represent potential low hanging fruit for improving overall system performance. Protein Ontology concept recognition remains a target for future work as system performances did not surpass an F-score of 0.55. In-

clusion of the extension classes generally resulted in improvement of performance when compared to runs using only the proper ontology concepts, possibly attributable to the labels and synonyms that were provided for the extension classes. One exception is for GO_MF_EXT where performance is expected to suffer with inclusion of the extension class annotations as the proper ontology class count was limited to a very small subset of the original ontology. Overall, however, performance on the CRAFT-CA task demonstrated state-of-the-art performance for ontological concept recognition in biomedical text.

5 Conclusion

The CRAFT-ST 2019 provides a platform to gauge performance on three fundamental NLP tasks, automated dependency parse construction, coreference resolution, and ontology concept annotation against a high quality, manually annotated corpus of full-text biomedical articles. Submitted runs from participating systems demonstrate promising results, particularly with respect to automated dependency parse construction and some aspects of ontological concept annotation. Clear needs for improved extraction of protein ontology concepts remain, while the neural approaches used have addressed long standing deficiencies in the recognition of biological process concepts in text. Coreference resolution system performances highlight the existing challenges of coreference resolution in general, and of coreference resolution over biomedical text in particular.

The approaches taken by participants in the CRAFT-ST 2019 mirror the current themes in AI and NLP today. Neural approaches are unsurprisingly the preferred methodology for addressing these NLP tasks. The CRAFT ST 2019 have provided new benchmarks for these fundamental NLP tasks, setting the stage for the next evolution of system development.

Acknowledgments

The authors would like to thank Kevin Cohen and Karin Verspoor for their input during the early planning stages for the CRAFT Shared Task 2019, and Tiffany Callahan for help with the coreference chain length statistics. The authors gratefully acknowledge their support from NIH grants R01LM009254, R01LM008111 and T15LM009451.

---

[14]`https://github.com/dmis-lab/biobert`

References

Daniel Andor, Chris Alberti, David Weiss, Aliaksei Severyn, Alessandro Presta, Kuzman Ganchev, Slav Petrov, and Michael Collins. 2016. Globally normalized transition-based neural networks. In *Proceedings of the 54th Annual Meeting of the Association for Computational Linguistics (Volume 1: Long Papers)*, pages 2442–2452, Berlin, Germany. Association for Computational Linguistics.

Michael Ashburner, Catherine A Ball, Judith A Blake, David Botstein, Heather Butler, J Michael Cherry, Allan P Davis, Kara Dolinski, Selina S Dwight, Janan T Eppig, et al. 2000. Gene ontology: tool for the unification of biology. *Nature genetics*, 25(1):25.

Michael Bada, Miriam Eckert, Donald Evans, Kristin Garcia, Krista Shipley, Dmitry Sitnikov, William A Baumgartner, K Bretonnel Cohen, Karin Verspoor, Judith A Blake, et al. 2012. Concept annotation in the craft corpus. *BMC bioinformatics*, 13(1):161.

Michael Bada, Nicole Vasilevsky, William A Baumgartner, Melissa Haendel, and Lawrence E Hunter. 2017. Gold-standard ontology-based anatomical annotation in the craft corpus. *Database*, 2017.

Jonathan Bard, Seung Y Rhee, and Michael Ashburner. 2005. An ontology for cell types. *Genome biology*, 6(2):R21.

Iz Beltagy, Arman Cohan, and Kyle Lo. 2019. Scibert: A pretrained language model for scientific text. In *Proceedings of the 2019 Conference on Empirical Methods in Natural Language Processing*, Hong Kong. Association for Computational Linguistics.

Steven Bird, Ewan Klein, and Edward Loper. 2009. *Natural language processing with Python: analyzing text with the natural language toolkit.* " O'Reilly Media, Inc.".

Robert Bossy, Wiktoria Golik, Zorana Ratkovic, Philippe Bessières, and Claire Nédellec. 2013. Bionlp shared task 2013–an overview of the bacteria biotope task. In *Proceedings of the BioNLP Shared Task 2013 Workshop*, pages 161–169.

Robert Bossy, Julien Jourde, Alain-Pierre Manine, Philippe Veber, Erick Alphonse, Maarten Van De Guchte, Philippe Bessières, and Claire Nédellec. 2012. Bionlp shared task-the bacteria track. In *BMC bioinformatics*, volume 13, page S3. BioMed Central.

Sabine Buchholz and Erwin Marsi. 2006. Conll-x shared task on multilingual dependency parsing. In *Proceedings of the tenth conference on computational natural language learning*, pages 149–164. Association for Computational Linguistics.

Billy Chiu, Gamal Crichton, Anna Korhonen, and Sampo Pyysalo. 2016. How to train good word embeddings for biomedical nlp. In *Proceedings of the 15th workshop on biomedical natural language processing*, pages 166–174.

Han-Cheol Cho, Naoaki Okazaki, Makoto Miwa, and Jun?ichi Tsujii. 2010. Nersuite: a named entity recognition toolkit. *Tsujii Laboratory, Department of Information Science, University of Tokyo, Tokyo, Japan.*

Jinho D Choi and Martha Palmer. 2012. Guidelines for the clear style constituent to dependency conversion. *Technical Report 01–12.*

Miji Choi, Haibin Liu, William Baumgartner, Justin Zobel, and Karin Verspoor. 2016. Coreference resolution improves extraction of biological expression language statements from texts. *Database*, 2016.

Andrew B Clegg and Adrian J Shepherd. 2007. Benchmarking natural-language parsers for biological applications using dependency graphs. *BMC bioinformatics*, 8(1):24.

K Bretonnel Cohen, Arrick Lanfranchi, Miji Joo-young Choi, Michael Bada, William A Baumgartner, Natalya Panteleyeva, Karin Verspoor, Martha Palmer, and Lawrence E Hunter. 2017. Coreference annotation and resolution in the colorado richly annotated full text (craft) corpus of biomedical journal articles. *BMC bioinformatics*, 18(1):372.

Kirill Degtyarenko, Paula De Matos, Marcus Ennis, Janna Hastings, Martin Zbinden, Alan McNaught, Rafael Alcántara, Michael Darsow, Mickaël Guedj, and Michael Ashburner. 2007. Chebi: a database and ontology for chemical entities of biological interest. *Nucleic acids research*, 36(suppl_1):D344–D350.

Greg Durrett and Dan Klein. 2013. Easy victories and uphill battles in coreference resolution. In *Proceedings of the 2013 Conference on Empirical Methods in Natural Language Processing*, pages 1971–1982.

Karen Eilbeck, Suzanna E Lewis, Christopher J Mungall, Mark Yandell, Lincoln Stein, Richard Durbin, and Michael Ashburner. 2005. The sequence ontology: a tool for the unification of genome annotations. *Genome biology*, 6(5):R44.

Scott Federhen. 2011. The ncbi taxonomy database. *Nucleic acids research*, 40(D1):D136–D143.

Christopher Funk, William Baumgartner, Benjamin Garcia, Christophe Roeder, Michael Bada, K Bretonnel Cohen, Lawrence E Hunter, and Karin Verspoor. 2014. Large-scale biomedical concept recognition: an evaluation of current automatic annotators and their parameters. *BMC bioinformatics*, 15(1):59.

Lenz Furrer, Anna Jancso, Nicola Colic, and Fabio Rinaldi. 2019. Oger++: hybrid multi-type entity recognition. *Journal of cheminformatics*, 11(1):7.

Anupama Gupta, Imon Banerjee, and Daniel L Rubin. 2018. Automatic information extraction from unstructured mammography reports using distributed semantics. *Journal of biomedical informatics*, 78:78–86.

Negacy Degefa Hailu. 2019. *Investigation of traditional and deep neural sequence models for biomedical concept recognition*. Ph.D. thesis, University of Colorado.

Lynette Hirschman, Alexander Yeh, Christian Blaschke, and Alfonso Valencia. 2005. Overview of biocreative: critical assessment of information extraction for biology.

Eduard Hovy, Mitchell Marcus, Martha Palmer, Lance Ramshaw, and Ralph Weischedel. 2006. Ontonotes: the 90% solution. In *Proceedings of the human language technology conference of the NAACL, Companion Volume: Short Papers*, pages 57–60.

Guillaume Klein, Yoon Kim, Yuntian Deng, Vincent Nguyen, Jean Senellart, and Alexander M Rush. 2018. Opennmt: Neural machine translation toolkit. *arXiv preprint arXiv:1805.11462*.

Martin Krallinger, Florian Leitner, Obdulia Rabal, Miguel Vazquez, Julen Oyarzabal, and Alfonso Valencia. 2013. Overview of the chemical compound and drug name recognition (chemdner) task. In *BioCreative challenge evaluation workshop*, volume 2, page 2. Citeseer.

Kenton Lee, Luheng He, Mike Lewis, and Luke Zettlemoyer. 2017. End-to-end neural coreference resolution. In *Proceedings of the 2017 Conference on Empirical Methods in Natural Language Processing*, pages 188–197, Copenhagen, Denmark. Association for Computational Linguistics.

Nafise Sadat Moosavi and Michael Strube. 2016. Which coreference evaluation metric do you trust? a proposal for a link-based entity aware metric. In *Proceedings of the 54th Annual Meeting of the Association for Computational Linguistics (Volume 1: Long Papers)*, pages 632–642.

Christopher J Mungall, Carlo Torniai, Georgios V Gkoutos, Suzanna E Lewis, and Melissa A Haendel. 2012. Uberon, an integrative multi-species anatomy ontology. *Genome biology*, 13(1):R5.

Darren A Natale, Cecilia N Arighi, Winona C Barker, Judith A Blake, Carol J Bult, Michael Caudy, Harold J Drabkin, Peter D?Eustachio, Alexei V Evsikov, Hongzhan Huang, et al. 2010. The protein ontology: a structured representation of protein forms and complexes. *Nucleic acids research*, 39(suppl_1):D539–D545.

Philip V Ogren, K Bretonnel Cohen, and Lawrence Hunter. 2005. Implications of compositionality in the gene ontology for its curation and usage. In *Biocomputing 2005*, pages 174–185. World Scientific.

Naoaki Okazaki. 2007. Crfsuite: a fast implementation of conditional random fields (crfs).

Sameer Pradhan, Xiaoqiang Luo, Marta Recasens, Eduard Hovy, Vincent Ng, and Michael Strube. 2014. Scoring coreference partitions of predicted mentions: A reference implementation. In *Proceedings of the conference. Association for Computational Linguistics. Meeting*, volume 2014, page 30. NIH Public Access.

Sameer Pradhan, Alessandro Moschitti, Nianwen Xue, Olga Uryupina, and Yuchen Zhang. 2012. Conll-2012 shared task: Modeling multilingual unrestricted coreference in ontonotes. In *Joint Conference on EMNLP and CoNLL-Shared Task*, pages 1–40. Association for Computational Linguistics.

Barry Smith, Michael Ashburner, Cornelius Rosse, Jonathan Bard, William Bug, Werner Ceusters, Louis J Goldberg, Karen Eilbeck, Amelia Ireland, Christopher J Mungall, et al. 2007. The obo foundry: coordinated evolution of ontologies to support biomedical data integration. *Nature biotechnology*, 25(11):1251.

Larry Smith, Lorraine K Tanabe, Rie Johnson nee Ando, Cheng-Ju Kuo, I-Fang Chung, Chun-Nan Hsu, Yu-Shi Lin, Roman Klinger, Christoph M Friedrich, Kuzman Ganchev, et al. 2008. Overview of biocreative ii gene mention recognition. *Genome biology*, 9(2):S2.

Larry H Smith and W John Wilbur. 2009. The value of parsing as feature generation for gene mention recognition. *Journal of biomedical informatics*, 42(5):895–904.

Dongdong Sun, Minghui Wang, and Ao Li. 2017. Mptm: A tool for mining protein post-translational modifications from literature. *Journal of bioinformatics and computational biology*, 15(05):1740005.

James Z Wang, Zhidian Du, Rapeeporn Payattakool, Philip S Yu, and Chin-Fu Chen. 2007. A new method to measure the semantic similarity of go terms. *Bioinformatics*, 23(10):1274–1281.

Daniel Zeman, Jan Hajič, Martin Popel, Martin Potthast, Milan Straka, Filip Ginter, Joakim Nivre, and Slav Petrov. 2018. Conll 2018 shared task: multilingual parsing from raw text to universal dependencies. In *Proceedings of the CoNLL 2018 Shared Task: Multilingual Parsing from Raw Text to Universal Dependencies*, pages 1–21.

UZH@CRAFT-ST: a Sequence-labeling Approach to Concept Recognition

Lenz Furrer[*†]   **Joseph Cornelius**[*]   **Fabio Rinaldi**[*†]

[*]University of Zurich, Institute of Computational Linguistics
[†]Swiss Institute of Bioinformatics
Andreasstr. 15, CH-8050 Zürich, Switzerland
{`lenz.furrer,joseph.cornelius,fabio.rinaldi`}@uzh.ch

Abstract

As our submission to the CRAFT shared task 2019, we present two neural approaches to concept recognition. We propose two different systems for joint named entity recognition (NER) and normalization (NEN), both of which model the task as a sequence labeling problem. Our first system is a BiLSTM network with two separate outputs for NER and NEN trained from scratch, whereas the second system is an instance of BioBERT fine-tuned on the concept-recognition task. We exploit two strategies for extending concept coverage, ontology pretraining and backoff with a dictionary lookup. Our results show that the backoff strategy effectively tackles the problem of unseen concepts, addressing a major limitation of the chosen design. In the cross-system comparison, BioBERT proves to be a strong basis for creating a concept-recognition system, although some entity types are predicted more accurately by the BiLSTM-based system.

1 Introduction

We describe our submission to the CRAFT shared task 2019. We participated in the concept annotation (CA) subtask, which comprises biomedical named entity recognition (NER) and normalization (NEN) for full-text scientific articles. We tested two different neural architectures, a BiLSTM-based network trained from scratch and a transformer system obtained by fine-tuning Bio-BERT. While NER+NEN tasks have often been approached with a pipeline architecture (NER output passed to NEN as input), we strove for tackling both tasks jointly in a single model.

In essence, we cast the task as a sequence-labeling problem, by directly predicting IDs as symbolic labels. This approach has the obvious drawback that the models will only ever predict IDs that were seen in the training data. In order to account for this limitation, we used different strategies to enrich the systems with information derived from terminology resources, such as ontology pretraining and combination with a rule-based dictionary-lookup system.

The source code of our systems is publicly available at `https://github.com/OntoGene/craft-st`.

2 Data

The CRAFT corpus (Bada et al., 2012; Cohen et al., 2017) is a collection of 97 full-text articles, of which 30 have been released only in the course of the present shared task. The documents were manually annotated with respect to 10 different entity types, linked to 8 manually curated ontologies of biomedical terminology:

CHEBI: chemicals/small molecules (Chemical Entities of Biological Interest)

CL: cell types (Cell Ontology)

GO_CC: cellular and extracellular components and regions (Gene Ontology)

GO_BP: biological processes (Gene Ontology)

GO_MF: molecular functionalities possessed by genes (Gene Ontology)

MOP: chemical reactions and other molecular processes (Molecular Process Ontology)

NCBITaxon: biological taxa and organisms (NCBI Taxonomy)

PR: proteins, genes, and transcripts (Protein Ontology)

SO: biomacromolecular entities, sequence features (Sequence Ontology)

UBERON: anatomical entities (UBERON)

In addition, the annotations are distributed in an extended variant, i.e. **CHEBI_EXT, CL_EXT** etc., resulting in a total of 20 annotation sets. For the extension annotations, the creators of the CRAFT corpus modified the given ontologies in

Proceedings of the 5th Workshop on BioNLP Open Shared Tasks, pages 185–195
Hong Kong, China, November 4, 2019. ©2019 Association for Computational Linguistics

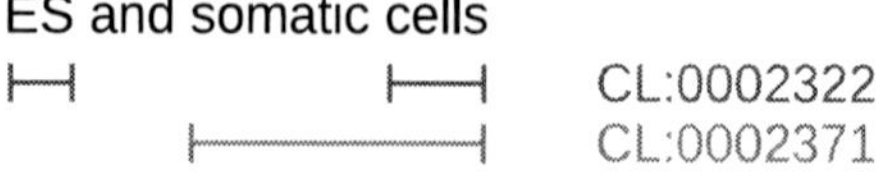

Figure 1: Example of discontinuous and overlapping annotations in an elliptical coordination construction.

a way to better represent actual usage of biomedical entities in scientific texts. In many cases, new concepts were added or existing ones were replaced; some concepts were merged across ontologies (e. g. CL_GO_EXT:cell, which refers to an unspecific cell).

The size of the ontologies varies considerably, ranging from 5 concepts for GO_MF to 1,167,358 concepts for NCBITaxon_EXT. The 67 articles released for training contain a total of 575,296 tokens and the 30 test articles contain 239,409 tokens. In the training set of the corpus, PR_EXT holds the most annotations (19,862 mentions of 1075 unique IDs) and MOP has the fewest (240 mentions of 16 unique IDs). The corpus includes 1264 discontinuous annotations, which are found most frequently among the GO_BP annotations with 493 occurrences. Of these, 788 annotations partially overlap with another annotation of the same type, sharing at least one token (cf. Figure 1).

Furthermore, the corpus contains 3362 annotations that overlap with an annotation of a different type. The three most common combinations are ⟨CL, UBERON⟩ (571), ⟨GO_BP, UBERON⟩ (500) and ⟨CL, GO_BP⟩ (349). The three most common terms with cross-type annotations are "gene expression" (161), "Mcm4/6/7" (107) and "Cln3" (97), whereby the ten most common terms account for 22.159% of the overlapping annotations.

For the present work, we treated each annotation set as a separate dataset independent of all others, resulting in 20 individual tasks. This is in accordance with how the evaluation is carried out.

2.1 Preprocessing

The CRAFT corpus is distributed with annotations in a stand-off format, i. e. separated from the text. The primary format is Knowtator XML, but a format-conversion suite is provided for producing BioNLP format, which is more easily processed and which is also required for the system predictions by the official evaluation suite.

The stand-off formats allow representing inter-laced annotations, such as discontinuous spans and overlapping concepts, which often occur together (cf. Figure 1). For sequence classification, however, two parallel sequences of tokens and labels with one-to-one correspondence are required, typically using IOB or IOBES tags. There is no straight-forward method to represent interlaced annotations in this format, even though potential solutions have been proposed (Metke-Jimenez and Karimi, 2016; Dai, 2018). Instead, we decided to use a lossy transformation which simplifies the annotations during the conversion. While this means that our systems cannot represent (and thus predict) all required types of annotations, we believe that the phenomenon is too rare to justify the increase in complexity (multi-class classification for overlaps, additional labels for discontinuity, more complex heuristics in postprocessing).

We used the *standoff2conll* suite[1] for converting the annotations from BioNLP to a CoNLL-like tab-separated format. We chose the "first-span" strategy for resolving discontinuous spans and "keep-longer" for overlapping concepts, the former of which we wrote ourselves in analogy to the existing "last-span" strategy. The *standoff2conll* suite also takes care of sentence splitting and tokenization, using rule-based approaches.

In addition, we applied abbreviation expansion using Ab3P (Sohn et al., 2008). We removed short-form candidates that were all-lowercase, consisted of only one character or had a P-precision (Ab3P's confidence metric) of less than 0.9. For each article, all occurrences of the remaining short forms were then replaced with their best-matching long-form (highest P-precision). Abbreviation expansion was only integrated in the BiLSTM system.

2.2 Postprocessing

Since our systems produce predictions in a CoNLL-like format, an additional conversion step was necessary to meet the requirements of the evaluation suite (BioNLP format). As another contribution to the *standoff2conll* tool, we wrote a converter for the inverted direction (CoNLL to stand-off). The converter is graceful with respect to invalid tag sequences (e. g. O – I – O) and makes use of existing functionality.

---

[1] https://github.com/spyysalo/standoff2conll

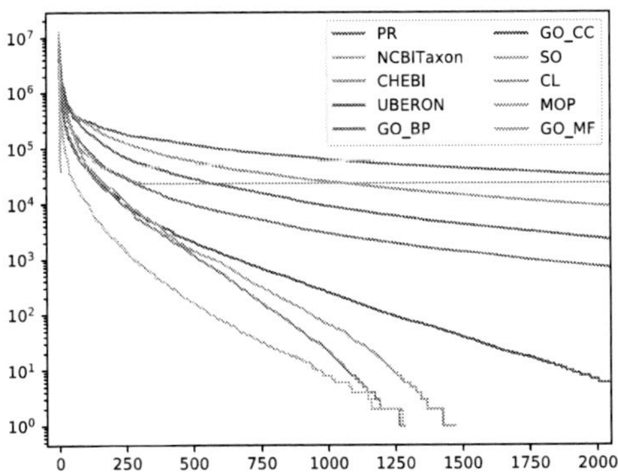

Figure 2: Occurrences of all concepts in the CRAFT ontologies, as annotated by OGER in a large subset of Medline+PMC, sorted by rank.

3 System Description

For the concept annotation task of the CRAFT shared task, we tested two different neural architectures, BiLSTM and transformer (BERT). In addition, we used a rule-based dictionary-lookup system (OGER), which served both as a baseline and as an auxiliary component in the machine-learning systems.

All three systems are applied to each of the annotation sets individually, i.e. each system performs 20 independent predictions. For the neural systems this means that we trained 20 separate models for each configuration; in the case of cross-validation, the number of models is multiplied by another factor.

In a supervised classification setup, an example-based model can only ever predict concepts that have been seen in the training phase. As the concept vocabularies are very large for most of the entity types, an annotated corpus with full coverage is out of reach. However, since the mentions of biomedical concepts resemble a Zipfian distribution (cf. Figure 2), it is often possible to achieve reasonable performance in terms of F-Score even with such a restricted label set. Yet a system that is limited to the concepts of a training corpus is undesirable in many application scenarios. For this reason, we searched for ways to combine the neural systems with the dictionary-based system OGER, which requires no training and can target the entire set of concepts from a given ontology.

Another common challenge of the neural systems, inherent to the sequence-labeling approach, is the classification of multi-word expressions, as each token is labeled individually. This is especially true for semantically weak tokens like stop words, single letters, or numbers (e. g. "I" in "Hexokinase I"). Correctly annotating these tokens is only possible in light of their context, which makes them exceedingly demanding with respect to generalization.

In contrast, OGER annotates multi-word expressions jointly with a single lookup for the entire span. As another difference, OGER can predict multiple concepts for the same span or even interleaved spans, whereas the sequence taggers can only assign one concept to each token.

3.1 Dictionary-based System

OGER (Basaldella et al., 2017; Furrer et al., 2019) is a fast, reliable concept-recognition system based on dictionary lookup. It is highly flexible in terms of matching rules (tokenization, spelling normalization) and supports a wide range of input/output formats. For the present work, we used the following spelling normalization rules: transliteration of Greek letter names, ise/ize conflation, and stemming. Based on the performance on the training set, we fine-tuned the configuration on a per-ontology basis; e.g. stemming was disabled for NCBITaxon and PR.

3.2 BiLSTM-based System

Architecture

Our first neural sequence tagger is a network with a bidirectional Long Short-Term Memory (BiLSTM) (Hochreiter and Schmidhuber, 1997; Schuster and Paliwal, 1997) layer at its core. Its architecture is illustrated in Figure 3. The input tokens x are represented using pretrained word embeddings (Chiu et al., 2016) and randomly initialized character embeddings, the latter of which are transformed into a token-level vector through a convolution and pooling operation (not shown in the figure). The token representation is concatenated with a dictionary feature x^O, which is a vector that encodes the predictions by OGER (using the same dimensionality as the NEN output vector over y^C, see below).

The subsequent layers are inspired by the work of Zhao et al. (2019), who propose a multi-task-learning framework to jointly tackle span detection (NER) and normalization (NEN). A key step to make NER and NEN compatible was to model NEN as a sequence-labeling problem, where IDs

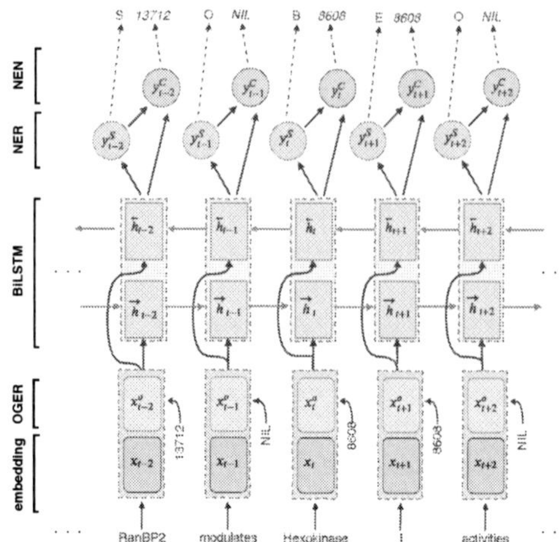

Figure 3: Architecture of the BiLSTM-based sequence tagger (simplified).

Figure 4: Example of labeling with IOBES tags (NER) and concept IDs (NEN).

are predicted for each token just like span tags in NER (cf. Figure 4). A BiLSTM layer consumes a sequence of token representations one sentence at a time. The sequence representation is then forked into two output layers with soft-max activation, which solve different tasks: The span-detection layer predicts one of the labels $y^S = \{I, O, B, E, S\}$, as in a classical single-type NER problem. The normalization layer predicts concept labels (IDs) from $y^C = y_T^C \cup y_P^C \cup y_O^C$, where y_T^C are all labels seen in the training corpus, y_P^C are the labels seen in ontology pretraining and y_O^C are all labels found by OGER. The label set y^C includes the NIL symbol, which denotes the absence of a concept annotation. In addition to the hidden states of the BiLSTM layer, the normalization layer takes the output of the span-detection layer as an input. In contrast to Zhao et al., there is no symmetric feedback between the two output layers, i. e. the span-detection layer does not "see" the output of the normalization layer. This allows training spans and concepts simultaneously.

Training a BiLSTM model for NER and NEN

Training is performed in two phases, ontology pretraining and main training. In the first phase, the model adapts to the domain of the respective entity type by means of terminology entries. At this stage, the model is trained on isolated names and synonyms extracted from the provided ontology files. Due to technical limitations, we restricted the pretraining data to the 1000 most com-

mon concepts of each ontology. As an approximation for determining the most commonly used concepts in the literature, we automatically annotated a large subset of Medline (26M abstracts) and PubMed Central (725k articles) with OGER. We sorted the annotated concepts by occurrence and manually removed high-frequency false positives. The model is then pretrained on the top 1000 concepts for a fixed number of 20 epochs.

In the main training phase, training continues with full sentences from the CRAFT corpus. At this stage, the model learns to predict concept mentions in real-world language usage, including contextual hints, frequency distribution, and challenges like rephrasing and non-standard spelling. While the main training is likely to override parts of the connections learnt during ontology pretraining, others may remain to form some kind of background knowledge. Main training is performed as 6-fold cross-validation, where the held-out set of each fold is used to determine when to stop training, using a patience value of 5 epochs. Thus, 6 models are trained for each entity type.

Agreement of NER and NEN Predictions

At prediction time, the softmax scores from all 6 models are averaged before the highest-ranking label for a particular token is determined. Also, when abbreviations have been expanded into multiple tokens during preprocessing, their scores are averaged prior to label selection. The outputs for NER and NEN are tested for agreement. Agreement means that both outputs see a given token t as either relevant or irrelevant, or formally:

$$(\hat{y}_t^S = O \wedge \hat{y}_t^C = \text{NIL}) \vee (\hat{y}_t^S \neq O \wedge \hat{y}_t^C \neq \text{NIL})$$

The labels $\hat{y}_t^S$ and $\hat{y}_t^C$ are chosen such that they satisfy the above requirement, while maximizing the overall score. In practice, we compare the score product of the irrelevant labels (O/NIL) to the score product of the top-ranking relevant labels

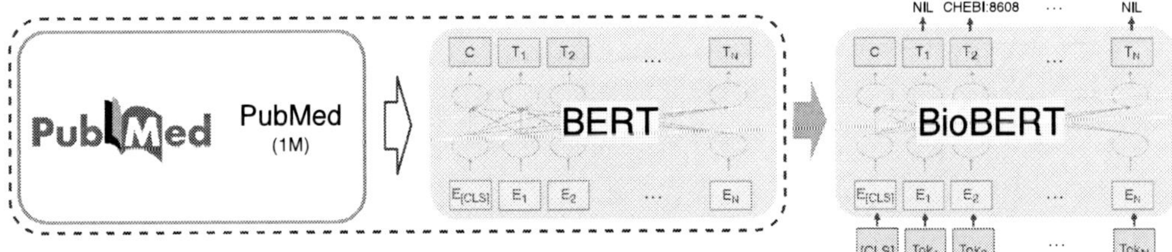

Figure 5: A symbolic illustration of the BioBERT model as a result of a BERT model pretrained on the PubMed corpus and fine-tuned for NEN on the CRAFT corpus.

of either output. This means that we might select a non-best-ranking label for one of the outputs.

3.3 BERT-based System

Background: BERT and BioBERT

The multi-layer BERT model (Devlin et al., 2019) is trained in an unsupervised setting to create bidirectional contextual representations of a token from unlabeled text conditioned on the left and the right context. Two tasks are used to train the BERT model: first, to predict whether two sentences follow each other, and second, to predict a randomly masked token. The resulting pretrained BERT model can be applied to a large number of tasks, such as question answering, next sentence prediction, or NER. Recently it has been shown that the use of pretrained BERT models is especially beneficial to NER tasks (Devlin et al., 2019). In contrast to traditional models used for NER tasks such as long short-term memory (LSTM) models and conditional random field (CRF) models (Habibi et al., 2017), which use context-independent word vector representations such as Word2vec (Mikolov et al., 2013) or GLOVE (Pennington et al., 2014), the BERT model learns context-dependent word vector representations.

A specialized variant of the BERT model for the biomedical domain is the BioBERT (Lee et al., 2019) model, which has been shown to produce state-of-the-art results for NER in the biomedical domain (Jin et al., 2019). The BioBERT model is initialized using the BERT model pretrained on general-domain data (Wikipedia, Bookcorpus) and is then pretrained an additional 200k steps on a corpus of one million PubMed abstracts.

Fine-tuning BioBERT for NER and NEN

For our second system in the CRAFT shared task, we used the readily pretrained BioBERT model available online.[2] We wrote a task-specific head for ID tagging and fine-tuned the model on the CRAFT corpus for another 55 epochs. Like the BiLSTM system, the model is trained to directly predict a sequence of concept IDs from a sequence of input tokens. Technically, we implemented this as an adaptation of an NER tagger by extending the tagset to all concept labels of the training set (cf. Figures 4 and 5).

As a variant, we fine-tuned another BioBERT model as a classical NER tagger over IOBES tags and combined the resulting predictions with annotations from OGER. Predictions were only kept if both OGER and BERT agreed, i. e. both produced a label different from O/NIL. This system, which resembles a traditional NER+NEN pipeline, combines the high recall of the dictionary-based system with the context-aware span detection of an example-based classifier.

Additionally, we combined the previous two systems into a third system. In this variant, the ID tagger takes precedence, whereas the span tagger serves as a backoff model. Whenever the first system does not predict an ID for a token, the backoff system gets a chance to provide an ID, thus joining the forces of two alternative approaches.

3.4 Related Work

Concept-recognition systems solve the task of detecting and linking textual mentions to terminology identifiers. In the past, this problem has often been approached with a pipeline combining an NER tagger with a dictionary-lookup module (e. g. Campos et al., 2013; Ghiasvand and Kate, 2014) or a rule-based system (D'Souza and Ng, 2015; Lee et al., 2016). Leaman et al. (2013) prepared the ground for machine-learning approaches to the normalization task, modeling it as a rank-

---

[2] `https://github.com/naver/biobert-pretrained`

<table>
<tr><td rowspan="2"></td><td rowspan="2"></td><td>CHE-
BI</td><td>CL</td><td>GO
BP</td><td>GO
CC</td><td>GO
MF</td><td>MOP</td><td>NCBI
Taxon</td><td>PR</td><td>SO</td><td>UBE-
RON</td></tr>
<tr><td>OGER (baseline)</td><td></td><td>0.5808</td><td>0.6657</td><td>0.2832</td><td>0.6838</td><td>0.8632</td><td>0.4459</td><td>0.5947</td><td>0.4581</td><td>0.5362</td><td>0.6561</td></tr>
<tr><td rowspan="3">BERT BiLSTM</td><td>no-pretraining</td><td>0.7293</td><td>0.5939</td><td>0.7293</td><td>0.7051</td><td>0.9764</td><td>0.7932</td><td>0.9609</td><td>0.3591</td><td>0.8824</td><td>0.7076</td></tr>
<tr><td>pretraining (Run 2a)</td><td>0.7412</td><td>0.5810</td><td>0.7455</td><td>0.7191</td><td>0.9670</td><td>0.0000</td><td>0.9584</td><td>0.3526</td><td>0.8909</td><td>0.7350</td></tr>
<tr><td>pick-best</td><td>0.7442</td><td>0.5990</td><td>0.7483</td><td>0.7149</td><td>0.9670</td><td>0.8014*</td><td>0.9611</td><td>0.3596</td><td>0.9027</td><td>0.7404</td></tr>
<tr><td rowspan="3">BERT BiLSTM</td><td>IDs (Run 1)</td><td>0.7555</td><td>0.6316</td><td>0.7966</td><td>0.7626</td><td>0.9221</td><td>0.8601</td><td>0.9669</td><td>0.4762</td><td>0.8933</td><td>0.7416</td></tr>
<tr><td>spans+OGER</td><td>0.6586</td><td>0.6522</td><td>0.2957</td><td>0.7603</td><td>0.9838</td><td>0.7683</td><td>0.8451</td><td>0.8026</td><td>0.8163</td><td>0.6784</td></tr>
<tr><td>IDs+spans+OGER (Run 3)</td><td>0.7700</td><td>0.6487</td><td>0.8037</td><td>0.7645</td><td>0.9561</td><td>0.8705</td><td>0.9694</td><td>0.5443</td><td>0.8954</td><td>0.7488</td></tr>
<tr><td rowspan="2"></td><td rowspan="2"></td><td>CHE-
BI
EXT</td><td>CL
EXT</td><td>GO
BP
EXT</td><td>GO
CC
EXT</td><td>GO
MF
EXT</td><td>MOP
EXT</td><td>NCBI
Taxon
EXT</td><td>PR
EXT</td><td>SO
EXT</td><td>UBE-
RON
EXT</td></tr>
<tr><td>OGER (baseline)</td><td></td><td>0.6797</td><td>0.7236</td><td>0.3644</td><td>0.8220</td><td>0.6731</td><td>0.4106</td><td>0.5994</td><td>0.4826</td><td>0.4604</td><td>0.6810</td></tr>
<tr><td rowspan="3">BERT BiLSTM</td><td>no-pretraining</td><td>0.8031</td><td>0.7263</td><td>0.7758</td><td>0.8674</td><td>0.7154</td><td>0.7996</td><td>0.9583</td><td>0.4004</td><td>0.9092</td><td>0.6900</td></tr>
<tr><td>pretraining (Run 2a)</td><td>0.8173</td><td>0.7199</td><td>0.7712</td><td>0.8725</td><td>0.7411</td><td>0.5630</td><td>0.9554</td><td>0.4005</td><td>0.9167</td><td>0.7312</td></tr>
<tr><td>pick-best</td><td>0.8168</td><td>0.7289</td><td>0.7755</td><td>0.8723</td><td>0.7438</td><td>0.7996*</td><td>0.9549</td><td>0.4122</td><td>0.9187</td><td>0.7458</td></tr>
<tr><td rowspan="3">BERT BiLSTM</td><td>IDs (Run 1)</td><td>0.8143</td><td>0.7375</td><td>0.8085</td><td>0.8918</td><td>0.6530</td><td>0.8240</td><td>0.9682</td><td>0.4706</td><td>0.9056</td><td>0.7654</td></tr>
<tr><td>spans+OGER</td><td>0.7180</td><td>0.7187</td><td>0.3799</td><td>0.8862</td><td>0.6715</td><td>0.4562</td><td>0.8351</td><td>0.8011</td><td>0.5640</td><td>0.7029</td></tr>
<tr><td>IDs+spans+OGER</td><td>0.8209</td><td>0.7484</td><td>0.8138</td><td>0.8936</td><td>0.6691</td><td>0.8437</td><td>0.9722</td><td>0.5516</td><td>0.9069</td><td>0.7714</td></tr>
</table>

*ontology pretraining disabled

Table 1: F-Score results of our experiments using the CRAFT corpus. Underlined numbers denote submitted results; other results were obtained in post-submission experiments. Bold figures mark the best result for each entity type (column).

ing problem. This approach has been adopted by many (Zhang et al., 2014; Cho et al., 2017), also using different neural architectures (Li et al., 2017; Liu and Xu, 2018; Tutubalina et al., 2018).

There have been continued efforts to jointly address NER and NEN, fighting the problem of error propagation inherent to pipeline architectures. Dictionary-based approaches can detect and normalize concept mentions in a single step (Tseytlin et al., 2016; Pafilis et al., 2013), even though postfiltering (Basaldella et al., 2017; Cuzzola et al., 2017) or other strategies are usually required to achieve good performance. Example-based approaches include probabilistic (Leaman and Lu, 2016) and graphical (Lou et al., 2017; ter Horst et al., 2017) systems for jointly learning NER+NEN in shared or interdependent models. Zhao et al. (2019) propose a multi-task-learning set-up for neural NER and NEN with bidirectional feedback, as mentioned earlier.

Recently, it has been shown that BERT models that are pretrained on biomedical and clinical datasets are beneficial for the NER task in the biomedical domain (Lee et al., 2019; Beltagy et al., 2019). To address the NEN task with BERT-based models, Kim et al. (2019) combined the BioBERT model with a rule-based approach to multi-type resolution and a dictionary lookup for the normalization.

4 Results

The results of our experiments are summarized in Tables 1 and 2. The tables contain both officially submitted results (printed with underline) and post-submission runs. The results were obtained by the official evaluation suite, which measures performance in terms of Slot Error Rate (SER) (Makhoul et al., 1999) and F-Score (F1). Both metrics are based on the counts of matches (true positives), insertions (false positives), deletions (false negatives) and substitutions (partial positives). The substitutions, as defined by Bossy et al. (2013), are a way to give partial credit to system predictions that are partially correct, e.g. when the correct ID was assigned to one token of a multi-word expression. While F1 is a measure of accurateness ranging from 1 (perfect) to 0 (no matching prediction at all), SER is a measure of errors ranging from 0 (perfect) to above 1 (more errors than ground-truth annotations). The rankings produced by the two metrics are not guaranteed to be identical; in fact, we report several cases where F1 and SER disagree on the question of which system performed best. For both metrics, the scores are micro-averaged across all 30 documents of the test set.

We used the plain dictionary-based system OGER as a baseline. For the BiLSTM system, we compared three different configurations: no-

	CHE-BI	CL	GO BP	GO CC	GO MF	MOP	NCBI Taxon	PR	SO	UBE-RON
OGER (baseline)	0.7873	**0.4862**	0.9826	0.6120	0.3032	1.6238	1.0122	1.9768	1.2617	0.5584
BiLSTM no-pretraining	0.4280	0.5628	0.4231	0.4829	0.0443	<u>0.3507</u>	0.0688	0.9017	0.1934	0.4379
pretraining (<u>Run 2a</u>)	<u>0.4089</u>	<u>0.5781</u>	<u>0.3991</u>	<u>0.4296</u>	<u>0.0638</u>	1.0000	<u>0.0733</u>	<u>0.8597</u>	<u>0.1786</u>	<u>0.3913</u>
pick-best	0.4038	0.5563	0.3956	0.4455	0.0638	0.3453*	0.0723	0.8501	**0.1593**	0.3864
BERT IDs (<u>Run 1</u>)	<u>0.3569</u>	<u>0.5780</u>	<u>0.3145</u>	<u>0.3848</u>	<u>0.1507</u>	<u>0.2882</u>	<u>0.0580</u>	<u>0.7612</u>	<u>0.1689</u>	**0.3752**
spans+OGER	0.5111	0.5000	0.8276	**0.3788**	**0.0319**	0.3762	0.2240	**0.3052**	0.2918	0.4749
IDs+spans+OGER (<u>Run 3</u>)	**0.3388**	0.5620	**0.3047**	0.3888	0.0869	**0.2684**	**0.0537**	0.6863	0.1680	<u>0.3770</u>

	CHE-BI EXT	CL EXT	GO BP EXT	GO CC EXT	GO MF EXT	MOP EXT	NCBI Taxon EXT	PR EXT	SO EXT	UBE-RON EXT
OGER (baseline)	0.6032	**0.3361**	0.8677	0.3493	0.5459	1.8108	0.9869	1.7056	1.1596	0.5210
BiLSTM no-pretraining	0.3152	0.3555	0.3398	0.2076	0.4266	<u>0.3445</u>	0.0744	0.8354	0.1398	0.4552
pretraining (<u>Run 2a</u>)	<u>0.3016</u>	<u>0.3547</u>	<u>0.3357</u>	<u>0.2032</u>	<u>0.3922</u>	0.5564	<u>0.0776</u>	<u>0.8047</u>	<u>0.1257</u>	<u>0.3943</u>
pick-best	0.3016	0.3497	0.3333	0.2051	**0.3881**	0.3445*	0.0784	0.7715	**0.1230**	0.3730
BERT IDs (<u>Run 1</u>)	<u>0.2664</u>	0.3667	<u>0.2867</u>	**0.1678**	<u>0.5081</u>	0.3440	0.0538	<u>0.7257</u>	<u>0.1475</u>	**0.3371**
spans+OGER	0.4224	0.3417	0.7419	0.1907	0.4676	0.6432	0.2353	**0.3030**	0.5566	0.4450
IDs+spans+OGER	**0.2571**	0.3583	**0.2786**	0.1681	0.4999	**0.3080**	**0.0466**	0.6464	0.1466	0.3384

*ontology pretraining disabled

Table 2: SER results of our experiments using the CRAFT corpus. For mark-up (underline/bold) see Table 1.

pretraining, pretraining, and pick-best. For the no-pretraining run, we skipped the pretraining phase over the ontology names. The pretraining run corresponds to the description in Section 3.2; we (unofficially[3]) submitted this run as Run 2a, except for MOP and MOP_EXT, where pretraining was disabled since it had an extraordinarily negative effect for this entity type in early experiments already. In the pick-best run, we trained each model two or three times and picked the one with the best performance on the held-out set in the cross-validation; again, ontology pretraining was disabled for MOP[_EXT] for this run.

For the transformer architecture, we also compared three systems: BERT-IDs, BERT-spans+OGER, and BERT-IDs+BERT-spans+OGER. BERT-IDs was trained to predict concept identifiers directly; we submitted these results as Run 1 (except for CL_EXT, GO_CC_EXT, MOP_EXT, NCBITaxon_EXT, and UBERON_EXT, which we analyzed only in post-submission experiments due to time constraints). BERT-spans+OGER combines IOBES predictions with annotations from OGER in a pipeline fashion. The last configuration combines the previous two in a backoff manner; this was submitted as Run 3 (extension types post-submission only).

For many entity types, the BERT systems beat the BiLSTM systems, which in turn clearly out-

performed the dictionary-based baseline. A notable exception to this pattern is CL, where no neural system was as accurate as OGER. However, the baseline is beaten by all other systems in many cases; this is particularly true for SER, where the baseline shows very poor performance for a number of entity types.

Among the BiLSTM systems, the effect of ontology pretraining is somewhat heterogeneous; while it clearly improved performance for some entity types (such as CHEBI[_EXT], UBERON[_EXT]), it had a marginal or even negative effect on others (e.g. NCBITaxon[_EXT]). As expected from the cross-validation results, ontology pretraining heavily decreased performance for MOP and MOP_EXT. The pick-best setting yielded modest improvements in most of the cases. In three cases (GO_MF_EXT, SO, SO_EXT), this configuration achieves the best overall scores.

Among the BERT-based systems, directly predicting IDs usually gave better results than joining span predictions with OGER annotations, and combining the two systems in a backoff manner yielded another improvement. However, the span detector coupled with OGER outperformed the two ID taggers in five cases (CL, GO_MF[_EXT], PR[_EXT]), three of which constitute best overall scores (GO_MF, PR[_EXT]). The most notable results are the ones for PR and PR_EXT, where BERT-spans+OGER beat all other systems by a margin of more than 0.25 F1/0.34 SER.

[3]after the deadline, but before the release of the ground-truth annotations

	ground-truth concepts		OGER		BiLSTM pretraining		BiLSTM pick-best		BERT-spans+ OGER		BERT-IDs+ BERT-spans+ OGER	
	unique	occ.	P	R	P	R	P	R	P	R	P	R
CHEBI	110	447	0.33	0.65	–		–		0.74	0.47	0.70	0.11
CHEBI_EXT	134	538	0.37	0.71	–		–		0.62	0.49	0.76	0.09
CL	52	484	0.72	0.31	–		–		0.88	0.22	0.59	0.04
CL_EXT	52	484	0.72	0.31	–		–		0.71	0.25	0.71	0.11
GO_BP	120	484	0.21	0.25	–		–		0.56	0.12	0.66	0.06
GO_BP_EXT	126	508	0.22	0.28	–		–		0.29	0.18	0.62	0.07
GO_CC	32	184	0.19	0.35	–		–		0.50	0.17	0.49	0.06
GO_CC_EXT	36	231	0.28	0.47	–		–		0.58	0.19	0.60	0.07
GO_MF	1	1	0.10	0.50	–		–		–		–	
GO_MF_EXT	73	416	0.38	0.22	–		–		0.57	0.15	0.54	0.04
MOP	2	2	0.08	1.00	–		–		–		–	
MOP_EXT	2	2	0.08	1.00	–		–		–		–	
NCBITaxon	40	87	0.02	0.50	–		–		0.40	0.34	0.75	0.22
NCBITaxon_EXT	44	95	0.02	0.54	–		–		0.43	0.35	0.85	0.25
PR	278	4782	0.26	0.86	0.61	4E-4	0.63	4E-4	0.81	0.74	0.69	0.15
PR_EXT	309	5156	0.27	0.84	0.22	3E-3	0.34	8E-3	0.84	0.73	0.65	0.20
SO	16	101	0.04	0.87	–		–		0.10	0.06	0.52	0.02
SO_EXT	25	123	0.05	0.78	–		–		0.28	0.47	0.85	0.41
UBERON	203	1297	0.47	0.33	0.74	2E-3	0.69	2E-3	0.74	0.25	0.59	0.06
UBERON_EXT	207	1308	0.47	0.33	0.76	2E-3	0.87	1E-3	0.78	0.27	0.60	0.06

Table 3: System performance for unseen concepts: precision (P) and recall (R) calculated over the subset of annotations and predictions of IDs that were absent from the training data. A dash (–) denotes that the system only predicted known IDs for the given entity type. The systems BiLSTM no-pretraining and BERT-IDs are omitted as they cannot predict unseen labels.

5 Discussion

The results show that, in general, neural sequence taggers can be successfully applied to biomedical concept recognition, using a single model for joint NER+NEN. Unfortunately, we cannot compare our results to other work, as no other team has submitted results to the concept-annotation task and no official baseline is available at the time of writing. Since the CRAFT test set has only been released in the course of the present shared task, it is not possible to directly benchmark our results against previous work (such as Funk et al., 2014; Tseytlin et al., 2016; Hailu, 2019) either. However, the tested systems allow for a comparison of different approaches.

The strategies for extending the concept coverage – a vital feature for many applications – show a mixed picture. Pretraining on ontology names has led to limited benefit only. While it has demonstrated a positive effect for many entity types, it has been able to increase the set of recognized concepts only occasionally. As can be seen in Table 3, ontology pretraining led to prediction of IDs outside the training data in four entity types (PR[_EXT], UBERON[_EXT]). Even though the majority of the predicted unseen IDs is correct, they only account for a fraction of the ground-truth annotations.

On the other hand, combining BERT span predictions with OGER annotations resulted in correct predictions of unseen IDs for almost all entity types – the exceptions being GO_MF, MOP, and MOP_EXT, which suffer from a small number of concepts or positive examples in the training data. The BERT-spans+OGER system is particularly strong for PR[_EXT], where recognizing unseen concepts is especially important due to the diversity and abundance of protein mentions in the literature. When this system is used as a backoff for BERT-IDs, the recall for unseen concepts drops due to the bias for existing knowledge inherent to the ID tagger. In some cases this bias is beneficial for precision, i.e. the ID tagger suppresses many false-positive predictions of OGER (e.g. CHEBI_EXT, NCBITaxon[_EXT], SO[_EXT]), while in other cases false positives of the ID tagger hide correct OGER predictions, leading to lower precision.

A few examples of correctly predicted IDs absent from the training corpus are given in context in the following. BERT-IDs+BERT-spans+OGER predicted CHEBI_PR_EXT:somatostatin in document 17503968 (two occurrences):

> However, the **somatostatin receptor 2 (SSTR-2)** antagonist PRL-2903 does

not interfere with the ability of glucose (at 3 and 7 mM) to inhibit glucagon secretion from mouse islets [47].

The same system predicted CHEBI:60004 in document 11604102:

> Adult mouse testes were homogenized in a buffer containing 20 mM Tris, pH 7.5, 100 mM KCl, 5 mM MgCl2, 0.3% NP-40, 40 U/ml of Rnasin ribonuclase inhibitor (Promega, Madison, WI), and a **mixture** of 10 protease inhibitors provided [...]

BiLSTM pick-best predicted PR:000008373 in document 16968134:

> Decreased Osteogenic Differentiation Correlates with Abnormal Distribution of **Cx43**

The creators of the CRAFT corpus have put great effort in building an annotated corpus with high quality and consistency across all entity types. However, the diversity of the different types requires a lot of engineering for tackling them all. A single approach is not sufficient to meet the differing needs of all entity types. The experiments with the test set have yielded a few surprising results, such as the comparatively good performance of the dictionary-based approach on CL or the outstanding scores for BERT-spans+OGER on PR[_EXT].

Of the two concept extension strategies, the NER+dictionary backoff has worked well, whereas the effect of ontology pretraining was not too conclusive. Since we tested each of the strategies with only one system architecture, it is not entirely clear which component contributed the most to the success – the network architecture or the extension strategy. Testing the inverse combinations, i. e. BERT with ontology pretraining and BiLSTM with OGER backoff, is left for future work.

Acknowledgments

The research activities of the OntoGene/BioMeXT group at the University of Zurich are supported by the Swiss National Science Foundation (grant CR30I1_162758). We would like to thank the organisers for this well-organised shared task with high-quality annotations and prompt support. Additional thanks to the anonymous reviewers who provided valuable feedback.

References

Michael Bada, Miriam Eckert, Donald Evans, Kristin Garcia, Krista Shipley, Dmitry Sitnikov, William A. Baumgartner, K. Bretonnel Cohen, Karin Verspoor, Judith A. Blake, and Lawrence E. Hunter. 2012. Concept annotation in the CRAFT corpus. *BMC Bioinformatics*, 13(1):1–20.

Marco Basaldella, Lenz Furrer, Carlo Tasso, and Fabio Rinaldi. 2017. Entity recognition in the biomedical domain using a hybrid approach. *Journal of Biomedical Semantics*, 8(1):51.

Iz Beltagy, Kyle Lo, and Arman Cohan. 2019. SciBERT: A pretrained language model for scientific text. In *Proceedings of the 2019 Conference on Empirical Methods in Natural Language Processing and 9th International Joint Conference on Natural Language Processing (EMNLP-IJCNLP)*.

Robert Bossy, Wiktoria Golik, Zorana Ratkovic, Philippe Bessières, and Claire Nédellec. 2013. BioNLP shared task 2013 – an overview of the bacteria biotope task. In *Proceedings of the BioNLP Shared Task 2013 Workshop*, pages 161–169, Sofia, Bulgaria. Association for Computational Linguistics.

David Campos, Sérgio Matos, and José Luís Oliveira. 2013. A modular framework for biomedical concept recognition. *BMC Bioinformatics*, 14(281).

Billy Chiu, Gamal Crichton, Anna Korhonen, and Sampo Pyysalo. 2016. How to train good word embeddings for biomedical NLP. In *Proceedings of the 15th Workshop on Biomedical Natural Language Processing*, pages 166–174, Berlin, Germany. Association for Computational Linguistics.

Hyejin Cho, Wonjun Choi, and Hyunju Lee. 2017. A method for named entity normalization in biomedical articles: application to diseases and plants. *BMC Bioinformatics*, 18(1):451.

K. Bretonnel Cohen, Karin Verspoor, Karën Fort, Christopher Funk, Michael Bada, Martha Palmer, and Lawrence E. Hunter. 2017. *The Colorado Richly Annotated Full Text (CRAFT) Corpus: Multi-Model Annotation in the Biomedical Domain*, pages 1379–1394. Springer Netherlands, Dordrecht.

John Cuzzola, Jelena Jovanović, and Ebrahim Bagheri. 2017. RysannMD: a biomedical semantic annotator balancing speed and accuracy. *Journal of Biomedical Informatics*, 71:91–109.

Xiang Dai. 2018. Recognizing complex entity mentions: A review and future directions. In *Proceedings of ACL 2018, Student Research Workshop*, pages 37–44, Melbourne, Australia. Association for Computational Linguistics.

Jacob Devlin, Ming-Wei Chang, Kenton Lee, and Kristina Toutanova. 2019. BERT: pre-training of deep bidirectional transformers for language understanding. In *Proceedings of NAACL-HLT 2019*, pages 4171–4186, Minneapolis, MN, USA.

Jennifer D'Souza and Vincent Ng. 2015. Sieve-based entity linking for the biomedical domain. In *Proceedings of the 53rd Annual Meeting of the Association for Computational Linguistics and the 7th International Joint Conference on Natural Language Processing*, pages 297–302, Beijing, China. Association for Computational Linguistics.

Christopher Funk, William Baumgartner, Benjamin Garcia, Christophe Roeder, Michael Bada, K. Bretonnel Cohen, Lawrence E. Hunter, and Karin Verspoor. 2014. Large-scale biomedical concept recognition: an evaluation of current automatic annotators and their parameters. *BMC Bioinformatics*, 15(1):1–29.

Lenz Furrer, Anna Jancso, Nicola Colic, and Fabio Rinaldi. 2019. OGER++: hybrid multi-type entity recognition. *Journal of Cheminformatics*, 11(1):7.

Omid Ghiasvand and Rohit J. Kate. 2014. UWM: disorder mention extraction from clinical text using CRFs and normalization using learned edit distance patterns. In *Proceedings of the 8th International Workshop on Semantic Evaluation (SemEval 2014)*, pages 828–832, Dublin, Ireland. Association for Computational Linguistics.

Maryam Habibi, Leon Weber, Mariana Neves, David Luis Wiegandt, and Ulf Leser. 2017. Deep learning with word embeddings improves biomedical named entity recognition. *Bioinformatics*, 33(14):i37–i48.

Negacy Degefa Hailu. 2019. *Investigation of traditional and deep neural sequence models for biomedical concept recognition*. Ph.D. thesis, University of Colorado at Denver, Anschutz Medical Campus.

Sepp Hochreiter and Jürgen Schmidhuber. 1997. Long short-term memory. *Neural Computation*, 9(8):1735–1780.

Hendrik ter Horst, Matthias Hartung, and Philipp Cimiano. 2017. *Joint Entity Recognition and Linking in Technical Domains Using Undirected Probabilistic Graphical Models*, volume 10318, pages 166–180. Springer.

Qiao Jin, Bhuwan Dhingra, William W. Cohen, and Xinghua Lu. 2019. Probing biomedical embeddings from language models. In *Proceedings of the 3rd Workshop on Evaluating Vector Space Representations for NLP*, pages 82–89, Minneapolis, MN, USA. Association for Computational Linguistics.

Donghyeon Kim, Jinhyuk Lee, Chan Ho So, Hwisang Jeon, Minbyul Jeong, Yonghwa Choi, Wonjin Yoon, Mujeen Sung, and Jaewoo Kang. 2019. A neural named entity recognition and multi-type normalization tool for biomedical text mining. *IEEE Access*, 7:73729–73740.

Robert Leaman, Rezarta Islamaj Doğan, and Zhiyong Lu. 2013. DNorm: disease name normalization with pairwise learning to rank. *Bioinformatics*, 29(22):2909–2917.

Robert Leaman and Zhiyong Lu. 2016. TaggerOne: joint named entity recognition and normalization with semi-Markov Models. *Bioinformatics*, 32(18):2839.

Hsin-Chun Lee, Yi-Yu Hsu, and Hung-Yu Kao. 2016. AuDis: an automatic CRF-enhanced disease normalization in biomedical text. *Database*, 2016:baw091.

Jinhyuk Lee, Wonjin Yoon, Sungdong Kim, Donghyeon Kim, Sunkyu Kim, Chan Ho So, and Jaewoo Kang. 2019. BioBERT: a pre-trained biomedical language representation model for biomedical text mining. *Bioinformatics*. Btz682.

Haodi Li, Qingcai Chen, Buzhou Tang, Xiaolong Wang, Hua Xu, Baohua Wang, and Dong Huang. 2017. CNN-based ranking for biomedical entity normalization. *BMC Bioinformatics*, 18(11):385.

Hongwei Liu and Yun Xu. 2018. A deep learning way for disease name representation and normalization. In Xuanjing Huang, Jing Jiang, Dongyan Zhao, Yansong Feng, and Yu Hong, editors, *Natural Language Processing and Chinese Computing*, pages 151–157. Springer International Publishing.

Yinxia Lou, Yue Zhang, Tao Qian, Fei Li, Shufeng Xiong, and Donghong Ji. 2017. A transition-based joint model for disease named entity recognition and normalization. *Bioinformatics*, 33(15):2363–2371.

John Makhoul, Francis Kubala, Richard Schwartz, and Ralph Weischedel. 1999. Performance measures for information extraction. In *Proceedings of DARPA broadcast news workshop*, pages 249–252.

Alejandro Metke-Jimenez and Sarvnaz Karimi. 2016. Concept identification and normalisation for adverse drug event discovery in medical forums. In *Proceedings of the First International Workshop on Biomedical Data Integration and Discovery (BMDID 2016)*.

Tomas Mikolov, Ilya Sutskever, Kai Chen, Greg S. Corrado, and Jeff Dean. 2013. Distributed representations of words and phrases and their compositionality. In C. J. C. Burges, L. Bottou, M. Welling, Z. Ghahramani, and K. Q. Weinberger, editors, *Advances in Neural Information Processing Systems 26 (NIPS 2013)*, pages 3111–3119. Curran Associates, Inc.

Evangelos Pafilis, Sune P. Frankild, Lucia Fanini, Sarah Faulwetter, Christina Pavloudi, Aikaterini Vasileiadou, Christos Arvanitidis, and Lars Juhl Jensen. 2013. The SPECIES and ORGANISMS resources for fast and accurate identification of taxonomic names in text. *PLOS ONE*, 8(6):1–6.

Jeffrey Pennington, Richard Socher, and Christopher D. Manning. 2014. GloVe: Global vectors for word representation. In *Proceedings of the 2014 Conference on Empirical Methods in Natural Language Processing (EMNLP)*, pages 1532–1543, Doha, Qatar. Association for Computational Linguistics.

Mike Schuster and Kuldip K. Paliwal. 1997. Bidirectional recurrent neural networks. *IEEE Transactions on Signal Processing*, 45(11):2673–2681.

Sunghwan Sohn, Donald C. Comeau, Won Kim, and W. John Wilbur. 2008. Abbreviation definition identification based on automatic precision estimates. *BMC Bioinformatics*, 9(1):402.

Eugene Tseytlin, Kevin Mitchell, Elizabeth Legowski, Julia Corrigan, Girish Chavan, and Rebecca S. Jacobson. 2016. NOBLE – Flexible concept recognition for large-scale biomedical natural language processing. *BMC Bioinformatics*, 17(1):1–15.

Elena Tutubalina, Zulfat Miftahutdinov, Sergey Nikolenko, and Valentin Malykh. 2018. Medical concept normalization in social media posts with recurrent neural networks. *Journal of Biomedical Informatics*, 84:93–102.

Yaoyun Zhang, Jingqi Wang, Buzhou Tang, Yonghui Wu, Min Jiang, Yukun Chen, and Hua Xu. 2014. UTH_CCB: a report for SemEval 2014 – Task 7 analysis of clinical text. In *Proceedings of the 8th International Workshop on Semantic Evaluation (SemEval 2014)*, pages 802–806, Dublin, Ireland. Association for Computational Linguistics.

Sendong Zhao, Ting Liu, Sicheng Zhao, and Fei Wang. 2019. A neural multi-task learning framework to jointly model medical named entity recognition and normalization. In *Proceedings of the Thirty-Third AAAI Conference on Artificial Intelligence (AAAI-19)*, pages 817–824.

Coreference Resolution in Full Text Articles with BERT and Syntax-based Mention Filtering

Hai-Long Trieu[1], Khoa N. A. Duong[1], Nhung T. H. Nguyen[2], Makoto Miwa[1,3],
Hiroya Takamura[1], Sophia Ananiadou[2]
[1]Artificial Intelligence Research Center (AIRC),
National Institute of Advanced Industrial Science and Technology (AIST), Japan
[2]National Centre for Text Mining, University of Manchester, United Kingdom
[3]Toyota Technological Institute, Japan
{long.trieu, khoa.duong, takamura.hiroya}@aist.go.jp,
makoto-miwa@toyota-ti.ac.jp,
{nhung.nguyen, Sophia.Ananiadou}@manchester.ac.uk

Abstract

This paper describes our system developed for the coreference resolution task of the CRAFT Shared Tasks 2019. The CRAFT corpus is more challenging than other existing corpora because it contains full text articles. We have employed an existing span-based state-of-the-art neural coreference resolution system as a baseline system. We enhance the system with two different techniques to capture long-distance coreferent pairs. Firstly, we filter noisy mentions based on parse trees with increasing the number of antecedent candidates. Secondly, instead of relying on the LSTMs, we integrate the highly expressive language model–BERT into our model. Experimental results show that our proposed systems significantly outperform the baseline. The best performing system obtained F-scores of 44%, 48%, 39%, 49%, 40%, and 57% on the test set with B^3, BLANC, CEAFE, CEAFM, LEA, and MUC metrics, respectively. Additionally, the proposed model is able to detect coreferent pairs in long distances, even with a distance of more than 200 sentences.

1 Introduction

Coreference resolution is important not only in general domains but also in the biomedical domain. The Colorado Richly Annotated Full Text (CRAFT) corpus (Cohen et al., 2017) was constructed with an aim of boosting the performance of the task in the biomedical literature. Unlike other corpora, CRAFT is comprised of full text articles or full papers, its coreferent chains are arbitrarily long; the mean length of coreferent chains is 4 while the longest chain is 186, which makes the resolution even more difficult than usual. The corpus has been fully released in the CRAFT Shared Task 2019. In this paper, we present our approach to address the coreference resolution task in this challenging corpus.

We employ the state-of-the-art end-to-end coreference system (Lee et al., 2017) as our baseline. The system generates all continuous sequences of words (or spans) in each sentence as mention candidates, which means the number of candidates increases linearly to the number of sentences. Such candidates may contain a large number of noisy spans, which are spans in a sentence that do not fit any noun phrases according to the corresponding parse tree. Such noisy spans are often wasteful when being included in the list of candidates for the coreference resolution step. Especially for the CRAFT corpus, of which the average number of sentences is more than 300, the number of noisy spans would be many and needs to be reduced. Also, our observations on the CRAFT corpus show that in many cases, a mention and its antecedent are far away, e.g., a mention can occur in the result section of a paper while its antecedent is in the abstract section.

To address these problems, we enhance the baseline system in two ways; we propose to filter noisy spans by using syntactic information and increase the number of antecedent candidates to capture such long-distance coreferent pairs. We further boost the system by replacing the underlying Long Short-Term Memory (LSTM) (Hochreiter and Schmidhuber, 1997) layer with the Bidirectional Encoder Representations from Transformer (BERT) model (Devlin et al., 2019)—a contextualized language model that can efficiently capture context in a wide range of NLP tasks.

We have evaluated our system on six common metrics for coreference resolution including B^3, BLANC, CEAFE, CEAFM, LEA, and MUC using the official evaluation script provided by the shared task organizers. By increasing the num-

Proceedings of the 5th Workshop on BioNLP Open Shared Tasks, pages 196–205
Hong Kong, China, November 4, 2019. ©2019 Association for Computational Linguistics

ber of antecedents and filtering noisy ones, we could boost the recall of mention detection, hence improving the performance of coreference resolution. When incorporating BERT into the system, we could attain better scores in both mention detection and coreference resolution at every metrics.

Our contributions are as follows.

- We proposed a new method to filter noisy spans, which is a weakness of the baseline system (Lee et al., 2017). Our filtering method based on syntactic trees reduced up to 90% noisy spans but still kept 93% of correct mentions on the development set. The method helps our model more computationally efficient than the baseline one, hence allowing us to increase the number of antecedent candidates to capture long-distance coreferent pairs.

- We successfully integrated the BERT model to replace the LSTM layers for coreference resolution task and obtained significant improvement.

- Although we only experimented our model with the CRAFT corpus, our proposed method is general enough to be applied to other corpora with long documents.

2 Methods

2.1 LSTM-based Baseline Model

Our model is based on the span-based end-to-end model (Lee et al., 2017). The model employs an exhaustive method to create any continuous sequences of words (spans) in each sentence. The representation of a span from the k-th word to the l-th word in a sentence is calculated by concatenating the information of the first word, last word, head word, and the span width feature as follows:

$$m_{k,l} = [h_k, h_l, \hat{w}_{k..l}, \phi(k, l)], \quad (1)$$

where h_k and h_l are embeddings of the first and last words calculated by a bidirectional LSTM; $\hat{w}_{k..l}$ is the weighted sum of the word vectors; and $\phi(k, l)$ encodes the size of this span.

Mention scores are calculated using a feedforward neural network given the span representation.

$$s_m(k, l) = w_m \cdot \text{FFNN}_m(m_{k,l}), \quad (2)$$

where w_m is a learnable weight vector; and FFNN denotes a feed-forward neural network.

Since the span-based model generates a large number of spans, a simple technique is used to rank and filter spans based on a λ ratio multiplied by the document size and choose the k best candidates.

To find an antecedent for each mention, we calculate the antecedent score as follows:

$$s_a(m_{k,l}, m_{u,v}) = w_a \cdot \text{FFNN}_a([m_{k,l}, m_{u,v}, \\ m_{k,l} \circ m_{u,v}, \phi((k, l), (u, v))]), \quad (3)$$

where w_a is a learnable weight vector; $\circ$ denotes an element-wise multiplication and $\phi((k, l), (u, v))$ represents the feature vector between the two mentions.

2.2 Coreference Resolution with BERT

Recently, BERT (Devlin et al., 2019) shows significant improvement on various tasks in comparison with other deep learning models including LSTMs. This highly expressive language model is able to capture contextual information effectively. We, therefore, aim at investigating whether this architecture can work effectively on coreference resolution in comparison with the previous LSTM-based models.

In the BERT model, contextual representations are assigned to sub-words in each word. We use the representation of the last subword in a word as the representation of the word and calculated the span representation using Equation 1. Since the pre-trained BERT model just supports sentences up to 512 sub-words, we utilize a sliding window technique with a window size of 512 and stride of 256 for longer sentences and then retrieve subword embeddings from windows so that each subword has maximum left and right context. We adapted the mention score and antecedent score functions as Equations 2 and 3.

2.3 Learning Parse Trees to Filter Mentions

A weakness of the span-based baseline model is that the greedy method generates a large number of noisy, mostly meaningless, spans. Although Lee et al. (2017) proposed to select k-best candidates but this strategy is problematic when working on long documents, in which a mention is probably far away from its true antecedents while there are a large number of noisy candidates between them.

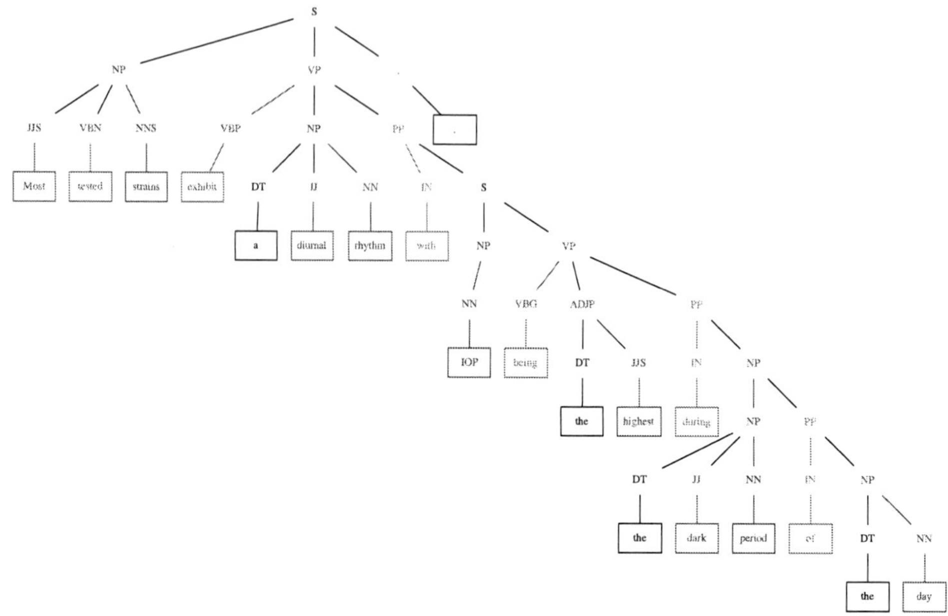

Figure 1: Three patterns corresponding to three gold mentions are extracted from the parse tree: "*a diurnal rhythm with IOP*" (pattern: **(NP,IN,NP)**), "*the dark period of the day*" (pattern: **(NP)**), "*the day*" (pattern: **(NP)**)

In order to overcome this issue, we propose to filter noisy spans based on their syntactic information. We observe that in the task like coreference resolution, mentions usually follow syntactic structures such as noun phrases. We therefore learn a syntactic parsing model to parse sentences and then extract patterns of gold mentions based on the resulting parse trees.

The end-to-end parsing model is trained jointly by the two following steps.

- **Part-of-speech (POS) classifier**: given raw sentences from the training set, words are split into sub-words with corresponding vectors from BERT embeddings. The last subword embedding of each word is used as the word embedding and passed through a linear layer to predict POS tags. The gold label POS tags are obtained from the CRAFT training set. Predicted POS tags and the raw texts will be used as the input for the parsing model.

- **Parser**: our model is based on the constituency parsing model (Kitaev and Klein, 2018), in which parse trees were built based on a self-attentive encoder and achieved state-of-the-art performance on the Penn

Treebank. Unlike their model, we replaced the self-attentive encoder by BERT.

Figure 1 presents an example of using a parse tree to extract patterns of gold mentions. In this example, three patterns corresponding to three gold mentions are extracted: (NP, IN, NP), (NP), and (NP).

In the coreference resolution model, generated spans that match with the learned patterns are fed into the span representation layer to create span embeddings, while unmatched spans are ignored.

3 Experimental Settings

3.1 Dataset

The organizer provided two subsets of the CRAFT corpus (Cohen et al., 2017): one for training and one for testing systems. To estimate our model before submitting testing results, we further divided the original training set into two subsets, namely training and development sets. Table 1 shows the statistics numbers of these three subsets.

3.2 Compared Models

In order to show the effect of our proposed methods, we compare the following models.

	Train	Dev	Test
No. documents	60	7	30
No. sentences	19,575	2,156	9,099
Avg. sentences/doc	326.25	308.0	303.3
No. mentions	69,413	8,342	33,749
No. discon. men.	4,041	444	1,845
No. coreference.	14,679	1,623	7,185

Table 1: Characteristics of the CRAFT corpus. (*discon. men.*: discontinuous mentions)

Syntactic Patterns	Frequency	Ratio (%)
NP	40,639	58.55
NN	10,746	15.48
NML	3,462	4.99
PRP$	1,012	1.46
NN, NN	1,012	1.46
NP, NN	678	0.98
LS	647	0.93

Table 2: The most frequent patterns of mentions in the training set. Please check Appendix A for the definition of relevant Penn Treebank labels.

- **LSTM**: this is the baseline model (Lee et al., 2017) based on LSTM. This model obtained the state-of-the-art performance on general domain. In this setting, all generated spans are used for calculating mention scores. The number of antecedent candidates was 250 and the $\lambda = 0.25$.

- **LSTM_filter**: this is the same as the LSTM baseline model, but we applied the filtering method and increased the antecedent number to 600 instead of 250.

- **BERT**: we employed the pre-trained SciBERT model (Beltagy et al., 2019) instead of using the LSTM as the baseline model. The number of antecedent candidates was 600.

- **BERT_filter**: we used the same settings as BERT but we combined it with the filtering method.

- **E2E_MetaMap** (Trieu et al., 2018): this model is based on the baseline model (Lee et al., 2017) but it particularly incorporated semantic type features extracted from the MetaMapLite (Demner-Fushman et al., 2017) to address biomedical documents. The maximum antecedent was 250.

The E2E_MetaMap implementation is based on the Tensorflow repository.[1] Meanwhile, the LSTM, LSTM_filter, and BERT_filter are based on the PyTorch repository.[2]

For the BERT model, we employed the PyTorch Pretrained BERT repository.[3] We trained the model with the Adam optimizer (Kingma and Ba, 2015). We included gradient clipping and dropout.

4 Results and Discussion

We firstly present the results of extracting patterns to filter mentions. We then report and discuss the performance of our models on the official test set. In order to deeply investigate the effect of the proposed method, we describe the intensive results of ablation tests on the development set. We finally conduct analysis to see how each model works on each group of sentence-level distance between mentions and antecedents.

4.1 Patterns of Gold Mentions

Table 2 reports some patterns[4] with the highest frequencies in the training set. In total, we extracted 1,561 unique patterns. To avoid low quality filtering, we kept patterns with a minimum frequency threshold of 5. The threshold was chosen from our experiments so that we could filter a large number of noisy spans but still kept a high recall on the development set. Specifically, this filtering method helps to reduce up to 90% noisy spans but still kept 93% of correct mentions on the development set.

4.2 Evaluation on the Test Set

The results on the official test set are presented in Table 3. In summary, our BERT_filter obtained the best performance on both mention and coreference detection in all metrics.

Mention detection For mention detection, most models obtained approximately the same precision of more than 70%. However, the recall

---

[1] https://github.com/kentonl/e2e-coref/tree/1f37582e68

[2] https://github.com/allenai/allennlp/tree/master/allennlp/models/coreference_resolution

[3] https://github.com/huggingface/pytorch-pretrained-BERT/tree/34cf67fd6c

[4] The tag set in our patterns follows Penn Treebank POS tags.

Metric	Model	Mention			Coreference		
		P	R	F	P	R	F
B^3	LSTM	0.7565	0.3578	0.4858	0.6177	0.1583	0.2520
	LSTM_filter	0.7292	0.4187	0.5320	0.5764	0.2524	0.3511
	BERT	0.7416	0.5603	0.6383	0.5151	0.3544	0.4199
	BERT_filter	0.7314	0.5778	**0.6456**	0.5166	0.3838	**0.4404**
	E2E_MetaMap	0.6713	0.5272	0.5906	0.5247	0.2791	0.3644
BLANC	LSTM	0.7565	0.3578	0.4858	0.6434	0.2153	0.3227
	LSTM_filter	0.7292	0.4187	0.5320	0.6471	0.3903	0.4869
	BERT	0.7416	0.5603	0.6383	0.5376	0.4350	0.4809
	BERT_filter	0.7314	0.5778	**0.6456**	0.5056	0.4731	**0.4888**
	E2E_MetaMap	0.6713	0.5272	0.5906	0.5297	0.4140	0.4648
CEAFE	LSTM	0.7565	0.3578	0.4858	0.3590	0.2076	0.2631
	LSTM_filter	0.7292	0.4187	0.5320	0.4100	0.2408	0.3034
	BERT	0.7416	0.5603	0.6383	0.4366	0.3305	0.3762
	BERT_filter	0.7314	0.5778	**0.6456**	0.4544	0.3537	**0.3978**
	E2E_MetaMap	0.6713	0.5272	0.5906	0.3545	0.3101	0.3308
CEAFM	LSTM	0.7565	0.3578	0.4858	0.5141	0.2431	0.3301
	LSTM_filter	0.7292	0.4187	0.5320	0.5847	0.3357	0.4265
	BERT	0.7416	0.5603	0.6383	0.5432	0.4104	0.4676
	BERT_filter	0.7314	0.5778	**0.6456**	0.5551	0.4385	**0.4900**
	E2E_MetaMap	0.6713	0.5272	0.5906	0.4662	0.3662	0.4102
LEA	LSTM	0.7565	0.3578	0.4858	0.5733	0.1331	0.2161
	LSTM_filter	0.7292	0.4187	0.5320	0.5415	0.2265	0.3194
	BERT	0.7416	0.5603	0.6383	0.4692	0.3135	0.3759
	BERT_filter	0.7314	0.5778	**0.6456**	0.4753	0.3454	**0.4000**
	E2E_MetaMap	0.6713	0.5272	0.5906	0.4864	0.2433	0.3244
MUC	LSTM	0.7565	0.3578	0.4858	0.6765	0.3007	0.4164
	LSTM_filter	0.7292	0.4187	0.5320	0.6656	0.3798	0.4837
	BERT	0.7416	0.5603	0.6383	0.6412	0.4842	0.5517
	BERT_filter	0.7314	0.5778	**0.6456**	0.6445	0.5111	**0.5701**
	E2E_MetaMap	0.6713	0.5272	0.5906	0.5995	0.4564	0.5182

Table 3: Results on the test set. The three official submissions of our team were BERT, BERT_filter and E2E_MetaMap. The non-coreference scores of BLANC are reported in Appendix B.

of the BERT_filter is much higher than those of the LSTM and LSTM_filter (57% vs. 35% and 41%, respectively). Consequently, the F-score of the BERT_filter is 16% and 11% points higher than the LSTM and LSTM_filter, respectively. The E2E_MetaMap is 5% points lower than the BERT_filter in F-score.

Coreference detection By obtaining the highest recall in mention detection, the BERT_filter could achieve the highest scores in coreference detection in all metrics. Using the mention filtering improved the baseline LSTM from 4-16% points in F-score varied by metrics. When replacing LSTM by BERT and combining with mention filtering, we obtained significant improvements: +19% points of B^3 and LEA; +16% points of MUC, BLANC and CEAFM; and +13% points of CEAFE in F-score.

The E2E_MetaMap performance is higher than the LSTM and LSTM_filter, but lower than the BERT_filter. As aforementioned, the LSTM model is based on the PyTorch implementation while the E2E_MetaMap is based on the Tensorflow repository. Therefore it is difficult to verify whether performance difference comes from using MetaMap features or from the implementation. Due to time constraint, we have not conducted experiments to clarify the reasons yet. We will leave this as our future work.

Metric	Model	Mention			Coreference		
		P	R	F	P	R	F
B^3	LSTM	0.7716	0.3782	0.5076	0.6387	0.1763	0.2764
	LSTM_filter	0.7193	0.4396	0.5457	0.5725	0.2572	0.3550
	BERT_250	0.7288	0.5675	0.6381	0.5258	0.32	0.3979
	BERT	0.7094	0.5742	0.6347	0.4807	0.3596	0.4115
	BERT_filter	0.7066	0.6014	**0.6498**	0.5021	0.3855	**0.4361**
BLANC	LSTM	0.7716	0.3782	0.5076	0.5665	0.1487	0.2356
	LSTM_filter	0.7193	0.4396	0.5457	0.5503	0.2235	0.3179
	BERT_250	0.7288	0.5675	0.6381	0.5129	0.3256	0.3983
	BERT	0.7094	0.5742	0.6347	0.4691	0.3168	0.3782
	BERT_filter	0.7066	0.6014	**0.6498**	0.5141	0.3757	**0.4341**
CEAFE	LSTM	0.7716	0.3782	0.5076	0.3659	0.2536	0.2996
	LSTM_filter	0.7193	0.4396	0.5457	0.4123	0.3252	0.3636
	BERT_250	0.7288	0.5675	0.6381	0.3888	0.3672	0.3777
	BERT	0.7094	0.5742	0.6347	0.4115	0.3674	0.3882
	BERT_filter	0.7066	0.6014	**0.6498**	0.4176	0.3993	**0.4083**
CEAFM	LSTM	0.7716	0.3782	0.5076	0.5285	0.2591	0.3477
	LSTM_filter	0.7193	0.4396	0.5457	0.5757	0.3518	0.4368
	BERT_250	0.7288	0.5675	0.6381	0.5073	0.3952	0.4443
	BERT	0.7094	0.5742	0.6347	0.5143	0.4163	0.4602
	BERT_filter	0.7066	0.6014	**0.6498**	0.5308	0.4518	**0.4881**
LEA	LSTM	0.7716	0.3782	0.5076	0.5974	0.1507	0.2407
	LSTM_filter	0.7193	0.4396	0.5457	0.5370	0.2276	0.3197
	BERT_250	0.7288	0.5675	0.6381	0.4811	0.2805	0.3544
	BERT	0.7094	0.5742	0.6347	0.4383	0.3196	0.3696
	BERT_filter	0.7066	0.6014	**0.6498**	0.4619	0.3464	**0.3959**
MUC	LSTM	0.7716	0.3782	0.5076	0.7065	0.3117	0.4325
	LSTM_filter	0.7193	0.4396	0.5457	0.6658	0.3783	0.4825
	BERT_250	0.7288	0.5675	0.6381	0.6418	0.4743	0.5455
	BERT	0.7094	0.5742	0.6347	0.6144	0.4850	0.5421
	BERT_filter	0.7066	0.6014	**0.6498**	0.6271	0.5179	**0.5673**

Table 4: Results on the development set.

4.3 Ablation Tests

We conducted experiments on the development set to show the effect of using mention filtering and BERT. In order to directly compare between BERT and LSTM, we also conducted an experiment with BERT and set a value of 250 to the number of antecedent candidates. We named it as BERT_250. Meanwhile, LSTM, LSTM_filter, BERT, BERT_filter have the same settings as described in Section 3.2. All of the results are reported in Table 4.

Mention Filtering When we used mention filtering, the mention detection precision dropped 6% points in the case of LSTM, but in the case of BERT it was almost the same. However, the filter-ing helped to improve recall in both cases, which is important to the coreference detection step. As a result, in the coreference resolution step, mention filtering improved 2-8% points of F-score in all metrics.

Using BERT Using BERT could significantly boost the performance of the baselines in both mention detection and coreference resolution. For mention detection, BERT produced almost the same precision with the LSTM but much higher recall (+17% points), which led to an increase of 10% points in F-score. For coreference detection, BERT-based models outperformed the LSTM-based ones from 4-14% points of F-scores in all metrics.

In summary, when combining both techniques

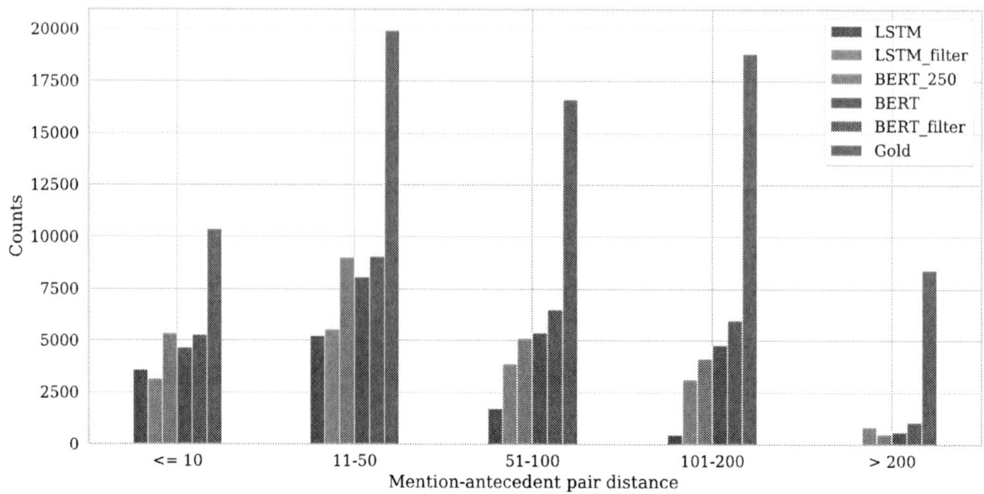

Figure 2: Numbers of correct predictions (true positives) grouped by sentence-level distance between a mention and its antecedent in the development set. The detail results are reported in Appendix C.

(BERT_filter vs. LSTM), we could make a significant increase of more than 14% points in F-score for mention detection and from 11% to 20% points in F-score in all metrics for coreference resolution on the development set.

4.4 Analysis

To investigate the effect of distance between a mention and its antecedent(s) on each model, we calculated the number of true positive coreference predictions in the development set and grouped them by sentence-level distance. Specifically, we divided the number of true positive predictions into five groups: $\leq$10, 11-50, 51-100, 101-200, and >200. The first two groups can be considered as short-distance coreference, e.g., abstract papers like the BioNLP dataset (Nguyen et al., 2011) with an average of nine sentences per document. Meanwhile, the other three groups can be considered as long-distance coreference like full papers in the CRAFT corpus.

Distribution of coreferent pairs in gold data As illustrated in Figure 2, only about 40.85% of the gold pairs are in the groups of short distance while the other 59.15% of them are in the groups of long distance. This means that if a model cannot deal with long distance coreference, pairs of mentions and antecedents in this region cannot be discovered. Among those long distance pairs, 47.8% are in between 51-200 sentences while the number of pairs whose distance is more than 200 is about 11.35%.

The effect of mention filtering The results in Figure 2 revealed that by using the filtering methods, we could effectively address long-distance coreferent pairs. It can be seen from the figure that the baseline model was good enough when working on short distance pairs, and the filtering may slightly harm the performance. However, for longer distances, the filtering contributed to increases of 5.46%, 55.26%, 84.60% and 100% for the groups of 11-50, 51-100, 101-200, and >200, respectively, in comparison with the baseline.

The effect of BERT Without using the filtering method, BERT itself could capture a fairly large number of long-distance pairs, which was even better than the LSTM_filter model.

Long-distance coreference When summing up the results of long-distance groups, i.e., three groups of 51-100, 101-200, and >200, we found that the LSTM_filter and BERT_filter models could predict 71.89% and 83.63% higher number of long-distance pairs than the LSTM one, which indicates the effect of our model in using the filtering method and BERT for long documents. Additionally, for the most tough case of detecting pairs in the distance of more than 200 sentences, our LSTM_filter model predicted 874 correct pairs (about 1.18% of the gold pairs), and the BERT_filter model predicted 1,081 correct pairs (about 1.46% of the gold pairs). Meanwhile, the LSTM model failed to detect pairs in this group.

Recall problem It is necessary to note that although our model improved the baseline LSTM and obtained promising results, the recall is still low in all groups of distance. For instance, the best performing model, i.e., BERT_filter could cover only 50.90%, 45.38%, 39.08%, 31.71%, and 12.82% of the gold pairs in the groups of $<=10$, 11-50, 51-100, 101-200, >200, respectively. This is an open issue that we will address in the future.

5 Conclusion

In this paper, we particularly address the challenge of coreference resolution in full text articles in the CRAFT Shared Task 2019. Specifically, we employ the span-based end-to-end model (Lee et al., 2017) and enhance the model by utilizing a syntax-based mention filtering method and BERT. To filter noisy mentions, we jointly train a parsing model with a POS classifier to obtain parse trees of sentences. We then generate syntactic patterns of gold mentions based on the resulting parse trees. Any mentions that satisfy the generated patterns will be fed into the coreference resolution model. We finally incorporate BERT into our model. Experimental results on the CRAFT corpus indicate that the proposed method is effective in capturing long-distance coreferences in long documents.

Acknowledgments

This paper is based on results obtained from a project commissioned by the New Energy and Industrial Technology Development Organization (NEDO). This work is also supported by PRISM (Public/Private R&D Investment Strategic Expansion PrograM).

References

Iz Beltagy, Arman Cohan, and Kyle Lo. 2019. Scibert: Pretrained contextualized embeddings for scientific text. *CoRR*, abs/1903.10676.

K. Bretonnel Cohen, Arrick Lanfranchi, Miji Jooyoung Choi, Michael Bada, William A. Baumgartner Jr., Natalya Panteleyeva, Karin Verspoor, Martha Palmer, and Lawrence E. Hunter. 2017. Coreference annotation and resolution in the colorado richly annotated full text (CRAFT) corpus of biomedical journal articles. *BMC Bioinformatics*, 18(1):372:1–372:14.

Dina Demner-Fushman, Willie J Rogers, and Alan R Aronson. 2017. MetaMap Lite: an evaluation of a new Java implementation of MetaMap. *Journal of the American Medical Informatics Association*, 24(4):841–844.

Jacob Devlin, Ming-Wei Chang, Kenton Lee, and Kristina Toutanova. 2019. BERT: Pre-training of deep bidirectional transformers for language understanding. In *Proceedings of the 2019 Conference of the North American Chapter of the Association for Computational Linguistics: Human Language Technologies, Volume 1 (Long and Short Papers)*, pages 4171–4186, Minneapolis, Minnesota. Association for Computational Linguistics.

Sepp Hochreiter and Jürgen Schmidhuber. 1997. Long short-term memory. *Neural computation*, 9(8):1735–1780.

Diederik P Kingma and Jimmy Lei Ba. 2015. Adam: A method for stochastic optimization. In *Proceedings of the 3rd International Conference on Learning Representations (ICLR2015)*.

Nikita Kitaev and Dan Klein. 2018. Constituency parsing with a self-attentive encoder. In *Proceedings of the 56th Annual Meeting of the Association for Computational Linguistics (Volume 1: Long Papers)*, pages 2676–2686, Melbourne, Australia. Association for Computational Linguistics.

Kenton Lee, Luheng He, Mike Lewis, and Luke Zettlemoyer. 2017. End-to-end neural coreference resolution. In *Proceedings of the 2017 Conference on Empirical Methods in Natural Language Processing*, pages 188–197, Copenhagen, Denmark. Association for Computational Linguistics.

Ngan Nguyen, Jin-Dong Kim, and Jun'ichi Tsujii. 2011. Overview of BioNLP 2011 protein coreference shared task. In *Proceedings of BioNLP Shared Task 2011 Workshop*, pages 74–82, Portland, Oregon, USA. Association for Computational Linguistics.

H-L. Trieu, N. T. H. Nguyen, M. Miwa, and S. Ananiadou. 2018. Investigating domain-specific information for neural coreference resolution on biomedical texts. In *Proceedings of the BioNLP 2018 workshop*, pages 183–188.

Colin Warner, Ann Bies, Christine Brisson, and Justin Mott. 2004. Addendum to the penn treebank ii style bracketing guidelines: Biomedical treebank annotation. *University of Pennsylvania*.

A Penn Treebank Labels

For the Penn Treebank labels in our syntactic patterns, we follow the BioMedical Treebank tagset definition (Warner et al., 2004). Please refer to Table 5 for the detail description.

B Non-Coreference Results

Unlike other metrics, the BLANC metric also contains non-coreference results. We report the results of the test set in Table 6.

C Results on (Mention-Antecedent) Pair Distance

We present the detail results of each model and the corresponding gold coreference grouped by the sentence-level distance of mention-antecedent pairs in Table 7. The results are calculated in five groups of distance: $<=10$, 11-50, 51-100, 101-200, >200.

Tag	Description
NP	noun phrase
NN	noun, singular or mass
NML	sub-NP nominal substrings
PRP$	possessive pronoun
LS	list item marker

Table 5: The definition of relevant Penn Treebank labels.

Metric	Model	Mention			Non-Coreference			
		P	R	F	P	R	F	BLANC-F
BLANC	LSTM	0.7565	0.3578	0.4858	0.5725	0.1314	0.2137	0.2682
	LSTM_filter	0.7292	0.4187	0.5320	0.5432	0.1796	0.2699	0.3784
	BERT	0.7416	0.5603	0.6383	0.5537	0.3172	0.4033	0.4421
	BERT_filter	0.7314	0.5778	0.6456	0.5383	0.3356	0.4135	0.4512
	E2E_MetaMap	0.6713	0.5272	0.5906	0.4488	0.2812	0.3457	0.4053

Table 6: Non-coreference results for the BLANC metric on the testing set.

Model	<=10			11-50			51-100		
	TP	G.R (%)	O.R (%)	TP	G.R (%)	O.R (%)	TP	G.R (%)	O.R (%)
LSTM	3,588	34.61	4.83	5,230	26.22	7.05	1,735	10.43	2.34
LSTM_filter	3,164	30.52	4.26	5,532	27.73	7.45	3,878	23.31	5.23
BERT_250	5,347	51.58	7.20	9,015	45.19	12.15	5,115	30.75	6.89
BERT	4,651	44.87	6.27	8,062	40.41	10.86	5,381	32.35	7.25
BERT_filter	5,276	50.90	7.11	9,053	45.38	12.20	6,502	39.08	8.76
Group Gold	10,366	100	13.97	19,949	100	26.88	16,636	100	22.41

Model	101-200			>200		
	TP	G.R (%)	O.R (%)	TP	G.R (%)	O.R (%)
LSTM	484	2.57	0.65	0	0.00	0.00
LSTM_filter	3,143	16.69	4.23	874	10.37	1.18
BERT_250	4,163	22.10	5.61	527	6.25	0.71
BERT	4,798	25.47	6.46	620	7.35	0.84
BERT_filter	5,974	31.71	8.05	1,081	12.82	1.46
Group Gold	18,837	100	25.38	8,431	100	11.36
Total Gold	74,219					

Table 7: Results of models on each distance group; **TP**: True Positive; **G.R**: Group Ratio = True Positive/Group Gold; **O.R**: Overall Ratio = True Positive/Total Gold

Neural Dependency Parsing of Biomedical Text: TurkuNLP entry in the CRAFT Structural Annotation Task

Thang Minh Ngo, Jenna Kanerva, Filip Ginter, and Sampo Pyysalo
Turku NLP Group, Department of Future Technologies,
University of Turku, Finland
`{first.last}@utu.fi`

Abstract

We present the approach taken by the TurkuNLP group in the CRAFT Structural Annotation task, a shared task on dependency parsing. Our approach builds primarily on the Turku neural parser, a native dependency parser that ranked among the best in the recent CoNLL tasks on parsing Universal Dependencies. To adapt the parser to the biomedical domain, we considered and evaluated a number of approaches, including the generation of custom word embeddings, combination with other in-domain resources, and the incorporation of information from named entity recognition. We achieved a labeled attachment score of 89.7%, the best result among task participants.

1 Introduction

Syntactic analysis (parsing) is a fundamental task in natural language processing (NLP) and a prerequisite for many related tasks. There is a long tradition of research in automatic parsing targeting both constituency (phrase structure) and dependency representations, with most work focusing on the analysis of English news texts (Marcus et al., 1994). Syntactic analyses are required also by many methods for the analysis of biomedical text; for example, information extraction methods commonly rely on the shortest path over syntactic dependencies to identify how entities mentioned in text are related (Airola et al., 2008; Björne et al., 2009; Liu et al., 2013; Luo et al., 2016). The performance of parsers is known to be domain-dependent: to create high-quality analyses of e.g. biomedical texts, the tools should be trained on annotated corpora reflecting the domain (Miwa et al., 2010). Syntactically annotated corpora of domain texts are thus required for much of biomedical NLP. These resources should also preferably follow the relevant standards in the representation of syntactic analyses to allow methods developed to these standards to be applied also for biomedical domain texts, thus allowing biomedical NLP to benefit from advances in parsing technology.

The CRAFT Structural Annotation (SA) task, organized in 2019 is a shared task on dependency parsing largely following the setting of the popular Conference on Computational Natural Language Learning (CoNLL) 2017 and 2018 shared tasks on dependency parsing (Zeman et al., 2017, 2018). These tasks emphasize real-world scenarios by casting the task as analyzing raw text (rather than e.g. pre-tokenized and tagged text) and applying universal, language-independent representations. The CRAFT SA task follows these tasks in providing only plain text as input, requiring participating systems to perform sentence segmentation, tokenization, part-of-speech tagging, lemmatization, and the identification of morphological features in addition to analyzing the syntactic structure of the input sentences. CRAFT SA also adopts the format and evaluation tools of the CoNLL tasks, and its representation matches the universal representation of these tasks in part. The CRAFT task is differentiated from the many corpus resources applied in the CoNLL tasks specifically in focusing on biomedical domain texts, and CRAFT is unique among syntactically annotated biomedical corpora in that its texts are drawn from full-text articles, rather than only article titles and abstracts.

We participated in the CRAFT SA task using an approach that builds primarily on the Turku neural parser (Kanerva et al., 2018), a native dependency parsing system that previously ranked among the best systems in the CoNLL 2018 task. As the parser is fully retrainable, designed to accept the format used for the CRAFT data, and agnostic to the details of the representation, it was possible to train it for the CRAFT task with little modification. Additionally, as the parser has not been de-

Proceedings of the 5th Workshop on BioNLP Open Shared Tasks, pages 206–215
Hong Kong, China, November 4, 2019. ©2019 Association for Computational Linguistics

veloped or previously applied to biomedical English, we consider a number of modifications and adaptions to improve on its performance, finding in particular that the strong baseline performance of the parser can be further improved through initialization with in-domain word vectors.

2 Background

Biomedical domain models have been available for a number of constituency parsers (e.g. Charniak and Johnson (2005), McClosky and Charniak (2008)) and have been widely applied in domain information extraction efforts, frequently in conjunction with heuristic conversions into dependency representations such as Stanford dependencies (De Marneffe and Manning, 2008). There have also been native dependency parsers available for the domain, such as Pro3Gres (Schneider and Rinaldi, 2004) and, later, GDep (Miyao et al., 2008), nevertheless the abovementioned McClosky-Charniak parser with Stanford dependencies conversion was the workhorse of biomedical dependency parsing for nearly a decade. Also the treebanks available for training the parsers in the biomedical domain have traditionally been constituency-based, for instance the Penn BioIE (Kulick et al., 2004) and especially the GENIA treebank (Tateisi et al., 2005). The BioInfer corpus (Pyysalo et al., 2007) was the first domain corpus to adopt Stanford Dependencies as the native annotation scheme, coinciding with a generally growing interest in dependency parsing and its applications.

The CoNLL 2006 and 2007 shared tasks (Buchholz and Marsi, 2006; Nivre et al., 2007) addressed multilingual dependency parsing, and while data was provided for different languages in the same format, the underlying representation (e.g. dependency types) was not standardized in these tasks. These tasks also included only prediction of syntactic trees, whereas tokenization and part-of-speech tags were given for the participants.

In recent years, there has been an increased interest in native dependency parsing, reflected in efforts such as Universal Dependencies (UD) (Nivre et al., 2016) and the CoNLL 2017 and 2018 shared tasks on multilingual parsing using UD data (Zeman et al., 2017, 2018). While these efforts have covered a wide range of languages, genres and text domains, and introduced end-to-end parsing from plain text as the objective, they have not specifi-

	Train	Test
Documents	67	30
Sentences	21 731	9 099
Tokens	561 032	232 619

Table 1: CRAFT Structural Annotation statistics

	Train	Devel	Eval
Documents	47	10	10
Sentences	15 007	3 421	3 303
Tokens	387 473	91 306	82 253

Table 2: CRAFT Train data split for development

cally involved scientific articles or biomedical domain texts.

3 Data

3.1 CRAFT data

The primary resource used for training systems for the task is the CRAFT corpus syntactic annotation provided by the task organizers. Table 1 summarizes the key statistics of the data.

The test annotations were only made available after participants had submitted their predictions, and no train/development split was defined for the provided data. For development purposes, we thus split the provided training dataset of 67 documents randomly into a set of 47 used for training, a devel set of 10 used for early stopping during training, and 10 used for evaluation during development. The statistics of this split are shown in Table 2

The original CRAFT corpus syntactic annotation uses a modified Penn Treebank (PTB) constituency formalism (Verspoor et al., 2012), and the dependency annotation provided for the task was automatically created by conversion from the constituency representation. The source data was first converted into the CoNLL-X format using the SD dependency representation and PTB POS tags using the approach of (Choi and Palmer, 2012), and this data was then further converted into the CoNLL-U format with custom scripts.

The resulting task dataset is in the UD *format* (CoNLL-U), but it only partially follows the UD standard in terms of its content. In particular, while the POS tags and morphological features conform to UD, the dependency representation – arguably the most important part of the data – does not, instead matching the SD representation of the CoNLL-X version of the data. Figure 1 shows SD

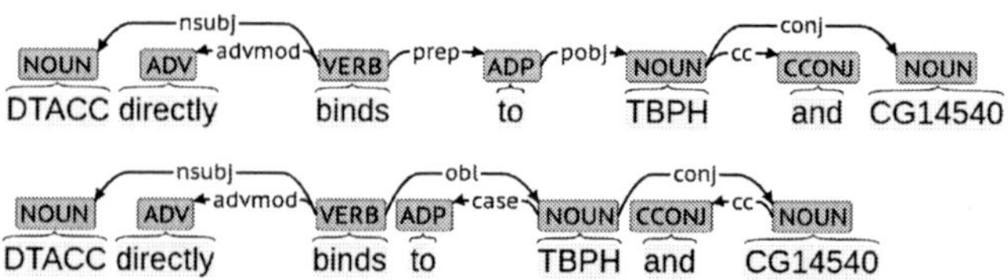

Figure 1: Illustration of Stanford Dependencies (top) and Universal Dependencies (bottom) analyses for an example sentence (from PMCID:15207008). The CRAFT dependency annotation follows the former representation.

and UD analyses for an example sentence from the CRAFT data. While a number of dependencies are identical between the two (e.g. `nsubj`), in UD dependencies primarily relate content words (e.g. verbs and nouns), with function words such as adpositions being dependents of content words rather than mediating their relations such as in SD (cf. *binds to TBPH* in Figure 1). There are also a number of minor differences such as the attachment of coordinating conjunctions to the first constituent in SD and the nearest to the right in UD.

While this discrepancy does not prevent the use of tools that are agnostic to the details of the representation (including many UD parsers), it does mean that the data is incompatible with existing UD resources and greatly complicates combination with other corpora, none of which are available in this particular hybrid SD/UD CoNLL-U representation. We expand on this issue below in Section 6.2.

3.2 Word vectors

We considered a number of previously released word vectors for initializing the parser. As a baseline we use the English word embeddings by Ginter et al. (2017) trained on general English extracted from Wikipedia and Internet crawls. These embeddings are trained using the word2vec (Mikolov et al., 2013) tool with lower-cased data, skip-gram algorithm, window size of 10 and 100 dimensions. The vectors were originally provided for the CoNLL 2017 and 2018 multilingual parsing shared task, and thus used by many of the participating systems in their English parsing models. We also considered a number of word vectors induced specifically on biomedical text for domain tasks, including those created by Pyysalo et al. (2013)[1] and Chiu et al. (2016)[2].

3.3 Unlabelled data

To induce new word vectors (Section 4.3) and conduct co-training experiments (Section 5.2), we used unlabelled texts from PubMed titles and abstracts and PubMed Central (PMC) full texts. The data was drawn from the PubMed 2017 baseline distribution and a 2017 download of the PMC Open Access subset.[3] The texts were segmented into sentences using the GENIA sentence splitter and then tokenized using the PTBTokenizer included in Stanford CoreNLP tools (Manning et al., 2014) and the tokenized sentences shuffled randomly. The resulting dataset consists of 12.5 billion tokens in 500 million sentences. As the text of the full-text articles of the CRAFT corpus contains characters outside of the basic ASCII character set, we created word vectors on the original extracted texts instead of first applying a mapping to ASCII characters as was done in a number of similar previous efforts (e.g. (Pyysalo et al., 2013)).

4 Methods

4.1 Turku Parser

Our primary parser used in all experiments is the Turku Neural Parser Pipeline[4] (Kanerva et al., 2018), a full parser pipeline meant for end-to-end analysis from raw text into UD. The pipeline includes sentence and word segmentation, part-of-speech and morphological tagging, syntactic parsing, and lemmatization.

The segmentation component in the Turku pipeline is built using UDPipe (Straka and Straková, 2017), where the token and sentence boundaries are jointly predicted using a single-layer bidirectional GRU network. Universal (UPOS) and language-specific (XPOS) part-of-speech tags, as well as morphological features

---

[1] `http://bio.nlplab.org/`
[2] `https://github.com/cambridgeltl/BioNLP-2016`

[3] We used 2017 data as we had a plain text version readily available from previous work.
[4] `https://turkunlp.org/Turku-neural-parser-pipeline/`

(FEATS) are predicted with a modified version of the one published by Dozat et al. (2017), a time-distributed classifier over tokens in a sentence embedded using bidirectional LSTM network. The tagger has two separate classification layers, one for universal part-of-speech and one originally used for language-specific part-of-speech tags. The bidirectional encoding is shared between both classifiers. In the modified version (Kanerva et al., 2018), the second classifier is used to jointly predict the language-specific POS tags together with morphological features by simply concatenating the two input columns into one. The syntactic analysis is based on a graph-based parser by Dozat et al. (2017), a biaffine classifier with MST decoder on top of a bidirectional LSTM network. The lemmatizer component by Kanerva et al. (2019) is a sequence-to-sequence model, where the lemma is generated one character at a time from the given input word form and morphological features.

In the Turku parser pipeline, all these components are wrapped into a single system. All components directly supports training with CoNNL-U formatted treebanks while being completely label agnostic, thus not requiring the treebank to actually follow the UD guidelines and label sets. Therefore, the parser can be trained on CRAFT corpus as is. The Turku Parser was ranked second on LAS and MLAS, and first on BLEX on the CoNLL-2018 Shared Task, making it highly competitive.

4.2 UDPipe

UDPipe[5] (Straka and Straková, 2017) is an easily trainable parsing pipeline including segmentation, morphological tagging, lemmatization and syntactic parsing. UDPipe has long been the "go-to" UD parser and has also served as the organizers' baseline in the 2017 and 2018 CoNLL Shared Tasks on Multilingual Parsing from Raw Text to Universal Dependencies. Tokenization and sentence segmentation is implemented jointly, using a single-layer GRU network, predicting for each character whether it is a sentence boundary, token boundary, or token-internal. The tagger is an averaged perceptron disambiguating from a set of candidate analyses generated based on the last four characters of the word. Lemmatization is carried out by generating a set of candidate lemma rules, each

of which produce a lemma by removing and possibly substituting characters from the word prefix and suffix. As in tagging, an averaged perceptron then disambiguates among the candidates. The dependency parser is a transition-based parser with a feed-forward neural network serving as the classifier that decides on the next transition taken by the parser.

4.3 Word vectors

For inducing new sets of word vectors, we used the word2vec[6] (Mikolov et al., 2013) and Fast-Text[7] (Joulin et al., 2016; Bojanowski et al., 2017) tools. In brief, these tools generate a vector representation for each token based on the similarity of the contexts in which they appear in a large corpus of unannotated text. Word vectors were induced on texts extracted from PubMed abstracts and PMC Open Access publications (Section 3.3) using both the skip-gram and continuous bag-of-words (CBOW) models implemented in both tools. Model parameters were primarily kept at their default values, but we performed a series of experiments with different values of the window parameter, which has been found to be particularly impactful in previous work (Chiu et al., 2016).

4.4 Evaluation

The CRAFT SA shared task adopted the evaluation metrics and evaluation implementation of the CoNLL'18 shared task. In particular, performance was evaluated in terms of the Labeled attachment score (LAS), Morphology-aware labeled attachment score (MLAS), and Bi-lexical dependency score (BLEX) metrics, defined as follows (Zeman et al., 2018):

LAS The percentage of nodes having correctly assigned parent token, as well as correct type of the dependency relation. All tokens are considered in the evaluation, including also punctuation.

MLAS Similar to LAS, but with an additional requirement of having also functional dependents and certain morphological features predicted correctly. In addition the metric is calculated only based on content bearing words discarding functional words and punctuation. Thus, MLAS measures the percentage of content words having correctly assigned parent token, relation type, func-

---

[5] http://ufal.mff.cuni.cz/udpipe

[6] https://github.com/tmikolov/word2vec
[7] https://fasttext.cc/

Parser	Word vectors	LAS
Turku	Bio, word2vec/CBOW (window 2)	89.86
Turku	Bio (CRAFT tokens), word2vec/CBOW (default parameters)	89.78
Turku	Bio (CRAFT tokens), word2vec/CBOW (win2)	89.69
Turku	Bio, word2vec/CBOW (default parameters)	89.55
Turku	Bio, word2vec/CBOW (window 20)	89.73
Turku	Bio, FastText/CBOW (default parameters)	89.50
Turku	Bio, word2vec/skipgram (default parameters)	89.63
Turku	CoNLL	89.27
UDPipe	Bio, word2vec/CBOW (window 2)	85.00
UDPipe	CoNLL	84.66
UDPipe	Bio, word2vec/CBOW (default parameters)	84.22

Table 3: Development set results with different word vectors. CoNLL = baseline CoNLL shared task word vectors, Bio = custom word vectors induced on PubMed and PMC articles, CRAFT tokens = input text tokenized with model trained on CRAFT data. (For details on the word2vec and FastText tools, their CBOW and skipgram models, and parameters, see Section 3.2)

tional dependents and certain morphological features.

BLEX The proportion of correct relations between two content bearing words with an additional requirement that the lemma of the dependent must be correct. Functional words and punctuation tokens are discarded.

As LAS is the best established and most frequently applied of these metrics, we focused on optimizing this metric during development and report results for experiments conducted during development in terms of LAS only. For the final three test set submissions, we provide results for the full set of metrics implemented in the CoNLL evaluation script. In addition to the three metrics above, this includes measures of token, sentence and word segmentation agreement with gold (**Tokens**, **Sentences** and **Words** metrics), agreement of the universal (**UPOS**) and language-specific (**XPOS**) part-of-speech tags and morphological features (**UFeats**), the three previous together (**AllTags**), and agreement on lemmas (**Lemmas**). We refer to Zeman et al. (2018) for further details on these additional metrics.

We note that in the CRAFT test set evaluation, performance for each metric was calculated as an average of the results for the 30 test set documents, rather than over the catenation of the documents as in the CoNLL evaluation.

5 Results

During the development of our system, we considered a number of approaches in an iterative and incremental process. In this section, we first present the strategies we found effective, namely the use of custom in-domain word vectors and data aug-

mentation. We then present the results from our three test set submissions and an analysis of these results using various additional metrics.

5.1 Word vectors

A simple but highly effective way to adapt machine learning systems that operate on vector representations of words to new domains is to initialize them with word embeddings induced on domain texts. We evaluated a variety of previously introduced and newly induced word embeddings in this way (see above) using both the Turku and UDPipe parsers, and summarize results for notable baseline vectors and selected in-domain word vectors in Table 3.

We find that using the general out-of-domain CoNLL word vectors, the parsers already achieve high baseline LAS scores, 84.66% for UDPipe and 89.27% for our primary, Turku system. In our limited experiments with UDPipe we found somewhat mixed results from the use of custom biomedical domain word vectors. For the Turku parser, a number of the in-domain word embeddings did prove effective, with the best-performing combination of data preprocessing, model and parameters achieving a LAS of 89.86%, a 5% relative reduction in LAS error from the CoNLL word vector baseline. Regarding the alternative settings for inducing word vectors, we broadly found CBOW to be more effective than the skipgram model and small windows to be more effective than either default parameters or large windows. We did not see an advantage of FastText over word2vec vectors and conducted the majority of our experiments with the latter tool.

Two of the runs submitted for the final evaluation used settings from these experiments, namely

Parser	Word vectors	Extra data (size, source)	LAS
Turku	Bio, word2vec/CBOW (window 2)	4k sentences, PMC	89.92
Turku	Bio, word2vec/CBOW (window 2)	10k sentences, PMC and PubMed	89.87
Turku	Bio, word2vec/CBOW (window 2)	6k sentences, PubMed	89.78
Turku	Bio, word2vec/CBOW (window 2)	10k sentences, PMC	89.84
Turku	Bio, word2vec/CBOW (window 2)	20k sentences, PMC	89.41

Table 4: Development set results with extra training data

Bio, word2vec/CBOW (window 2) and CRAFT tokenized, word2vec/CBOW (default parameters).

5.2 Training data augmentation

After identifying the word vectors that achieved the highest LAS score for this data, we implemented and evaluated multiple techniques to increase the number of training examples beyond the given training data. Most of the approaches we considered failed to improve on performance, largely due to incompatibilities in annotation (see Section 6.1), but we found limited success with a co-training approach (Blum and Mitchell, 1998).

Specifically, we first used the best Turku and UDPipe parser models introduced in our previous experiments to analyze a large sample of unannotated text from PubMed abstracts and PMC full text articles. We then compared the results to identify sentences that are identically segmented and tokenized and given identical syntactic analyses (heads and dependency relations) by the two systems. We then created random samples of varying sources and sizes from this data, generating comparatively high-quality automatically annotated additional training data. This data was combined with the original CRAFT training data to create an extended training set that was then used to create a new model with the Turku parser. We present a selection of development results from this setting in Table 4.

While we achieved some minor improvements in some of the experiments, the co-training approach did not improve the performance as system as much as could be hoped based on e.g. the effectiveness of self-training for parsing (McClosky et al., 2006). There may be a number of reasons for the limited effectiveness of our approach, potentially including sub-domain mismatch between our unlimited samples of PubMed and PMC documents and the comparatively narrow and focused domain of CRAFT texts. We nevertheless chose to include the model with the best result in these experiments with **4k sentences, PMC** as extra training data to include in our final submissions.

5.3 Test set results

The properties of the three runs we submitted to the task are summarized in Table 5 together with their development and test set LAS scores. We find that test set performance closely follows the results of development experiments, producing the same ranking of the three runs as well as results within 0.3% points of the development results in all three cases.

As expected on the basis of the development experiments, the two runs without extra training data are highly competitive, and augmenting the training data via co-training while keeping the word vectors constant provides only a modest benefit. Nevertheless, the run that combined custom indomain word vectors and co-training to adapt the Turku parser to biomedical text achieved the highest performance not only among our runs but also out of all six runs submitted to the task.

5.4 Analysis of final results

Table 6 provides a detailed look at the performance of our three final submissions using all metrics implemented in the CoNLL 2018 shared task evaluation script (see Section 4.4). All of the metrics are averaged F1 scores across all 30 test files.

We find very similar results across all three runs. Segmentation performance is acceptable for sentence splitting (over 97.5%) and very high for tokenization (over 99.5%), indicating limited remaining benefit from further focus on identifying sentence and token boundaries. Part-of-speech tags (UPOS and XPOS) as well as morphological features are each assigned at a high level of consistency (approx. 98% each), and lemmas are correctly identified in approx. 99% of cases, indicating that the parser is well adapted to the challenges of specialized biomedical domain terminology. The only metrics showing notable remaining room for improvement are dependency-based (last five rows in Table 6). The relatively close results for the unlabeled and labeled attachment score metrics (UAS and LAS) indicate that the identification of the correct dependency relation is not a key factor

211

Parser	Word vectors	Extra data	LAS(dev)	LAS(test)
Turku	Bio, word2vec/CBOW (window 2)	4k sentences, pmcoa articles	89.92	89.695
Turku	Bio, word2vec/CBOW (window 2)	No	89.86	89.650
Turku	Bio (CRAFT tokens), word2vec/CBOW (defaults)	No	89.78	89.536

Table 5: Final submission results on test data

Metrics	Run 1	Run 2	Run 3
Tokens	99.593	99.555	99.593
Sentences	97.590	97.621	97.590
Words	99.593	99.555	99.593
UPOS	98.221	98.179	98.184
XPOS	97.806	97.758	97.789
UFeats	98.282	98.233	98.265
AllTags	97.752	97.718	97.729
Lemmas	98.999	98.981	99.048
UAS	90.942	90.882	90.794
LAS	89.695	89.650	89.536
CLAS	87.373	87.294	87.201
MLAS	85.549	85.441	85.318
BLEX	86.630	86.595	86.544

Table 6: Final submission test results for all metrics

limiting the performance of the parser, and that the remaining challenges for substantially advancing the performance of the system lie specifically in more accurately recovering the dependency structure of the sentences.

6 Discussion

In the following, we briefly discuss a number of ideas we considered that failed to improve on the performance of the parser and address the relationship between the CRAFT SA task data and Universal Dependencies.

6.1 What did not work

During the relatively brief development period for participating in the shared task, we considered a number of variants and potential extensions of our approach that failed to improve on the performance of the system. Although these were not developed and evaluated with the rigor required to report full experimental results, we summarize some of these ideas here in the hope that they may help others in their work.

Corpus combinations As the CRAFT dependency annotations were created by automatic conversion from PTB source, we considered the possibility of combining the task training data with additional similarly converted annotations. We performed several preliminary experiments converting the PTB Wall Street Journal section (Marcus et al., 1994) and the original GENIA treebank data (Tateisi et al., 2005) as well as a version of the GENIA treebank that as previously converted using the Stanford Dependency Converter.[8] The results of these experiments were disappointing; initial single-corpus experiments using the converted data failed to reach the expected level of performance, and all combinations of this data with CRAFT data resulted in decreased performance. We also initially considered attempting combinations with English corpora from the Universal Dependencies collection, but abandoned this idea due to incompatibilities in the representations (see below).

Entity mentions As the CRAFT corpus annotation integrates not only syntactic but also entity mention (or concept) annotation, there is an opportunity to integrate information on named entities and related concepts into the parsing process.[9] Briefly, the intuition is that a model that has information on which tokens are e.g. part of chemical or species names could better parse their mentions and associated text. To explore this idea, we converted the CRAFT concept annotation into a token-level begin-in-out (BIO) representation using custom tools, and appended these annotations into the XPOS column of the CoNLL-U data, creating merged POS and entity tags. We then trained on this data, creating joint models that integrate dependency parsing and entity mention information. However, the performance of these models was mixed, with minimal improvements in few cases and a reduction in LAS in others, and we chose not to pursue the idea further.

Previously introduced in-domain word embeddings Throughout development, we evaluated many word vectors, including both previously introduced and newly induced as well as biomedical domain and out-of-domain embeddings. The

---

[8] https://github.com/allenai/
genia-dependency-trees
[9] This idea was also advanced by the organizers in the CRAFT SA task description.

general pattern we found was the vectors introduced for the CoNLL shared task represented a very strong baseline, and many in-domain word vectors previously made available by the biomedical NLP community (including ones previously introduced by some of the authors) failed to improve on the results achieved with these vectors. We were only consistently able to improve over the CoNLL word vector baseline by newly inducing custom in-domain word vectors for the parser. We attribute some of this effect to the differences in the dimensionality of previously introduced vectors: although the parser can be configured to accept vectors of any size, some part of its development may have specifically optimized for the 100-dimensional CoNLL word vectors. It is also likely that part of the effect is explained by the presence of non-ASCII characters in the CRAFT data, as many in-domain word vectors were created on texts specifically mapped to ASCII as a pre-processing step.

6.2 CRAFT and Universal Dependencies

Universal Dependencies have become the *de facto* standard representation for computational dependency parsing, and the UD repository[10], containing over 100 UD treebanks covering more than 70 languages as of this writing, is a key interface connecting corpus creators and researchers working on parsing technology. There are several potential benefits to a biomedical domain UD corpus, especially the potential for combining existing English resources and domain transfer techniques. However, the CRAFT Structural Annotation shared task dataset differs from UD standards and conventions on a number of points, hindering its adoption as a UD resource.

Most obviously, despite being provided in the CoNLL-U format, the CRAFT data does not fully adopt UD types and annotation conventions. As noted above, the dependency relation types are drawn from a predecessor of UD, Stanford dependencies (SD), and the dependency annotation similarly follows SD rather than UD conventions. While the SD and UD representations are quite similar in many ways, they differ systematically in particular in that UD emphasizes content words over function words (see also Figure 1) and diverge in many details of the representation.

We also noted that the lemmas in the CRAFT

data don't always correspond to the canonical (or base) forms of the words. In addition to numbers expressed as digits all having the lemma value "0", spelled-out cardinal numbers (e.g. "one") have the value "#crd#" in place of a lemma, ordinal numbers (e.g. "first") have "#ord#", and hyperlinks (e.g. "http://www.ncbi.nlm.nih.gov/") have "#hlink#". These exceptions are not part of UD and contrary to the representation of lemmas in existing English UD resources.

Based on our experience with the SD and UD representations and in creating UD corpora by conversion from other formats, we believe it should be possible to automatically convert the present CRAFT corpus annotations into a full UD representation using a combination of existing tools and some deterministic mappings addressing issues specific to this data. Such conversion would allow the inclusion of the corpus in the UD repository, increasing the availability of biomedical English training data to the parsing community.

7 Conclusions

In this paper, we have presented the approach of the TurkuNLP team to the CRAFT SA dependency parsing shared task. Building on the basis of the Turku neural parser and UDPipe, we considered a number of modifications and adaptations to better address the full-text biomedical domain articles of the task, including the induction of custom word vectors and the extension of the training data with additional automatically parsed data. Experiments showed the Turku parser to clearly outperform the UDPipe baseline at the task and demonstrated that initializing the parser with custom in-domain word vectors could further improve on its strong off-the-shelf performance. Our adapted version of the Turku parser achieved the highest result on the test set of the shared task with a labeled attachment score of 89.7%.

All of the tools and resources applied in this work, as well as the newly trained parsing models, are made available under open licenses.

Acknowledgments

We thank the task organizers for their help and responsiveness to feedback during the development period. We are grateful to CSC – IT Center for Science for computational resources used to train our models.

---

[10] https://universaldependencies.org/

References

Antti Airola, Sampo Pyysalo, Jari Björne, Tapio Pahikkala, Filip Ginter, and Tapio Salakoski. 2008. All-paths graph kernel for protein-protein interaction extraction with evaluation of cross-corpus learning. *BMC bioinformatics*, 9(11):S2.

Jari Björne, Juho Heimonen, Filip Ginter, Antti Airola, Tapio Pahikkala, and Tapio Salakoski. 2009. Extracting complex biological events with rich graph-based feature sets. In *Proceedings of BioNLP Shared Task*, pages 10–18.

Avrim Blum and Tom Mitchell. 1998. Combining labeled and unlabeled data with co-training. In *Proceedings of the eleventh annual conference on Computational learning theory*, pages 92–100.

Piotr Bojanowski, Edouard Grave, Armand Joulin, and Tomas Mikolov. 2017. Enriching word vectors with subword information. *Transactions of the Association for Computational Linguistics*, 5:135–146.

Sabine Buchholz and Erwin Marsi. 2006. Conll-x shared task on multilingual dependency parsing. In *Proceedings of the tenth conference on computational natural language learning*, pages 149–164. Association for Computational Linguistics.

Eugene Charniak and Mark Johnson. 2005. Coarse-to-fine n-best parsing and maxent discriminative reranking. In *Proceedings of the 43rd annual meeting on association for computational linguistics*, pages 173–180. Association for Computational Linguistics.

Billy Chiu, Gamal Crichton, Anna Korhonen, and Sampo Pyysalo. 2016. How to train good word embeddings for biomedical nlp. In *Proceedings of the 15th workshop on biomedical natural language processing*, pages 166–174.

Jinho D Choi and Martha Palmer. 2012. Guidelines for the clear style constituent to dependency conversion. *Technical Report 01–12*.

Marie-Catherine De Marneffe and Christopher D Manning. 2008. The stanford typed dependencies representation. In *Coling 2008: proceedings of the workshop on cross-framework and cross-domain parser evaluation*, pages 1–8. Association for Computational Linguistics.

Timothy Dozat, Peng Qi, and Christopher D Manning. 2017. Stanfords graph-based neural dependency parser at the conll 2017 shared task. In *Proceedings of the CoNLL 2017 Shared Task: Multilingual Parsing from Raw Text to Universal Dependencies*, pages 20–30.

Filip Ginter, Jan Hajič, Juhani Luotolahti, Milan Straka, and Daniel Zeman. 2017. CoNLL 2017 shared task - automatically annotated raw texts and word embeddings. LINDAT/CLARIN digital library at the Institute of Formal and Applied Linguistics (ÚFAL), Faculty of Mathematics and Physics, Charles University.

Armand Joulin, Edouard Grave, Piotr Bojanowski, and Tomas Mikolov. 2016. Bag of tricks for efficient text classification. *arXiv preprint arXiv:1607.01759*.

Jenna Kanerva, Filip Ginter, Niko Miekka, Akseli Leino, and Tapio Salakoski. 2018. Turku Neural Parser Pipeline: An end-to-end system for the CoNLL 2018 shared task. In *Proceedings of the CoNLL 2018 Shared Task: Multilingual Parsing from Raw Text to Universal Dependencies*.

Jenna Kanerva, Filip Ginter, and Tapio Salakoski. 2019. Universal lemmatizer: A sequence to sequence model for lemmatizing universal dependencies treebanks. *arXiv preprint arXiv:1902.00972*.

Seth Kulick, Ann Bies, Mark Liberman, Mark Mandel, Ryan McDonald, Martha Palmer, Andrew Schein, Lyle Ungar, Scott Winters, and Pete White. 2004. Integrated annotation for biomedical information extraction. In *HLT-NAACL 2004 Workshop: Linking Biological Literature, Ontologies and Databases*, pages 61–68.

Haibin Liu, Lawrence Hunter, Vlado Kešelj, and Karin Verspoor. 2013. Approximate subgraph matching-based literature mining for biomedical events and relations. *PloS one*, 8(4):e60954.

Yuan Luo, Özlem Uzuner, and Peter Szolovits. 2016. Bridging semantics and syntax with graph algorithmsstate-of-the-art of extracting biomedical relations. *Briefings in bioinformatics*, 18(1):160–178.

Christopher D. Manning, Mihai Surdeanu, John Bauer, Jenny Finkel, Steven J. Bethard, and David McClosky. 2014. The Stanford CoreNLP natural language processing toolkit. In *Association for Computational Linguistics (ACL) System Demonstrations*, pages 55–60.

Mitchell Marcus, Beatrice Santorini, and Mary Ann Marcinkiewicz. 1994. Building a large annotated corpus of english: The penn treebank. *Computational Linguistics*.

David McClosky and Eugene Charniak. 2008. Self-training for biomedical parsing. In *Proceedings of the 46th Annual Meeting of the Association for Computational Linguistics on Human Language Technologies: Short Papers*, pages 101–104. Association for Computational Linguistics.

David McClosky, Eugene Charniak, and Mark Johnson. 2006. Effective self-training for parsing. In *Proceedings of NAACL*, pages 152–159.

Tomas Mikolov, Kai Chen, Greg Corrado, and Jeffrey Dean. 2013. Efficient estimation of word representations in vector space. *arXiv preprint arXiv:1301.3781*.

Makoto Miwa, Sampo Pyysalo, Tadayoshi Hara, and Jun'ichi Tsujii. 2010. A comparative study of syntactic parsers for event extraction. In *Proceedings of the 2010 Workshop on Biomedical Natural Language Processing*, pages 37–45. Association for Computational Linguistics.

Yusuke Miyao, Rune Sætre, Kenji Sagae, Takuya Matsuzaki, and Junichi Tsujii. 2008. Task-oriented evaluation of syntactic parsers and their representations. In *Proceedings of ACL*, pages 46–54.

Joakim Nivre, Marie-Catherine De Marneffe, Filip Ginter, Yoav Goldberg, Jan Hajic, Christopher D Manning, Ryan McDonald, Slav Petrov, Sampo Pyysalo, Natalia Silveira, et al. 2016. Universal dependencies v1: A multilingual treebank collection. In *Proceedings of the Tenth International Conference on Language Resources and Evaluation (LREC 2016)*, pages 1659–1666.

Joakim Nivre, Johan Hall, Sandra Kübler, Ryan McDonald, Jens Nilsson, Sebastian Riedel, and Deniz Yuret. 2007. The conll 2007 shared task on dependency parsing. In *Proceedings of the 2007 Joint Conference on Empirical Methods in Natural Language Processing and Computational Natural Language Learning (EMNLP-CoNLL)*.

Sampo Pyysalo, Filip Ginter, Veronika Laippala, Katri Haverinen, Juho Heimonen, and Tapio Salakoski. 2007. On the unification of syntactic annotations under the stanford dependency scheme: A case study on BioInfer and GENIA. In *ACL'07 workshop on Biological, translational, and clinical language processing (BioNLP'07)*, pages 25–32. Association for Computational Linguistics.

Sampo Pyysalo, Filip Ginter, Hand Moen, Sophia Ananiadou, and Tapio Salakoski. 2013. Distributional semantics resources for biomedical text processing. *Proceedings of LBM*, pages 39–44.

Gerold Schneider and James Rinaldi, Fabio; Dowdall. 2004. Fast, deep-linguistic statistical minimalist dependency parsing. In *COLING-2004 Recent Advances in Dependency Grammars*, pages 33–40.

Milan Straka and Jana Straková. 2017. Tokenizing, pos tagging, lemmatizing and parsing ud 2.0 with udpipe. In *Proceedings of the CoNLL 2017 Shared Task: Multilingual Parsing from Raw Text to Universal Dependencies*, pages 88–99, Vancouver, Canada. Association for Computational Linguistics.

Yuka Tateisi, Akane Yakushiji, Tomoko Ohta, and Junichi Tsujii. 2005. Syntax annotation for the genia corpus. In *Proceedings of IJCNLP'05*.

Karin Verspoor, Kevin Bretonnel Cohen, Arrick Lanfranchi, Colin Warner, Helen L Johnson, Christophe Roeder, Jinho D Choi, Christopher Funk, Yuriy Malenkiy, Miriam Eckert, et al. 2012. A corpus of full-text journal articles is a robust evaluation tool for revealing differences in performance of biomedical natural language processing tools. *BMC bioinformatics*, 13(1):207.

Daniel Zeman, Jan Hajič, Martin Popel, Martin Potthast, Milan Straka, Filip Ginter, Joakim Nivre, and Slav Petrov. 2018. Conll 2018 shared task: multilingual parsing from raw text to universal dependencies. In *Proceedings of the CoNLL 2018 Shared Task: Multilingual Parsing from Raw Text to Universal Dependencies*, pages 1–21.

Daniel Zeman, Martin Popel, Milan Straka, Jan Hajič, Joakim Nivre, Filip Ginter, Juhani Luotolahti, Sampo Pyysalo, Slav Petrov, Martin Potthast, et al. 2017. Conll 2017 shared task. *Proceedings of the CoNLL 2017 Shared Task Multilingual Parsing from Raw Text to Universal Dependencies*.

RDoC Task at BioNLP-OST 2019: A Mental Health Informatics Task with Research Domain Criteria

Mohammad Anani[1*], Nazmul Kazi[1*], Matthew Kuntz[23] and Indika Kahanda[1]

[1] Gianforte School of Computing, Montana State University, MT, USA
[2] National Alliance of Mental Illness (NAMI) Montana, Helena, MT, USA
[3] Center for Mental Health Research and Recovery, Montana State University, MT, USA

mohammad.anani@student.montana.edu
{kazinazmul.hasan,matthew.kuntz,indika.kahanda}@montana.edu

* The authors wish it to be known that Mohammad Anani and
Nazmul Kazi should be regarded as joint first authors.

Abstract

BioNLP Open Shared Tasks (BioNLP-OST) is an international competition organized to facilitate development and sharing of computational tasks of biomedical text mining and solutions to them. For BioNLP-OST 2019, we introduced a new mental health informatics task called "RDoC Task", which is composed of two subtasks: information retrieval and sentence extraction through National Institutes of Mental Health's Research Domain Criteria framework. Five and four teams around the world participated in the two tasks, respectively. According to the performance on the two tasks, we observe that there is room for improvement for text mining on brain research and mental illness.

1 Introduction and Motivation

The breadth of brain research is too expansive to be effectively curated without computational tools especially involving machine learning models. For example, a Pubmed search for "Brain" on August 12, 2019, revealed 854,612 articles[1]. More specifically, an August 12, 2019 search for the single mental illness diagnosis of "depression" revealed 530,519 articles[2]. And a search for anxiety revealed 224,305 articles[3]. It is not possible for researchers to functionally analyze all of the critical data patterns both within a single diagnosis or across diagnoses that could be revealed by those articles.

The challenge of curating brain research has been further complicated by the National Institute of Mental Health's adoption of the Research Domain Criteria (RDoC) [6]. Since 1952, the Diagnostic and Statistical Manual of Mental Disorders and International Classification of Diseases [5] (popularly known as DSM and ICD, respectively), have "reigned supreme" as the single "overarching model of psychiatric classification" [14]. That supremacy began to crumble in 2010 when the National Institute of Mental Health launched the RDoC initiative, an alternate framework to conceptually organize and direct biological research on mental disorders [1]. The RDoC initiative intends "to foster integration not only of psychological and biological measures but also of the psychological and biological constructs those measures measure" [13].

The RDoC initiative has fostered significant debate among brain health researchers. It has also created a significant categorization challenge - specifically how to curate articles completed under the DSM-ICD criteria so their data can be incorporated into the RDoC model. Brain science cannot afford to lose critical insights from the numerous articles on different sides of the categorization divide. Hence, it is vital that all existing and future biomedical literature related to brain research is correctly categorized with respect to the RDoC terminology in addition to DSM-ICD models.

However, manual curation of brain research articles using RDoC terminology by human annotators can be highly resource-consuming due to several reasons. RDoC framework is comprehensive and complex. It is made up six major *domains* of human functioning, which is further broken down to multiple *constructs* that comprise different aspects of the overall range of functions[4]. The RDoC matrix helps describe these constructs using several *units of analysis* such as molecules and circuits. On top of this, the rate of publication of biomedical literature (and by extension brain re-

---

[1] Pubmed search for Brain conducted on August 12, 2019
[2] Pubmed search for depression conducted on August 12, 2019
[3] Pubmed search for anxiety conducted on August 12, 2019

[4] https://www.nimh.nih.gov/research/research-funded-by-nimh/rdoc/definitions-of-the-rdoc-domains-and-constructs.shtml

Proceedings of the 5th Workshop on BioNLP Open Shared Tasks, pages 216–226
Hong Kong, China, November 4, 2019. ©2019 Association for Computational Linguistics

search related literature) is growing at an exponential rate [10]. This means that the gap between annotated versus unannotated articles will continue to grow at an alarming rate unless more efficient means of automated annotation is developed soon.

In order to invite text mining teams around the world to develop informatics models for RDoC, we introduced the RDoC Task[5] at this years' BioNLP-OST 2019 workshop[6]. RDoC task is a combination of two subtasks focusing on a subset of RDoC constructs: (a) Task 1 (RDoC-IR) - retrieving PubMed Abstracts related to RDoC constructs, and (b) Task 2 (RDoC-SE) - extracting the most relevant sentence for a given RDoC construct from a known relevant abstract. Both these tasks represent two very important steps of the typical triage process [10], which are finding the articles related to RDoC constructs and then extracting a specific snippet of information that is useful for curation or downstream tasks such as automatic text summarization [15].

There have been several shared tasks on text mining from biomedical literature and clinical notes in the last decade [19, 12] as well as a few shared tasks related to mental health topics ([4, 18, 22, 21, 30]). CLPsych 2015 Shared Task [4] focused on identifying depression and PTSD users from twitter data, while the same task from the following year (i.e. CLPsych 2016 Shared Task [18]) revolved around classifying the severity of peer support forum posts. One of the i2b2[7] challenges from 2011 focused on the sentiment analysis of suicide notes [22, 21].

In 2017, Uzuner et al. introduced the "The RDoC for Psychiatry" challenge, which was composed of three tracks: de-identification of mental health records [28], determination of symptom severity from a psychiatric evaluation of a patient) related to one of the RDoC domains) [9], and the use of mental health records released through the challenge for answering novel questions [32, 29, 7]. In contrast, the RDoC task is a combination of information retrieval and sentence extraction from Biomedical literature related to RDoC constructs.

To generate benchmark data for the RDoC task, three annotators were used to curate the gold-standard datasets. The registration for the RDoC

[5]https://sites.google.com/view/rdoc-task/home
[6]http://2019.bionlp-ost.org
[7]https://www.i2b2.org/

Task opened in March of 2019. Over 30 teams around the world registered for the two tasks. Training data in two batches were released in the month of April. Test data, again in two batches, were released in June. The participants were asked to submit their final predictions by June 19. Eventually, 4 and 5 groups each competed in Tasks 1 and 2, respectively. The final results were made public immediately after the submission deadline.

Two (out of four) and four (out of five) teams each outperformed the baseline methods in task 1 and 2, respectively. The increase in performance over the baselines were more noticeable in task 2 suggesting that information retrieval for RDoC task may be more challenging. There was quite a lot of variation across the several RDoC constructs used for the tasks suggesting that the complexity of different constructs may hinder certain models and construct-specific methods or models may be a requirement in the future. Overall observations from the RDoC Task highlights the need for more sophisticated method development.

The rest of the paper is organized as follows. Section 2 describes the benchmark or gold-standard data preparation process, development of training and test sets, submission requirements, baseline methods used by the organizers, and the performance measures used for the evaluation. Section 3 presents and discusses the overall results for the two tasks. Finally, Section 4 summarizes the task findings as well as describes the potential future work.

2 RDoC Task setup

RDoC Task is a combination of two subtasks. Participants were allowed to choose to participate in one or both tasks. Task 1 is on retrieving PubMed Abstracts related to RDoC constructs, while Task 2 is on extracting the most relevant sentences for an RDoC construct from an already relevant abstract.

In task 1, participants are given a set of PubMed abstracts and they are required to rank abstracts according to relevance for various RDoC constructs. In task 2, participants are given a set of PubMed abstracts relevant for an RDoC construct, and they are required to extract the most relevant sentence from each abstract for the corresponding RDoC construct.

2.1 Timeline

The RDoC Task was organized in two main phases (a) *Training* phase (8 weeks, from April-June 2019), and (b) *Evaluation* phase (1 week in mid-June). At the beginning of the training phase, participants were provided with labeled data (i.e. Training data) and they were expected to develop and fine-tune their models using these known labels. At the beginning of the Evaluation phase, unlabeled data (i.e. Test data) was made available to the participants. They were required to predict labels for this data and submit the predictions to the organizers at the end of the Evaluation phase. Finally, the organizers used the (with-held) labels of the test data for evaluating the accuracy of submissions.

2.2 The benchmark preparation

For the RDoC Task, 8 RDoC constructs out of 25 total constructs from the latest version of the RDoC matrix[8] were used. The motivation was to restrict ourselves to a subset of RDoC framework for which benchmark data can be gathered within a reasonable time-frame. However, these 8 constructs completely cover two of the six domains in the RDoC framework – namely *Negative Valence Systems* and *Arousal and Regulatory* Systems as shown in Table 1.

Table 1: Subset of RDoC constructs used for this task and their domain.

Domain	Construct
Negative Valence Systems	Acute Threat (Fear)
	Potential Threat (Anxiety)
	Frustrative Nonreward
	Sustained Threat
	Loss
Arousal/Regulatory Systems	Arousal
	Circadian Rhythms
	Sleep and Wakefulness

Under the guidance of the Subject Matter Experts from the National Alliance of Mental Illness (NAMI) Montana, the RDoC task benchmark was created by using Entrez e-search utility [26] to search the PubMed database to collect abstracts related to RDoC constructs. That is, we start by using the RDoC construct name as the only keyword to retrieve relevant articles.

If such an approach does not generate the desired number of articles or is too ambiguous on its own (e.g., *Loss* construct), we have utilized terms from the *Behaviors* unit of the RDoC matrix in addition to the construct name.

For example, the The query used for *Loss* construct was "Loss""Amotivation" or "Loss""Anhedonia" or "Loss""Crying" or "Loss""Guilt" or "Loss""Rumination" or "Loss""Sadness" or "Loss""Shame" or "Loss""Withdrawal" or "Loss""Worry". This retrieves about 315 articles, whereas using only "Loss" as the sole query retrieves too many articles (approximately one million articles).

Other queries follow a similar format as Loss when very few ($<$200) or too many ($>$10,000) articles were retrieved with the RDoC construct name as the only keyword. 200 abstracts was the desired minimum number of abstracts per construct that we were planning to send to each annotator. So, if the initial search retrieved less articles, it was deemed too narrow for our objective, and we added terms from the *Behavior* elements belonging to that construct to retrieve more than 200 articles. For example, for the construct Frustrative Nonreward, a PubMed search with the construct name only returns 52 abstracts (retrieved on 09/30/2019)[9]. The RDoC page for Frustrative Nonreward contains one element under the Behavior unit: "physical and relational aggression"[10]. Then, using this term, the search query becomes: "Frustrative Nonreward" or "physical aggression" or "relational aggression", which returns 736 abstracts.

10,000 was a rough estimation of an excessively inclusive search term as determined by our Subject Matter Expert. In other words, the construct name on its own (construct Loss, for example) has a very general definition, resulting in retrieving a large heterogeneous set of articles. Therefore, in these situations, other more specific terms describing the construct were used to limit the scope. Upon generating a search query that retrieves a satisfactory number of articles, we sort them by relevance to the query used.

Then the above-retrieved articles were provided

to three annotators for curation (an example of the annotation guidelines used is available online[11]). For each construct, they were asked to read the title and the abstract and determine whether it provides enough evidence that the abstract was related to the construct. If it was related it was annotated as "positive" (or "negative" otherwise). In addition, they were asked to identify up to 3 most relevant sentences to the abstract (i.e. the sentences that provide most evidence that the abstract is related to the said construct). The inter-annotator agreements are given in Table 2. Example annotation of an abstract is depicted in Figure 1.

While acknowledging we generated a *closed* set of articles for the information retrieval task, we emphasize that this complete process was guided by NAMI experts. They typically use keyword search for first finding relevant articles. Then they use manual curation to remove false positives. Hence, our benchmark datasets are developed using this approach. We wanted the RDoC Task to resemble how a typical curator would find information in this domain.

Table 2: Inter-annotator agreement of Task 1 and Task 2. κ_{free}: Free-Marginal Multirater Kappa [24] computed online[12]

RDoC Construct	κ_{free} Task 1	κ_{free} Task 2
Acute Threat	0.37	0.24
Potential Threat	0.45	0.27
Frustrative Nonreward	0.24	0.20
Sustained Threat	0.18	0.14
Loss	0.25	0.29
Arousal	0.64	0.35
Circadian Rhythms	0.95	0.35
Sleep & Wakefulness	0.97	0.51

We consolidated the labels from the three annotators using the majority vote (i.e. if at least 2 annotators agreed on a label, that was used as the final label for the abstract). In addition, we collected all the most relevant sentences by the three annotators (i.e. set union) as the final set of sentences. This means each abstract could have up to 9 most relevant sentences. In our dataset, at most 6 sentences were observed. This consolidated data was used to create training and test sets as described below.

We believe that the task of identifying the most relevant sentence was more challenging for the annotators than the task of identifying whether a given abstract was related to an RDoC construct or not (for the latter task, annotators were choosing between two labels while for the former, they were choosing from k sentences in the abstract). Therefore, it was possible that there would be more variability in annotations for the former task. So, we used the set union to allow for more flexibility.

2.3 Train, Test and Submission data

In the context of the RDoC task, training data refers to the labeled data sets initially provided to the participants for developing their models. Test sets refer to the unlabeled (i.e. with withheld labels) data sets for which they were asked to submit predictions. All the datasets are available online[13].

For each construct, two separate sets of articles (referred to as Set 1 and Set 2) were annotated. Data from the Set 1 and Set 2 were allocated for training and test data, respectively. Annotators were not aware of this distinction. Set 1 and Set 2 splits were randomly performed per each construct separately before annotation. Therefore, explicit stratified sampling was not applicable.

For each construct, a random subset of positive examples from Set 1 was used as the training examples for both Task 1 and 2 (negative examples were not provided). 80% of random abstracts from Set 2 were used as the test set for Task 1 (this included both positive and negative examples). The subset of positive examples in the rest of the Set 2 (i.e. 20%) was used as the test set for Task 2 (negative examples were not used).

2.3.1 Train data

As mentioned above, we provided the participants of the RDoC task with training examples for each of these 8 RDoC constructs. For task 1, the training examples are randomly selected subsets of positive abstracts for each of the RDoC constructs as shown in Table 3. For task 2, we provided up to 6 most relevant sentences for each of the abstract provided as part of Task 1 train data. In other words, the same set of PubMed IDs were used for training data of both tasks. The distribution of the training examples across the eight constructs is

---

[11]https://montana.box.com/s/kh0hmyn1jcj5ajvr2nibq4iwwgiv3led

[13]https://www.cs.montana.edu/rdoc-task/data/

Title: Characteristics of Physical Aggression in Children of Immigrant Mothers and Non-immigrant Mothers: A Cross-Sectional Analysis of the Survey of Young Canadians.

Abstract: Physical aggression (PA) is important to regulate as early as the preschool years in order to ensure healthy development of children. This study aims to determine the prevalence and characteristics of PA in children of immigrant and non-immigrant mothers. Bivariate and multivariable logistic regression was performed, with the outcome, PA, and covariates including maternal, child, household and neighbourhood characteristics. Twenty percent of children of non-immigrant mothers and 16% of children of immigrant mothers reported PA. The characteristics of PA differ between children of immigrant versus non-immigrant mothers therefore healthcare providers, policy makers, and researchers should be mindful to address PA in these two groups separately, and find ways to tailor current recommended coping strategies and teach children alternative ways to solve problems based on their needs.

RDoC Construct: Sustained Threat

Figure 1: An example of annotating an abstract for both Task 1 and Task 2. The abstract is annotated positive for *Sustained Threat* (Task 1; highlighted in purple) and the most relevant sentence in the abstract is identified (Task 2; highlighted in yellow).

provided in the Table 3 and the distribution of the number of most relevant sentences per construct is shown in Table 4.

Table 3: The number of training examples (positively labeled abstracts) provided for Tasks 1 and 2 across constructs.

RDoC construct	# Abstracts	%
Acute Threat (Fear)	39	14.7
Potential Threat (Anxiety)	27	10.2
Frustrative Nonreward	21	7.9
Sustained Threat	18	6.8
Loss	28	10.5
Arousal	38	14.3
Circadian Rhythms	47	17.7
Sleep and Wakefulness	48	18.1
Total	266	100.0

2.3.2 Test data

The Task 1 test set provided the participants with a random list of 999 relevant (positive) and irrelevant articles (negative) for each of the RDoC constructs (but without the actual labels). The label distribution is given in Table 5. The task 2 test set provided the participants with a list of relevant articles from which they had to extract a relevant sentence with respect to the given RDoC category. The set of abstracts used for test sets of task 1 and

2 were mutually independent for obvious reasons. The distribution of the test set for task 2 across constructs is shown in Table 6 and the distribution of the number of most relevant sentences per construct is provided in Table 4.

2.3.3 Participant Submissions

For task 1, participants were required to submit scores for each abstract in the test set. Scores should correspond to the predicted relevance of the abstract to the given construct. For task 2, participants were required to submit sentences from each abstract that is predicted as the most relevant sentence to the given construct. Submitting a score was not required.

Participants uploaded their submissions through an online web application[14]. We designed the web system to validate the content format of each submission before uploading the file(s) in the server. Upon finding a line that is not properly formatted, the system alerts the participant with an error message including the ill-formatted line number. If the file(s) are properly formatted, the system uploads the submission in the server, automatically analyzes the submission using python scripts and immediately reports the scores of two selected constructs, *Acute Threat (Fear)* and *Loss*, back to the participant.

The participants were allowed to make an un-

---

[14]https://www.cs.montana.edu/rdoc-task/

Table 4: Distribution of the number of most relevant (gold-standard) sentences in abstracts for each construct in the training data. #x: the percentage of abstracts with x relevant sentences.

RDoC Construct	Train Data						Test Data			
	#1	#2	#3	#4	#5	#6	#1	#2	#3	#4
Acute Threat (Fear)	0.0	15.4	35.9	35.9	10.3	2.6	15.8	31.6	42.1	10.5
Potential Threat (Anxiety)	11.1	33.3	55.6	0.0	0.0	0.0	38.2	35.3	20.6	5.9
Frustrative Nonreward	4.8	47.6	47.6	0.0	0.0	0.0	54.3	37.1	8.6	0.0
Sustained Threat	5.6	61.1	33.3	0.0	0.0	0.0	38.9	41.7	16.7	2.8
Loss	10.7	25.0	42.9	21.4	0.0	0.0	61.8	32.4	5.9	0.0
Arousal	7.9	63.2	28.9	0.0	0.0	0.0	23.1	53.8	15.4	7.7
Circadian Rhythms	2.1	51.1	46.8	0.0	0.0	0.0	20.0	40.0	26.7	13.3
Sleep and Wakefulness	10.4	62.5	27.1	0.0	0.0	0.0	26.7	36.7	30.0	6.7

Table 5: The number of abstracts in test set for task 1. Pos and %: number of positively labeled abstracts and their percentages, and Neg: number of negatively labeled abstracts.

RDoC construct	# Pos	%	# Neg
Acute Threat (Fear)	53	67.1	26
Potential Threat (Anxiety)	124	89.2	15
Frustrative Nonreward	96	66.7	48
Sustained Threat	82	56.2	64
Loss	90	65.2	48
Arousal	97	89.8	11
Circadian Rhythms	123	100.0	0
Sleep and Wakefulness	121	99.2	1
Total	786	78.7	213

Table 6: The number of abstracts and their percentages in test set for task 2.

RDoC construct	# Abstracts	%
Acute Threat (Fear)	19	7.8
Potential Threat (Anxiety)	34	13.9
Frustrative Nonreward	35	14.3
Sustained Threat	36	14.8
Loss	34	13.9
Arousal	26	10.7
Circadian Rhythms	30	12.3
Sleep and Wakefulness	30	12.3
Total	244	100.0

limited number of submissions and the scores from past submissions were discarded upon a new submission. This meant they could re-submit until they achieved a satisfactory performance for the above two constructs. The performance scores for all the constructs were made available immediately after the submission deadline. The older scores were only discarded for the purposes of the final evaluation. However, these scores are retained for potential future research.

2.4 Baseline methods

We used TF-IDF [23] with smooth IDF weights and cosine similarity [27] to calculate the similarity score for each document against a query and used these scores to rank the documents by relevance. Regardless of the task, we used the corresponding construct name concatenated with its definition as the query string. We used the definitions of constructs as defined by the National Institute of Mental Health listed online[15].

For task 1, each document is the title concatenated with the corresponding abstract and the similarity scores are used to rank the articles for each construct. For task 2, documents are the sentences of the abstracts and the top-ranked sentence per abstract was returned based on the similarity scores. All the baseline models were implemented using the Scikit-learn Python library [20]. No preprocessing techniques were applied to the abstract text. In addition to the above TFIDF-based baseline, we also used BM25 [25] as a baseline. But due to its comparatively lower performance on both tasks 1 and 2, BM25 values are not reported in this paper.

[15]https://www.nimh.nih.gov/research/research-funded-by-nimh/rdoc/definitions-of-the-rdoc-domains-and-constructs.shtml

2.5 Metrics used for evaluation

For task 1, we use Mean Average Precision (MAP) [16] as the performance measure because it is one of the most frequently used measures for IR [31, 8, 11]. First, we compute the Average Precision (AP) for each construct independently and macro-average across the constructs to compute the Mean Average Precision. For task 2, due to the non-applicability of utilizing popular standard measures such as precision and recall [3], we define the *Accuracy* as the percentage of abstracts with correctly predicted most relevant sentence. If at least one of the gold-standard sentences match the predicted sentence, it is counted as 1 and 0 otherwise (therefore, note that this measure is not the same as the typical accuracy measure used in Natural Language Processing and Machine Learning. We average across constructs to get the *Macro Average Accuracy*.

It should be pointed out that, technically, there is no "negative" class for the task 2 (in the traditional sense used for predictive models). Participants are given abstracts already known to be relevant to a construct. They are asked to submit just one sentence that they think is the most relevant (or that helps them the most for finding the relevance between the given abstract and the construct). Hence the participants are unable to gain undue advantages due to any class imbalances even though the above-defined performance measure may closely resemble the typical "Accuracy". Also, since we did not collect confidence scores for task 2, we did not compute threshold independent measures such as AUROC (area under the ROC curve).

3 Results and Discussion

Inter-annotator agreements for many of the constructs in both tasks 1 and 2 are relatively low (see table 2). According to the annotators, there were several reasons why information retrieval and sentence extraction with RDoC was reasonably challenging. The very generalized nature of the RDoC constructs, as well as ambiguity in the language stating the purpose/hypothesis/results of the experiment, made it difficult to find the relevance of a given abstract to an RDoC construct. The way the abstracts were written, made it seem such that it could be potentially tied to/or not, to various RDoC sentences.

Annotators reported that they had difficulties

with the 'Sustained Threat' and the 'Frustrative Non-Reward' constructs. For example, some annotators felt that every abstract that they read was related to Frustrative Non-Reward construct because many of the abstracts specifically studied the relational and physical aggressive behaviors. Although a lot of the studies tested these behaviors, it was challenging to figure out if they were "directly" related to Frustrative Non-Reward or not. For instance, several studies comparatively tested relational and physical aggression between genders (2 behaviors of Frustrative Non-Reward), but the abstracts didn't explicitly mention "withdrawal or prevention" of a reward (the definition). Therefore, when annotating, if they've felt that the research would benefit or help further understand Frustrative Non-Reward and its associated behaviors, they've annotated it as related (this included environmental, social, and biological factors influencing relational and physical aggression).

Over thirty teams registered to participate in at least one of the RDoC tasks. Eventually, 5 teams submitted their predictions; four teams submitted for both tasks and one team for only task 1. In the following analysis, we will be using the unique team identifiers (assigned during the task registration[16]) for referring to the 5 teams. Note that these team identifiers bear no significance other than identifying different teams.

3.1 Task 1: Information Retrieval

Four teams submitted their predictions for this task and their scores are reported in Table 7. Bold entries indicate the highest score for the corresponding construct. Although included in Table 7, we excluded the two constructs, *Circadian Rhythms* and *Sleep and Wakefulness*, from the final analysis since these constructs contain one and zero negative articles, respectively, leading to perfect performance (see Table 5). Team 30 achieved the highest mean average precision (0.86) among all teams. Though Team 10 achieved the second-highest mean average precision (0.85) that is very close to the highest, we found a statistically significant difference between the scores of these two teams (paired t-test, p=0.005, $\alpha = 0.05$). Team 30 achieved the highest scores for *Frustrative Nonreward, Loss* and *Potential Threat (Anxiety)* whereas Team 10 achieved the highest scores for the other three constructs. Though it seems the

---

[16]https://sites.google.com/view/rdoc-task/registration

scores achieved by the Team 10 and 30 is close to the baseline, we found these scores to be statistically significantly higher from the baseline for both Team 10 (paired t-test, $p=0.022$) and Team 30 (paired t-test, $p=0.043$) using $\alpha = 0.05$.

The last column in Table 7 reports the average score for the corresponding construct. It is seemingly easier to rank the relevant articles for *Arousal* and *Potential Threat (Anxiety)* whereas it is moderately difficult for *Sustained Threat*. Sustained Threat being more challenging for IR may be explained by the fact that the annotators also found it to be the most challenging construct for task 1 annotation.

3.2 Task 2: Sentence Extraction

Five teams submitted their predictions for this task and their scores are reported in Table 8. Bold entries indicate the highest score for the corresponding construct. Team 30 again achieved the highest macro average accuracy (0.58) among all the teams and the highest score for five out of eight constructs. Team 7 achieved the highest score for the rest of the three constructs with significant improvement over Team 30. Construct-wise highest scores of *Sustained Threat, Arousal* and *Circadian Rhythms*, achieved by either Team 7 or Team 30, are higher by about 0.27 compared to the baseline performance. In addition, the highest scores for other constructs are also higher by more than 0.17 compared to the baseline performance.

Frustrative Nonreward has the lowest average score (0.31) among all the constructs. Moreover, its highest score (0.43) is also the lowest among all the highest scores. So, extracting the most relevant sentences for *Frustrative Nonreward* is seemingly more difficult compared to the other constructs.

Typically, participating teams performed relatively better on shorter abstracts (see Table 9), which is intuitive due to that fact the models have a higher chance of finding the most similar sentences for shorter abstracts. Similarly, they performed well for abstracts with more gold-standard sentences (see Table 10). This is also intuitive because when there are more gold-standard sentences, there is a higher chance of matching one of them.

4 Conclusion and Future work

We introduced a novel mental health informatics task called RDoC task at this years BioNLP-OST 2019 workshop. RDoC task is a combination of two subtasks on information retrieval and sentence extraction using the RDoC framework. Originally, over 30 teams registered, highlighting a significant interested in mental health informatics and/ or RDoC. Eventually, four and five teams participated in the information retrieval and sentence extraction tasks, respectively.

Overall results show that the top-performing team was able to easily outperform the baseline models for most of the constructs. On the other hand, the baseline methods outperform at least one system (often more). This is surprising given that the baseline models are not sophisticated. One reason could be that the baseline methods do not utilize training data, while the participating methods may have been overfitted to the training data. Another reason could be, these simple baselines perform better than (most likely more complex) participating models due to working with shorter documents (i.e. abstracts). If the full texts were made available, models primarily depended on TFIDF may struggle to achieve good performance. Regardless, this calls for more sophisticated methods for both tasks because any other sophisticated method (such as Lucene [17] or MetaMap [2]) used a baseline may have outperformed even more participating teams.

The publicly made available gold-standard data should serve as a valuable resource for the brain research/ mental health and RDoC researchers and curators going forward. In the future iterations of the RDoC task, we would like to incorporate either all available or a well-representative set of RDoC constructs covering all domains. We plan to improve the quality of benchmark data using "reconciliation" instead of "majority voting" as well as using improved search that uses MeSH and/ or other vocabularies.

And equally important aspect would be to explore information extraction tasks such as extracting various entities under different RDoC units of analysis, which is likely more useful for the curators. This would also mean an exploration of incorporating full text in addition to abstracts will be required due to the abundance of entities existing in the full articles compared to just the abstract. Last but not least, exploring clever ways to maintain the enthusiasm of the registered teams would be highly valuable to the overall success of the future iterations of the RDoC task .

Table 7: Performance of retrieving PubMed Abstracts related to the corresponding RDoC construct (Task 1). Four teams participated (T10, T21, T22, and T30). IQR: inter-quartile range. Bolded scores are the highest across all teams per the construct.

RDoC construct	Baseline	T10	T21	T22	T30	Avg	IQR
Acute Threat (Fear)	0.74	**0.89**	0.83	0.67	0.85	0.81	0.17
Potential Threat (Anxiety)	0.90	0.87	0.89	0.81	**0.94**	0.88	0.10
Frustrative Nonreward	0.70	0.69	0.67	0.61	**0.73**	0.68	0.10
Sustained Threat	0.64	**0.64**	**0.64**	0.41	0.63	0.58	0.18
Loss	0.77	0.74	0.71	0.61	**0.78**	0.71	0.14
Arousal	0.95	**0.93**	0.91	0.84	0.92	0.90	0.07
Circadian Rhythms	1.00	1.00	1.00	1.00	1.00	1.00	0.00
Sleep and Wakefulness	1.00	1.00	1.00	0.98	1.00	1.00	0.02
Mean Average Precision	0.84	0.85	0.83	0.74	**0.86**	–	–

Table 8: Performance of extracting the most relevant sentence from each abstract related to the corresponding RDoC construct (Task 2). Five teams participated (T7, T10, T21, T22, and T30). IQR: inter-quartile range.

RDoC construct	Baseline	T7	T10	T21	T22	T30	Avg	IQR
Acute Threat (Fear)	0.53	0.58	0.68	0.37	0.47	**0.74**	0.57	0.29
Potential Threat (Anxiety)	0.41	0.41	0.32	0.15	0.38	**0.59**	0.37	0.27
Frustrative Nonreward	0.23	**0.43**	0.34	0.11	0.29	0.37	0.31	0.20
Sustained Threat	0.19	**0.47**	0.36	0.14	**0.47**	0.42	0.37	0.22
Loss	0.53	0.26	0.56	0.26	0.62	**0.74**	0.49	0.42
Arousal	0.46	0.46	0.62	0.12	0.42	**0.73**	0.47	0.41
Circadian Rhythms	0.43	**0.70**	0.47	0.10	0.60	0.47	0.47	0.37
Sleep and Wakefulness	0.43	0.33	0.50	0.17	0.57	**0.60**	0.43	0.34
Macro Average Accuracy	0.40	0.46	0.48	0.18	0.48	**0.58**	–	–

Table 9: Variation of Accuracy over various size of abstract. #m-n: abstracts with m to n sentences.

RDoC construct	#3-8	#9-14	#15-20
Acute Threat	0.60	0.64	0.40
Potential Threat	0.47	0.39	–
Frustrative Nonreward	0.28	0.25	0.50
Sustained Threat	0.39	0.32	0.40
Loss	0.62	0.60	0.31
Arousal	0.53	0.39	–
Circadian Rhythms	0.38	0.54	0.00
Sleep & Wakefulness	0.58	0.42	–

Table 10: Variation of Accuracy over the number of most relevant (gold-standard) sentences in abstracts. #x: abstracts with x relevant (gold-standard) sentences.

RDoC construct	#1	#2	#3	#4
Acute Threat	0.29	0.60	0.69	0.80
Potential Threat	0.31	0.47	0.57	0.75
Frustrative Nonreward	0.18	0.35	0.54	–
Sustained Threat	0.29	0.42	0.37	0.25
Loss	0.56	0.65	0.50	–
Arousal	0.47	0.45	0.65	0.40
Circadian Rhythms	0.17	0.41	0.60	0.61
Sleep & Wakefulness	0.23	0.56	0.67	0.80

Acknowledgments

This work was partially funded by The Center for Mental Health Research and Recovery (CMHRR) at Montana State University (MSU). We would like to thank Robell Basset, Lenin Lewis, Ninoo De Silva, and Hannah Reiser (from the Department of Psychology, MSU), and Soumilee Chaudhuri (from the Department of Cell Biology & Neuroscience, MSU) for assisting the curation process.

References

[1] Dean Carcone and Anthony C Ruocco. Six years of research on the National Institute of Mental Healths Research Domain Criteria (RDoC) initiative: a systematic review. *Frontiers in cellular neuroscience*, 11:46, 2017.

[2] K Bretonnel Cohen, Tom Christiansen, and Lawrence E Hunter. Metamap is a superior baseline to a standard document retrieval engine for the task of finding patient cohorts in clinical free text. In *TREC*. Citeseer, 2011.

[3] Kevin Bretonnel Cohen and Dina Demner-Fushman. *Biomedical natural language processing*, volume 11. John Benjamins Publishing Company, 2014.

[4] Glen Coppersmith, Mark Dredze, Craig Harman, Kristy Hollingshead, and Margaret Mitchell. Clpsych 2015 shared task: Depression and ptsd on twitter. In *Proceedings of the 2nd Workshop on Computational Linguistics and Clinical Psychology: From Linguistic Signal to Clinical Reality*, pages 31–39, 2015.

[5] Bruce N Cuthbert. The rdoc framework: facilitating transition from icd/dsm to dimensional approaches that integrate neuroscience and psychopathology. *World Psychiatry*, 13(1):28–35, 2014.

[6] Bruce N Cuthbert and Thomas R Insel. Toward the future of psychiatric diagnosis: the seven pillars of rdoc. *BMC medicine*, 11(1):126, 2013.

[7] Hong-Jie Dai, Emily Chia-Yu Su, Mohy Uddin, Jitendra Jonnagaddala, Chi-Shin Wu, and Shabbir Syed-Abdul. Exploring associations of clinical and social parameters with violent behaviors among psychiatric patients. *Journal of biomedical informatics*, 75:S149–S159, 2017.

[8] Daniel Dopp, Adam Morrone, and Indika Kahanda. KinDER: A biocuration tool for extracting kinase knowledge from biomedical literature. *Proceedings of the BioCreative VI Workshop*, Oct 2017.

[9] Michele Filannino, Amber Stubbs, and Özlem Uzuner. Symptom severity prediction from neuropsychiatric clinical records: Overview of 2016 cegs n-grid shared tasks track 2. *Journal of biomedical informatics*, 75:S62–S70, 2017.

[10] International Society for Biocuration. Biocuration: Distilling data into knowledge. *PLOS Biology*, 16(4):1–8, 04 2018.

[11] Julien Gobeill, Pascale Gaudet, Daniel Dopp, Adam Morrone, Indika Kahanda, Yi-Yu Hsu, Chih-Hsuan Wei, Zhiyong Lu, and Patrick Ruch. Overview of the biocreative vi text-mining services for kinome curation track. *Database*, 2018(1):bay104, 2018.

[12] Chung-Chi Huang and Zhiyong Lu. Community challenges in biomedical text mining over 10 years: success, failure and the future. *Briefings in bioinformatics*, 17(1):132–144, 2015.

[13] Jessica I Lake, Cindy M Yee, and Gregory A Miller. Misunderstanding rdoc. *Zeitschrift für Psychologie*, 2017.

[14] Scott O Lilienfeld and Michael T Treadway. Clashing diagnostic approaches: Dsm-icd versus rdoc. *Annual review of clinical psychology*, 12:435–463, 2016.

[15] Inderjeet Mani. *Advances in automatic text summarization*. MIT press, 1999.

[16] Christopher D. Manning, Prabhakar Raghavan, and Hinrich Schtze. *Evaluation in information retrieval*, page 139161. Cambridge University Press, 2008.

[17] Michael McCandless, Erik Hatcher, and Otis Gospodnetic. *Lucene in action: covers Apache Lucene 3.0*. Manning Publications Co., 2010.

[18] David N Milne, Glen Pink, Ben Hachey, and Rafael A Calvo. Clpsych 2016 shared task: Triaging content in online peer-support forums. In *Proceedings of the Third Workshop on Computational Linguistics and Clinical Psychology*, pages 118–127, 2016.

[19] Malvina Nissim, Lasha Abzianidze, Kilian Evang, Rob van der Goot, Hessel Haagsma, Barbara Plank, and Martijn Wieling. Sharing is caring: The future of shared tasks. *Computational Linguistics*, 43(4):897–904, 2017.

[20] F. Pedregosa, G. Varoquaux, A. Gramfort, V. Michel, B. Thirion, et al. Scikit-learn: Machine learning in Python. *Journal of Machine Learning Research*, 12:2825–2830, 2011.

[21] John P Pestian, Pawel Matykiewicz, and Michelle Linn-Gust. What's in a note: construction of a suicide note corpus. *Biomedical informatics insights*, 5:BII–S10213, 2012.

[22] John P Pestian, Pawel Matykiewicz, Michelle Linn-Gust, Brett South, Ozlem Uzuner, Jan Wiebe, K Bretonnel Cohen, John Hurdle, and Christopher Brew. Sentiment analysis of suicide notes: A shared task. *Biomedical informatics insights*, 5:BII–S9042, 2012.

[23] Anand Rajaraman and Jeffrey David Ullman. *Data Mining*, page 117. Cambridge University Press, 2011.

[24] Justus J Randolph. Free-marginal multirater kappa (multirater k [free]): An alternative to fleiss' fixed-marginal multirater kappa. *Online submission*, 2005.

[25] Stephen Robertson and Hugo Zaragoza. The probabilistic relevance framework: Bm25 and beyond. *Found. Trends Inf. Retr.*, 3(4):333–389, April 2009.

[26] Eric W. Sayers, Tanya Barrett, Dennis A. Benson, et al. Database resources of the National Center for Biotechnology Information. *Nucleic Acids Research*, 38(suppl_1):D5–D16, 1 2010.

[27] Amit Singhal et al. Modern information retrieval: A brief overview. *IEEE Data Eng. Bull.*, 24(4):35–43, 2001.

[28] Amber Stubbs, Michele Filannino, and Özlem Uzuner. De-identification of psychiatric intake records: Overview of 2016 cegs n-grid shared tasks track 1. *Journal of biomedical informatics*, 75:S4–S18, 2017.

[29] Tung Tran and Ramakanth Kavuluru. Predicting mental conditions based on history of present illness in psychiatric notes with deep neural networks. *Journal of biomedical informatics*, 75:S138–S148, 2017.

[30] Özlem Uzuner, Amber Stubbs, and Michele Filannino. A natural language processing challenge for clinical records: Research domains criteria (RDoC) for psychiatry. *Journal of biomedical informatics*, 75:S1–S3, 2017.

[31] Ellen M Voorhees, Donna K Harman, et al. *TREC: Experiment and evaluation in information retrieval*, volume 63. MIT press Cambridge, 2005.

[32] Yaoyun Zhang, Olivia Zhang, Yonghui Wu, Hee-Jin Lee, Jun Xu, Hua Xu, and Kirk Roberts. Psychiatric symptom recognition without labeled data using distributional representations of phrases and on-line knowledge. *Journal of biomedical informatics*, 75:S129–S137, 2017.

BioNLP-OST 2019 RDoC Tasks: Multi-grain Neural Relevance Ranking Using Topics and Attention Based Query-Document-Sentence Interactions

***Yatin Chaudhary[1,2], *Pankaj Gupta[1,2], Hinrich Schütze[2]**

[1]Corporate Technology, Machine-Intelligence (MIC-DE), Siemens AG Munich, Germany
[2]CIS, University of Munich (LMU) Munich, Germany
`{yatin.chaudhary, pankaj.gupta}@siemens.com`

Abstract

This paper presents our system details and results of participation in the RDoC Tasks of BioNLP-OST 2019. Research Domain Criteria (RDoC) construct is a multi-dimensional and broad framework to describe mental health disorders by combining knowledge from *genomics* to *behaviour*. Non-availability of RDoC labelled dataset and tedious labelling process hinders the use of RDoC framework to reach its full potential in Biomedical research community and Healthcare industry. Therefore, *Task-1* aims at retrieval and ranking of PubMed abstracts relevant to a given RDoC construct and *Task-2* aims at extraction of the most relevant sentence from a given PubMed abstract. We investigate (1) attention based supervised neural topic model and SVM for retrieval and ranking of PubMed abstracts and, further utilize BM25 and other relevance measures for re-ranking, (2) supervised and unsupervised sentence ranking models utilizing *multi-view* representations comprising of *query-aware attention-based sentence representation (QAR)*, bag-of-words (BoW) and TF-IDF. Our best systems achieved 1st rank and scored 0.86 mAP and 0.58 macro average accuracy in Task-1 and Task-2 respectively.

1 Introduction

The scientific research output of the biomedical community is becoming more sub-domain specialized and increasing at a faster pace. Most of the biomedical domain knowledge is in the form of unstructured text data. Natural Language Processing (NLP) techniques such as relation extraction and information retrieval have enabled us to effectively mine relevant information from a large corpus. These techniques have significantly reduced the time and effort required for knowledge mining and information extraction from past scientific studies and electronic health reports (EHR).

Information Retrieval (IR) is the process of retrieving relevant information from an unstructured text corpus, which satisfies a given query/requirement, for example Google search, email search, database search etc. This is generally achieved by converting the query and the document collection into an external representation which by preserving the important semantical information can reduce the IR processing time. This external representation can be generated using either statistical approach i.e., word counts or distributed semantical approach i.e., word embeddings. Therefore, there is a motivation to develop such IR system which can understand the specialized sub-domain language and domain-specific jargon of biomedical domain and assist researchers and medical professionals by effectively and efficiently retrieving most relevant information given a query.

RDoC Tasks aims at exploring information retrieval (IR) and information extraction (IE) tasks on selected abstracts from PubMed dataset. While Task-1 aims to rank abstracts i.e., *coarse granularity*, Task-2 aims to rank sentences i.e., *fine granularity* and hence the term *multi-grain*. An RDoC construct combines information from multiple sources like genomics, symptoms, behaviour etc. and therefore, is a much broader way of describing mental health disorders than symptoms based approach. Table 1 shows the association between PubMed abstracts and RDoC constructs depending on the semantic knowledge of the highlighted content words. Both of these tasks aim in the direction of *ease of accessibility* of PubMed abstracts labelled with diverse RDoC constructs so that this information can reach its full potential and can be of help to biomedical researchers and healthcare professionals.

---

* : Equal Contribution

Proceedings of the 5th Workshop on BioNLP Open Shared Tasks, pages 227–236
Hong Kong, China, November 4, 2019. ©2019 Association for Computational Linguistics

PMID	RDoC Construct	PubMed Abstract
14998902	Acute_Threat_Fear	*Title*: Mother lowers glucocorticoid levels of preweaning rats after acute threat. *Abstract*: Exposure to a deadly threat, an adult male rat, induced the release of corticosterone in 14-day-old rat pups. The endocrine stress response was decreased when the pups were reunited with their mother immediately after exposure. These findings demonstrate that social variables can reduce the consequences of an aversive experience.
21950094	Sleep_Wakefulness	*Title*: Central mechanisms of sleep-wakefulness cycle *Abstract*: Brief anatomical, physiological and neurochemical basics of the regulation of wakefulness, slow wave (NREM) sleep and paradoxical (REM) sleep are regarded as representing by the end of the first decade of the second millennium.

Table 1: RDoC construct - This table shows two PubMed abstracts labelled with two different RDoC construct and PubMed ID (PMID). Highlighted words (blue and red) in each abstract shows content words which together provide the semantic understanding of the corresponding RDoC constructs.

2 Task Description and Contributions

RDoc-IR Task-1: The task aims at retrieving and ranking the PubMed abstracts (within each of the eight clusters) that are relevant for the RDoC construct (i.e, a query) related to the cluster in the abstract appears. The training data consists of abstracts (title + sentences) each annotated with one or more RDoC constructs. Test data consists of abstracts without annotation and the goal is to submit a ranked lists of relevant articles for each medical domain RDoC construct.

RDoc-IE Task-2 The task aims at extracting the most relevant sentence from each PubMed abstract for the corresponding RDoC construct. The input consists of an abstract (title t and sentences s) for an RDoC construct q. The training data consists of abstracts each annotated with one RDoC construct and the most relevant sentence. Test data contains abstracts relevant for RDoC constructs and the goal is to submit a list of predicted most relevant sentence for each abstract.

Our Contributions: Following are our multi-fold contributions in this paper:

(1) **RDoC-IR Task-1**: We perform document (or abstract) ranking in two steps, first using supervised neural topic model and SVM. Moreover, we have introduced attentions in supervised neural topic model, along with pre-trained word embeddings from several sources. Then, we re-rank documents using BM25 and similarity scores between query and query-aware attention-based document representation.

Comparing with other participating systems in the shared task, our submission is ranked 1st with a mAP score of 0.86.

(2) **RDoC-IE Task-2**: We have addressed the sentence ranking task by introducing unsupervised and supervised sentence ranking schemes. Moreover, we have employed multi-view representations consisting of bag-of-words, TF-IDF and query-aware attention-based sentence representation via enhanced query-sentence interactions. We have also investigated relevance of title with the sentences and coined ways to incorporate both query-sentence and title-sentence relevance scores in ranking sentences with an abstract.

Comparing with other participating systems in the shared task, our submission is ranked 1st with a macro average accuracy of 0.58. Our code is available at `https://github.com/YatinChaudhary/RDoC_Task`.

3 Methodology

In this section, we first describe representing a query, sentence and document using local and distributed representation schemes. We further describe enhanced query-document (query-title and query-content) and query-sentence interactions to compute query-aware document or sentence representations for Task-1 and Task-2, respectively. Finally, we discuss the application of supervised neural topic modeling in ranking documents for task 1 and introduce unsupervised and supervised sentence rankers for Task-2.

3.1 Query, Sentence and Document Vectors

In this paper, we deal with texts of different lengths in form of query, sentence and document. In this section, we describe the way we represent the different texts.

Bag-of-words (BoW) and **Term frequency-inverse document frequency (TF-IDF)**: We use two the local representation schemes: BoW and

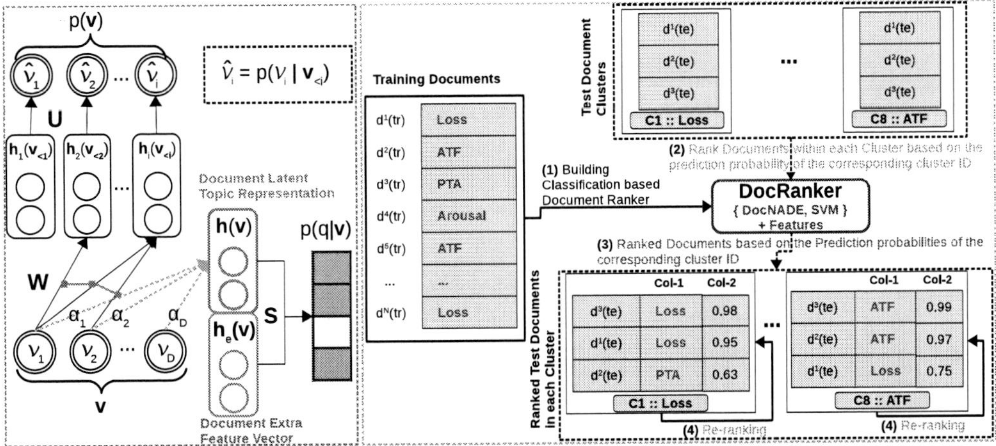

Figure 1: (Left) DocNADE Topic Model: Blue colored lines signify parameter sharing, α attention weights are used to compute *latent document representation* $\mathbf{h}(\mathbf{v})$; (Right) RDoC Task-1 system architecture, where the numbered arrow (1) denotes the flow. **Col-1** indicates "predicted label" by the DocRanker and **Col-2** indicates "prediction probability" ($p(q|\mathbf{v})$). "Features" inside DocRanker indicates *FastText* and *word2vec* pretrained embeddings.

TF-IDF (Manning et al., 2008) to compute sentence/document vectors.

Embedding Sum Representation (ESR): Word embeddings (Mikolov et al., 2013; Pennington et al., 2014) have been successfully used in computing distributed representation of text snippets (short or long). In ESR scheme, we employ the pre-trained word embeddings from FastText (Bojanowski et al., 2017) and word2vec (Mikolov et al., 2013). To represent a text (query, sentence or document), we compute the sum of (pre-trained) word vectors of each word in the text. E.g., ESR for a document d with D words can be computed as:

$$ESR(d) = \widetilde{\mathbf{d}} = \sum_{i=1}^{D} \mathbf{e}(d_i)$$ where, $\mathbf{e} \in \mathbb{R}^E$ is the pre-trained embedding vector of dimension E for the word d_i.

Query-aware Attention-based Representation (QAR) for Documents and Sentences: Unlike ESR, we reward the maximum matches between a query and document by computing density of matches between them, similar to McDonald et al. (2018). In doing so, we introduce a weighted sum of word vectors from pre-trained embeddings and therefore, incorporate importance/attention of certain words in document (or sentence) that appear in the query text.

For an enhanced query-aware attention based document (or sentence) representation, we first compute an histogram $\mathbf{a}_i(d) \in \mathbb{R}^D$ of attention weights for each word k in the document d (or sentence s) relative to the ith query word q_i, using cosine similarity:

$$\mathbf{a}_i(d) = [a_{i,k}]_{k=1}^{D} \text{ where, } a_{i,k} = \frac{\mathbf{e}(q_i)^T \mathbf{e}(d_k)}{||\mathbf{e}(q_i)|| \, ||\mathbf{e}(d_k)||}$$

for each kth word in the document d. Here, $\mathbf{e}(w)$ refers to an embedding vector of the word w.

We then compute an query-aware attention-based representation $\Phi_i(d)$ of document d from the viewpoint of ith query word by summing the word vectors of the document, weighted by their attention scores $\mathbf{a}_i(d)$:

$$\Phi_i(d) = \sum_{k=1}^{D} a_{i,k}(d) \, \mathbf{e}(d_k) = \mathbf{a}_i(d) \odot [\mathbf{e}(d_k)]_{k=1}^{D}$$

where $\odot$ is an element-wise multiplication operator.

Next, we compute density of matches between several words in query and the document by summing each of the attention histograms a_i for all the query terms i. Therefore, the *query-aware document representation* for a document (or sentence) relative to all query words in q is given by:

$$QAR(d) = \Phi_q(d) = \sum_{i}^{|q|} \Phi_i(d) \qquad (1)$$

Similarly, a *query-aware sentence representation* $\Phi_q(s)$ and *query-aware title representation* $\Phi_q(t)$ can be computed for the sentence s and document title t, respectively.

For query representation, we use ESR scheme as $\widetilde{\mathbf{q}} = \sum_{i=1}^{|q|} \mathbf{e}(w_i)$.

Figure 2 illustrates the computation of *query-aware attention-based sentence representation*.

3.2 Document Neural Topic Models

Topic models (TMs) (Blei et al., 2003) have shown to capture thematic structures, i.e., topics appearing within the document collection. Beyond interpretability, topic models can extract latent document representation that is used to perform document retrieval. Recently, Gupta et al. (2019a) and Gupta et al. (2019b) have shown that the neural network-based topic models (NTM) outperform LDA-based topic models (Blei et al., 2003; Srivastava and Sutton, 2017) in terms of generalization, interpretability and document retrieval.

In order to perform document classification and retrieval, we have employed supervised version of neural topic model with extra features and further introduced word-level attention in a neural topic model, i.e. in DocNADE (Larochelle and Lauly, 2012; Gupta et al., 2019a).

Supervised NTM (SupDocNADE): Document Neural Autoregressive Distribution Estimator (DocNADE) is a neural network based topic model that works on bag-of-words (BoW) representation to model a document collection in a language modeling fashion.

Consider a document d, represented as $\mathbf{v} = [v_1, ..., v_i, ..., v_D]$ of size D, where $v_i \in \{1, ..., Z\}$ is the index of ith word in the vocabulary and Z is the vocabulary size. DocNADE models the joint distribution $p(\mathbf{v})$ of document $\mathbf{v}$ by decomposing $p(\mathbf{v})$ into autoregressive conditional of each word v_i in the document, i.e., $p(\mathbf{v}) = \sum_{i=1}^{D} p(v_i|\mathbf{v}_{<i})$, where $\mathbf{v}_{<i} \in \{v_1, ..., v_{i-1}\}$.

As shown in Figure 1 (left), DocNADE computes each autoregressive conditional $p(v_i|\mathbf{v}_{<i})$ using a feed forward neural network for $i \in \{1, ..., D\}$ as,

$$p(v_i = w|\mathbf{v}_{<i}) = \frac{exp(b_w + \mathbf{U}_{w,:}\mathbf{h}(\mathbf{v}_{<i}))}{\sum_{w'} exp(b_{w'} + \mathbf{U}_{w',:}\mathbf{h}(\mathbf{v}_{<i}))}$$

$$\mathbf{h}_i(\mathbf{v}_{<i}) = f(\mathbf{c} + \sum_{j<i} \mathbf{W}_{:,v_j})$$

where, $f(\cdot)$ is a non-linear activation function, $\mathbf{W} \in \mathbb{R}^{H \times Z}$ and $\mathbf{U} \in \mathbb{R}^{Z \times H}$ are encoding and decoding matrices, $\mathbf{c} \in \mathbb{R}^{H}$ and $\mathbf{b} \in \mathbb{R}^{Z}$ are encoding and decoding biases, H is the number of units in latent representation $\mathbf{h}_i(\mathbf{v}_{<i})$. Here,

$\mathbf{h}_i(\mathbf{v}_{<i})$ contains information of words preceding the word v_i. For a document $\mathbf{v}$, the log-likelihood $\mathcal{L}(\mathbf{v})$ and latent representation $\mathbf{h}(\mathbf{v})$ are given as,

$$\mathcal{L}^{unsup}(\mathbf{v}) = \sum_{i=1}^{D} \log p(v_i|\mathbf{v}_{<i}) \tag{2}$$

$$\mathbf{h}(\mathbf{v}) = f(\mathbf{c} + \sum_{i=1}^{D} \mathbf{W}_{:,v_i}) \tag{3}$$

Here, $\mathcal{L}(\mathbf{v})$ is used to optimize the topic model in unsupervised fashion and $\mathbf{h}(\mathbf{v})$ encodes the topic proportion. See Gupta et al. (2019a) for further details on training unsupervised DocNADE.

Here, we extend the unsupervised version to DocNADE with a hybrid cost $\mathcal{L}^{hybrid}(\mathbf{v})$, consisting of a (supervised) discriminative training cost $p(y = q|\mathbf{v})$ along with an unsupervised generative cost $p(\mathbf{v})$ for a given query q and associated document $\mathbf{v}$:

$$\mathcal{L}^{hybrid}(\mathbf{v}) = \mathcal{L}^{sup}(\mathbf{v}) + \lambda \cdot \mathcal{L}^{unsup}(\mathbf{v}) \tag{4}$$

where $\lambda \in [0, 1]$. The supervised cost is given by:

$$\mathcal{L}^{sup}(\mathbf{v}) = p(y = q|\mathbf{v}) = \text{softmax}(\mathbf{d} + \mathbf{S}\,\mathbf{h}(\mathbf{v}))$$

Here, $\mathbf{S} \in \mathbb{R}^{L \times H}$ and $\mathbf{d} \in \mathbb{R}^{L}$ are output matrix and bias, L is the total number of unique RDoC constructs (i.e., unique query labels).

Supervised Attention-based NTM (a-SupDocNADE): Observe in equation 3 that the DocNADE computes document representation $\mathbf{h}(v)$ via aggregation of word embedding vectors without considering attention over certain words. However, certain content words own high important, especially in classification task. Therefore, we have introduced attention-based embedding aggregation in supDocNADE (Figure 1, left):

$$\mathbf{h}(\mathbf{v}) = f(\mathbf{c} + \sum_{i=1}^{D} \alpha_i \mathbf{W}_{:,v_i}) \tag{5}$$

Here, α_i is an attention score of each word i in the document $\mathbf{v}$, learned via supervised training.

Additionally, we incorporate extra word features, such as pre-trained word embeddings from several sources: FastText ($\mathbf{E}^{fast}$) (Bojanowski et al., 2017) and word2vec ($\mathbf{E}^{word2vec}$) (Mikolov et al., 2013). We introduce these features by concatenating $\mathbf{h}_e(\mathbf{v})$ with $\mathbf{h}(\mathbf{v})$ in the supervised portion of the a-supDocNADE model, as

$$\mathbf{h}_e(\mathbf{v}) = f\left(\mathbf{c} + \sum_{i=1}^{D} \alpha_i (\mathbf{E}^{fast}_{:,v_i} + \mathbf{E}^{word2vec}_{:,v_i})\right) \tag{6}$$

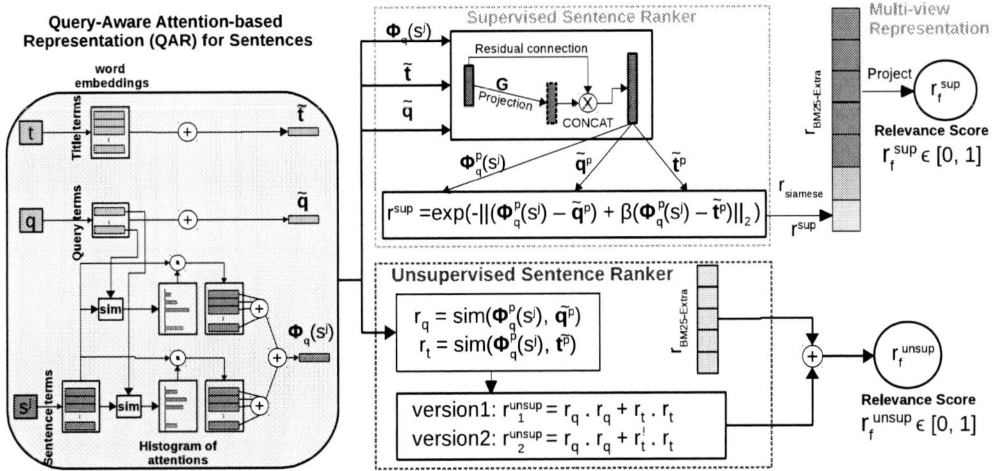

Figure 2: RDoC Task-2 System Architecture for Supervised and Unsupervised sentence ranking, consisting of: Query-aware Representation, Supervised and Unsupervised sentence rankers for computing the final relevance scores r_f^{sup} and r_f^{unsup}, respectively. Here, sim refers to cosine similarity.

Therefore, the classification portion of a-supDocNADE with additional features is given by:

$$p(q|\mathbf{v}) = \text{softmax}(\mathbf{d} + \mathbf{S}' \cdot \text{concat}(\mathbf{h}(\mathbf{v}), \mathbf{h}_e(\mathbf{v})))$$

where, $\mathbf{S}' \in \mathbb{R}^{H' \times L}$ and $H' = H + E^{fast} + E^{word2vec}$.

3.3 Traditional Methods for IR

BM25: A ranking function proposed by Robertson and Zaragoza (2009) is used to estimate the relevance of a document for a given query.

BM25-Extra: The relevance score of BM-25 is combined with four extra features: (1) percentage of query words with exact match in the document, (2) percentage of query words bigrams matched in the document, (3) IDF weighted document vector for feature #1, and (4) IDF weighted document vector for feature #2. Therefore, BM25-Extra returns a vector of 5 scores.

3.4 System Description for RDoC Task-1

RDoC Task-1 aims at retrieving and ranking of PubMed abstracts (title and content) that are relevant for 8 RDoC constructs. Participants are provided with 8 clusters, each with a RDoC construct label and required to rank abstracts within each cluster based on their relevance to the corresponding cluster label. Each cluster contains abstracts relevant to its RDoC construct, while some (or most) of the abstracts are noisy in the sense that

they belong to a different RDoC construct. Ideally, the participants are required to rank abstracts in each of the clusters by determining their relevance with the RDoC construct of the cluster in which they appear.

To address the RDoc Task-1, we learn a mapping function between latent representation $\mathbf{h}(\mathbf{v})$ of a document (i.e.., abstract) $\mathbf{v}$ and its RDoC construct, i.e., query words q in a supervised fashion. In doing so, we have employed supervised classifiers, especially supervised neural topic model **a-supDocNADE** (section 3.2) for document ranking. We treat q as label and maximize $p(q|\mathbf{v})$ leading to maximize $\mathcal{L}^{hybrid}(\mathbf{v})$ in *a-supDocNADE* model.

As demonstrated in Figure 1 (right), we perform document ranking in two steps:

(1) **Document Relevance Ranking**: We build a supervised classifier using all the training documents and their corresponding labels (RDoC constructs), provided with the training set. At the test time, we compute prediction probability score $p(CID = q|\mathbf{v}^{test}(CID)))$ of the label=CID for each test document $\mathbf{v}^{test}(CID)$ in the cluster, CID. This prediction probability (or confidence score) is treated as a relevance score of the document for the RDoC construct of the cluster. Figure 1(right) shows that we perform document ranking using the probability scores (col-2) of the RDoC construct (e.g. *loss*) within the cluster $C1$. Observe that a test document with least confidence

for a cluster are ranked lower within the cluster and thus, improving mean average precision (mAP). Additionally, we also show the predicted RDoC construct in col-1 by the supervised classifier.

(2) **Document Relevance Re-ranking**: Secondly, we re-ranked each document **v** (title+abstract) within each cluster (with label q) using *unsupervised ranking*, where the relevance scores are computed as: (a) **reRank(BM25-Extra)**: sum each of the 5 relevance scores to get the final relevance, and (b) **reRank(QAR)**: cosine-similarity(QAR(**v**), $\widetilde{q}$).

3.5 System Description for RDoC Task-2

The RDoC Task-2 aims at extracting the most relevant sentence from each of the PubMed abstract for the corresponding RDoC construct. Each abstract consists of title t and sentences s with an RDoC construct q.

To address RDoc Task-2, we first compute *multi-view* representation: BoW, TF-IDF and QAR (i.e., $\Phi_q(s^j)$) for each sentence s^j in an abstract d. On other hand, we compute ESR representation for RDoC construct (query q) and title t of the abstract d to obtain $\widetilde{q}$ and $\widetilde{t}$, respectively. Figure 2 and section 3.1 describe the computation of these representations. We then use the representations ($\Phi_q(s^j)$, $\widetilde{t}$ and $\widetilde{q}$) to compute a relevance scores of a sentence s_j relative to q and/or t via *unsupervised* and *supervised* ranking schemes, discussed in the following section.

3.5.1 Unsupervised Sentence Ranker

As shown in Figure 2, we first extract representations: $\Phi_q(s^j)$, $\widetilde{t}$ and $\widetilde{q}$ for the sentence s^j query q and title t. During ranking sentences within an abstract for the given RDoC construct q, we also consider title t in computing the relevance score for each sentence relative to q and t. It is inspired from the fact that the title often contains relevant terms (or words) appearing in sentence(s) of the document (or abstract). On top, we observe that q is a very short text and non-descriptive, leading to minimal text overlap with s.

We compute two relevance scores: r_q and r_t for a sentence s^j with respect to a query q and title t, respectively.

$$r_q = sim(\widetilde{q}, \Phi_q(s^j)) \text{ and } r_t = sim(\widetilde{t}, \Phi_q(s^j))$$

Now, we devise two ways to combine the relevance scores r_q and r_t in unsupervised paradigm:

$$\text{version1: } r_1^{unsup} = r_q \cdot r_q + r_t \cdot r_t$$

Observe that the relevance scores are weighted by itself. However, the task-2 expects a higher importance to the relevance score r_q over q_t. Therefore, we coin the following weighting scheme to give higher importance to r_q only if it is higher than r_t otherwise we compute a weight factor r'_t for r_t.

$$\text{version2: } r_2^{unsup} = r_q \cdot r_q + r'_t \cdot r_t$$

where r'_t is compute as:

$$r'_t = (r_t > r_q)|r_t - r_q|$$

The relevance score r_2^{unsup} is effective in ranking sentences when a query and sentence does not overlap. In such a scenario, a sentence is scored by title, penalized by a factor of $|r_t - r_q|$.

At the end, we obtain a final relevance score r_f^{unsup} for a sentence s^j by summing the relevance scores of BM25-Extra and r_1^{unsup} or r_2^{unsup}.

3.5.2 Supervised Sentence Ranker

Beyond unsupervised ranking, we further investigate sentence ranking in supervised paradigm by introducing a distance metric between the query (or title) and sentence vectors.

Figure 2 describes the computation of relevance score for a sentence s^j using a supervised sentence ranker scheme. Like the unsupervised ranker (section 3.5.1), the supervised ranker also employs vector representations: $\Phi_q(s^j)$, $\widetilde{t}$ and $\widetilde{q}$. Using the projection matrix **G**, we then apply a projection to each of the representation to obtain $\Phi_q^p(s^j)$, $\widetilde{t}^p$ and $\widetilde{q}^p$. Here, the operator $\otimes$ performs concatenation of the projected vector with its input via residual connection. Next, we apply a Manhattan distance metric to compute similarity (or relevance) scores, following Gupta et al. (2018):

$$r^{sup} = \exp\left(-||(\Phi_q^p(s^j), \widetilde{q}^p) + \beta\,(\Phi_q^p(s^j), \widetilde{t}^p)||_2\right)$$

where $\beta \in [0, 1]$ controls the relevance of title, determined by cross-validation. A final relevance score $r_f^{sup} \in [0, 1]$ is computed by feeding a vector $[r^{sup}, r_{siamese}^{sup}, \text{BM25-extra}]$ into a supervised linear regression, which is trained end-to-end by minimizing mean squared error between the r_f^{sup} and $\{0, 1\}$, i.e., 1 when the sentence s^j is relevant to query q. Here, $r_{siamese}^{sup}$ refers to a relevance

	L1	L2	L3	L4	L5	L6	L7	L8	Total
All data	39	38	47	21	28	27	48	18	266
Train set	31	30	37	16	22	21	38	14	209
Dev set	8	8	10	5	6	6	10	4	57
Test set (Task1)	79	108	123	144	138	139	122	146	999
Test set (Task2)	19	26	30	35	34	34	30	36	244

Table 2: Data statistics - # of PubMed abstracts belonging to each RDoC construct in different data partitions. (L1: "Acute Threat Fear"; L2: "Arousal"; L3: "Circadian Rhythms"; L4: "Frustrative Nonreward"; L5: "Loss"; L6: "Potential Threat Anxiety"; L7: "Sleep Wakefulness"; L8: "Sustained Threat")

Model	Feature	Classification Accuracy	Ranking mAP
SVM	BoW	0.947	**0.992**
Re-ranking	reRank #1	-	**0.992**
with	reRank #2	-	**0.992**
cluster label	reRank #3	-	**0.992**
a-supDocNADE	Random init	0.912	0.930
	+ FastText	0.947	0.949
	+ BioNLP	0.965	0.983
Re-ranking	reRank #1	-	0.985
with	reRank #2	-	**0.994**
cluster label	reRank #3	-	**0.994**

Table 3: RDoC Task-1 results (on development set): Classification accuracy and mean Average Precision (mAP) of a-supDocNADE and SVM models. Each model's *classification accuracy* and *ranking mAP* (using prediction probabilities) are shown together. Furthermore, each model's ranked clusters are re-ranked using different re-ranking algorithms. Best mAP score for each model is marked in **bold**. (*reRank #1*: "reRank(BM25-Extra)"; *reRank #2*: "reRank(QAR)"; *reRank #3*: "reRank(BM25-Extra) + reRank(QAR)")

score computed between q and s^j via Siamese-LSTM (Gupta et al., 2018).

To perform sentence ranking within an abstract for a given RDoC construct q, the relevance score r_f^{sup} (or r_f^{unsup}) is computed for all the sentences and a sentence with the highest score is extracted.

4 Experiments and Results

4.1 Data Statistics and Experimental Setup

Dataset Description: Dataset for RDoC Tasks contains a total of 266 PubMed abstracts labelled with 8 RDoC constructs in a single label fashion. Number of abstracts for each RDoC construct is described in Table 2, where *first* row describes the statistics for all abstracts and *second & third* row shows the split of those abstracts into *training* and *development* sets maintaining a 80-20 ratio for each RDoC construct. For Task-1, each PubMed abstract contains its associated title, PubMed ID (PMID) and label (RDoC construct). In addition for Task-2, each PubMed abstract also contains a list of most relevant sentences from that abstract. Final evaluation test data for Task-1 & Task-2 contains 999 & 244 abstracts respectively.

We use "RegexpTokenizer" from scikit-learn to tokenize abstracts and lower-cased all tokens. After this, we remove those tokens which occur in less than 3 abstracts and also remove stopwords (using nltk). For computing BM25-Extra relevance score, we use unprocessed raw text of sentences and titles.

Experimental Setup: As the training dataset labelled with RDoC constructs is very small, we use an external source of semantical knowledge by incorporating pretrained distributional word embeddings (Zhang et al., 2019) from FastText model (Bojanowski et al., 2017) trained on the entire corpus of PubMed and MIMIC III Clinical notes (Johnson et al., 2016). Similarly, we also use pretrained word embeddings (Moen and Ananiadou, 2013) from word2vec model (Mikolov et al., 2013) trained on PubMed and PMC abstracts. We create 3 folds* of train/dev splits for cross-validation.

RDoC Task-1: For DocNADE topic model, we use latent representation of size 50. We use pretrained FastText embeddings of size 300 and pretrained word2vec embeddings of size 200. For SVM, we use Bag-of-words (BoW) representation of abstracts with radial basis kernel function. PubMed abstracts are provided in eight different clusters, one for each RDoC construct, for final test set evaluation.

RDoC Task-2: We use pretrained FastText embeddings to compute *query-aware sentence representation* of a sentence ($\Phi_q(s^j)$), title ($\widetilde{t}$) and query ($\widetilde{q}$) representations. We also train Replicated-Siamese-LSTM (Gupta et al., 2018) model with input as sentence and query pair i.e., (s^j, q) and label as 1 if s^j is relevant otherwise 0. We use $\beta \in \{0, 1\}$.

4.2 Results: RDoC Task-1

Table 3 shows the performance of supervised Document Ranker models i.e, **a-supDocNADE** and **SVM**, for Task-1. *SVM* achieves a classification accuracy of 0.947 and mean average precision

---

*we only report results on $fold_1$ because of best scores on partial test dataset

Ranking (with Prediction Probability)			Re-ranking (with BM25-Extra)	
PMID	Pred Prob	Gold Label	PMID	Gold Label
22906122	0.90	PTA	26005838	PTA
24286750	0.77	PTA	22906122	PTA
17598732	0.61	PTA	28828218	PTA
26005838	0.56	PTA	26773206	PTA
28316567	0.46	*Loss*	24286750	PTA
28828218	0.45	PTA	17598732	PTA
26773206	0.41	PTA	28316567	*Loss*

Table 4: RDoC Task-1 analysis: Ranking of PubMed abstracts within "Potential Threat Anxiety (PTA)" cluster using supervised prediction probabilities ($p(q|\mathbf{v})$). It shows that an intruder/noisy abstract (Gold Label: *Loss*) is assigned higher probability than the abstracts with same Gold Label as the cluster. But, using re-ranking with BM25-Extra (`reRank(BM25-Extra)`) relevance score assigns lowest relevance to the intruder abstract.

(mAP) of 0.992 by ranking the abstracts in their respective clusters using the supervised prediction probabilities ($p(q|\mathbf{v})$). After that, we use three different relevance scores: (1) *reRank(BM25-Extra)*, (2) *reRank(QAR)* and (3) *reRank(BM25-Extra) + reRank(QAR)*, for re-ranking of the abstracts in their respective clusters. It is to be noted that the *ranking mAP* of the clusters using prediction probabilities is already the best possible i.e., the intruder abstracts (abstracts with different label (RDoC construct) than the cluster label) are at the bottom of the ranked clusters. Therefore, re-ranking of these clusters would not achieve a better score. Similarly, we train *a-supDocNADE* model with three different settings: (1) random weight initialization, (2) incorporating FastText embeddings ($\mathbf{h_e}(\mathbf{v})$) and (3) incorporating Fast-Text and word2vec embeddings ($\mathbf{h_e}(\mathbf{v})$). By using the pretrained embeddings, the classification accuracy increases from 0.912 to 0.965, this shows that distributional pretrained embeddings carry significant semantic knowledge. Furthermore, re-ranking using *reRank(BM25-Extra)* and *reRank(QAR)* further results in the improvement of mAP score (0.994 vs 0.983) by shifting the intruder documents at the bottom of each impure cluster.

4.3 Analysis: RDoC Task-1

Table 4 shows an impure "Potential Threat Anxiety" cluster of abstracts containing an intruder abstract with label (RDoC construct) "Loss". When this cluster is ranked on the basis of predic-

Model	Feature	Recall	F1	Macro-Average Accuracy
Unsupervised	reRank(BM25-Extra) [#1]	0.316	0.387	0.631
	version1 [#2]	**0.351**	**0.412**	**0.701**
	version2 [#3]	0.263	0.345	0.526
Supervised	$r_f^{sup}(\beta = 0)$ [#4]	**0.386**	**0.436**	**0.772**
	$r_f^{sup}(\beta = 1)$ [#5]	0.368	0.424	0.737
Ensemble	{#1, #2, #4}	**0.395**	**0.441**	**0.789**
	{#1, #3, #4}	0.316	0.387	0.631
	{#2, #4, #5}	**0.395**	**0.441**	**0.789**
	{#1, #2, #3, #4, #5}	0.368	0.424	0.737

Table 5: RDoC Task-2 results (on development set): Performance of unsupervised and supervised sentence rankers (Figure 2) under different configurations. Best scores for each model is marked in **bold**.

tion probabilities ($p(q|\mathbf{v})$), then "Loss" abstract is ranked third from the bottom and it degrades the mAP score of the retrieval system. But after re-ranking this cluster using *reRank(BM25-Extra)* relevance score, the "Loss" abstract is ranked at the bottom, thus maximizing the mAP score. Therefore, re-ranking with BM25-Extra on top of ranking with $p(q|\mathbf{v})$ is, evidently, a robust abstract/document ranking technique.

4.4 Results: RDoC Task-2

Table 5 shows results for Task-2 using *three* unsupervised and *two* supervised sentence ranker models. For unsupervised model, using *reRank(BM25-Extra)* relevance score between a query (q), label (RDoC construct) of an abstract, and all the sentences (s^j) in an abstract, we get an macro-average accuracy (MAA) of 0.631. However, using *version1* and *version2* models (see Fig 2), we achieve a MAA score of 0.701 and 0.526 respectively. Higher accuracy of *version1* model suggests that title (t) of an abstract also contains the essential information regarding the most relevant sentence. For supervised model, we get an MAA score of 0.772 and 0.737 by setting $\beta = 0$ & 1 in supervised relevance score (r_f^{sup}) equation in section 3.5.2. Hence, for supervised sentence ranker model, title (t) is playing a negative influence in correctly identifying the relevance (r_f^{sup}) of different sentences. Furthermore, we combine the knowledge of unsupervised and supervised sentence rankers by creating multiple ensembles (majority voting) of the predictions from different models. We achieve the highest MAA score of 0.789 by combining the predictions of (1) *reRank(BM25-Extra)*, (2) *version1*, and (3) r_f^{sup} with $\beta = 0$. Notice that all the proposed supervised and unsupervised sentence ranking mod-

Team	RDoC Task-1 (Official Results)									RDoC Task-2 (Official Results)								
	L1	L2	L3	L4	L5	L6	L7	L8	mAP	L1	L2	L3	L4	L5	L6	L7	L8	MAA
MIC-CIS	0.85	0.92	**1.00**	**0.73**	**0.78**	**0.94**	**1.00**	0.63	**0.86**	**0.74**	**0.73**	0.47	0.37	**0.74**	**0.59**	**0.60**	0.42	**0.58**
Javad Rafiei Asl	**0.89**	**0.93**	1.00	0.69	0.74	0.87	**1.00**	**0.64**	0.85	0.68	0.62	0.47	0.34	0.56	0.32	0.50	0.36	0.48
Ramya Tekumalla	0.83	0.91	1.00	0.67	0.71	0.89	**1.00**	**0.64**	0.83	0.37	0.12	0.10	0.11	0.26	0.15	0.17	0.14	0.18
Daniel Laden	0.67	0.84	1.00	0.61	0.61	0.81	**0.98**	0.41	0.74	-	-	-	-	-	-	-	-	-
Shyaman Jayasundara	-	-	-	-	-	-	-	-	-	0.47	0.42	0.60	0.29	0.62	0.38	0.57	**0.47**	0.48
Fei Li	-	-	-	-	-	-	-	-	-	0.58	0.46	**0.70**	**0.43**	0.26	0.41	0.33	**0.47**	0.46

Table 6: RDoC Tasks official results - performance on test set of different competing systems. Best score in each column is marked in **bold**. (Refer to Table 2 for header notations) (mAP: "Mean Average Precision"; MAA: Macro-Average Accuracy)

PubMed Abstract
(PMID: "23386529"; RDoC construct: "Loss")

Most Relevant Sentence (using `reRank(BM25-Extra)`)	Sentence ID	Gold Label
Nurses are expected to care for grieving women and families suffering from perinatal loss.	#1	Not relevant
Most Relevant Sentence (using `Ensemble {#1, #2, #4}`)	-	-
We found that nurses experience a grieving process similar to those directly suffering from perinatal loss.	#6	Relevant

Table 7: RDoC Task-2 analysis: This table shows that the most relevant sentence predicted using `reRank(BM25-Extra)` is actually not a relevant sentence, but `Ensemble {#1, #2, #4}` (Table 5) predicts the correct sentence as the most relevant.

els (except [#3]) outperform tranditional ranking models, e.g., *reRank(BM25-Extra)* in terms of query-document relevance score.

4.5 Analysis: RDoC Task-2

Table 7 shows that the most relevant sentence predicted by *reRank(BM25-Extra)* is actually a non-relevant sentence. But an ensemble of predictions from both unsupervised and supervised ranker models correctly predicts the relevant sentence. This suggests that complementary knowledge of different models is able to capture the relevance of sentences on different scales and majority voting among them is, evidently, a robust sentence ranking technique.

4.6 Results: RDoC Task 1 & 2 on Test set

Table 6 shows the final evaluation scores of different competing systems for both the RDoC Task-1 & Task-2 on final test set. Observe that our submission (MIC-CIS) scored a mAP score of 0.86 and MAA of 0.58 in Task-1 and Task-2, respectively. Notice that we outperform the second best

system by 20.83% (0.58 vs 0.48) margin in Task2.

5 Conclusion

In conclusion, both supervised neural topic model and SVM can effectively perform ranking of PubMed abstracts in a given cluster based on the prediction probabilities. However, a further re-ranking using *BM25-Extra* or *query-aware sentence representation (QAR)* has proven to maximize the mAP score by correctly assigning the lowest relevance score to the intruder abstracts. Also, unsupervised and supervised sentence ranker models using query-title-sentence interactions outperform the traditional BM25-Extra based ranking model by a significant margin.

In future, we would like to introduce complementary feature representation via hidden vectors of LSTM jointly with topic models and would like to further investigate the interpretability (Gupta et al., 2015; Gupta and Schütze, 2018) of the proposed neural ranking models in the sense that one can extract salient patterns determining relationship between query and text. Another promising direction would be introduce abstract information, such as part-of-speech and named entity tags (Lample et al., 2016; Gupta et al., 2016) to augment information retrieval (IR).

Acknowledgment

This research was supported by Bundeswirtschaftsministerium (bmwi.de), grant 01MD19003E (PLASS (plass.io)) at Siemens AG - CT Machine Intelligence, Munich Germany.

References

David M. Blei, Andrew Y. Ng, and Michael I. Jordan. 2003. Latent dirichlet allocation. *J. Mach. Learn. Res.*, 3:993–1022.

Piotr Bojanowski, Edouard Grave, Armand Joulin, and Tomas Mikolov. 2017. Enriching word vectors with subword information. *TACL*, 5:135–146.

Pankaj Gupta, Bernt Andrassy, and Hinrich Schütze. 2018. Replicated siamese LSTM in ticketing system for similarity learning and retrieval in asymmetric texts. In *Proceedings of the Third Workshop on Semantic Deep Learning, SemDeep@COLING 2018, Santa Fe, New Mexico, USA, August 20, 2018*, pages 1–11.

Pankaj Gupta, Yatin Chaudhary, Florian Buettner, and Hinrich Schütze. 2019a. Document informed neural autoregressive topic models with distributional prior. In *The Thirty-Third AAAI Conference on Artificial Intelligence, AAAI 2019, The Thirty-First Innovative Applications of Artificial Intelligence Conference, IAAI 2019, The Ninth AAAI Symposium on Educational Advances in Artificial Intelligence, EAAI 2019, Honolulu, Hawaii, USA, January 27 - February 1, 2019.*, pages 6505–6512.

Pankaj Gupta, Yatin Chaudhary, Florian Buettner, and Hinrich Schütze. 2019b. Texttovec: Deep contextualized neural autoregressive topic models of language with distributed compositional prior. In *7th International Conference on Learning Representations, ICLR 2019, New Orleans, LA, USA, May 6-9, 2019*.

Pankaj Gupta, Thomas Runkler, Heike Adel, Bernt Andrassy, Hans-Georg Zimmermann, and Hinrich Schütze. 2015. *Deep Learning Methods for the Extraction of Relations in Natural Language Text*. Master Thesis.

Pankaj Gupta and Hinrich Schütze. 2018. LISA: explaining recurrent neural network judgments via layer-wise semantic accumulation and example to pattern transformation. In *Proceedings of the Workshop: Analyzing and Interpreting Neural Networks for NLP, BlackboxNLP@EMNLP 2018, Brussels, Belgium, November 1, 2018*, pages 154–164.

Pankaj Gupta, Hinrich Schütze, and Bernt Andrassy. 2016. Table filling multi-task recurrent neural network for joint entity and relation extraction. In *COLING 2016, 26th International Conference on Computational Linguistics, Proceedings of the Conference: Technical Papers, December 11-16, 2016, Osaka, Japan*, pages 2537–2547.

Alistair EW Johnson, Tom J Pollard, Lu Shen, H Lehman Li-wei, Mengling Feng, Mohammad Ghassemi, Benjamin Moody, Peter Szolovits, Leo Anthony Celi, and Roger G Mark. 2016. Mimic-iii, a freely accessible critical care database. *Scientific data*, 3:160035.

Guillaume Lample, Miguel Ballesteros, Sandeep Subramanian, Kazuya Kawakami, and Chris Dyer. 2016. Neural architectures for named entity recognition. In *NAACL HLT 2016, The 2016 Conference of the North American Chapter of the Association for Computational Linguistics: Human Language Technologies, San Diego California, USA, June 12-17, 2016*, pages 260–270.

Hugo Larochelle and Stanislas Lauly. 2012. A neural autoregressive topic model. In *Advances in Neural Information Processing Systems 25: 26th Annual Conference on Neural Information Processing Systems 2012. Proceedings of a meeting held December 3-6, 2012, Lake Tahoe, Nevada, United States.*, pages 2717–2725.

Christopher D. Manning, Prabhakar Raghavan, and Hinrich Schütze. 2008. *Introduction to information retrieval*. Cambridge University Press.

Ryan McDonald, George Brokos, and Ion Androutsopoulos. 2018. Deep relevance ranking using enhanced document-query interactions. In *Proceedings of the 2018 Conference on Empirical Methods in Natural Language Processing, Brussels, Belgium, October 31 - November 4, 2018*, pages 1849–1860.

Tomas Mikolov, Ilya Sutskever, Kai Chen, Gregory S. Corrado, and Jeffrey Dean. 2013. Distributed representations of words and phrases and their compositionality. In *Advances in Neural Information Processing Systems 26: 27th Annual Conference on Neural Information Processing Systems 2013. Proceedings of a meeting held December 5-8, 2013, Lake Tahoe, Nevada, United States.*, pages 3111–3119.

SPFGH Moen and Tapio Salakoski Sophia Ananiadou. 2013. Distributional semantics resources for biomedical text processing. *Proceedings of LBM*, pages 39–44.

Jeffrey Pennington, Richard Socher, and Christopher D. Manning. 2014. Glove: Global vectors for word representation. In *Proceedings of the 2014 Conference on Empirical Methods in Natural Language Processing, EMNLP 2014, October 25-29, 2014, Doha, Qatar, A meeting of SIGDAT, a Special Interest Group of the ACL*, pages 1532–1543.

Stephen E. Robertson and Hugo Zaragoza. 2009. The probabilistic relevance framework: BM25 and beyond. *Foundations and Trends in Information Retrieval*, 3(4):333–389.

Akash Srivastava and Charles A. Sutton. 2017. Autoencoding variational inference for topic models. In *5th International Conference on Learning Representations, ICLR 2017, Toulon, France, April 24-26, 2017, Conference Track Proceedings*.

Yijia Zhang, Qingyu Chen, Zhihao Yang, Hongfei Lin, and Zhiyong Lu. 2019. Biowordvec, improving biomedical word embeddings with subword information and mesh. *Scientific data*, 6(1):52.

Author Index

Association for Computational Linguistics
209 N. Eighth Street
Stroudsburg, Pennsylvania 18360

ISBN 978-1-7138-0076-7